Integrating Digital Innovation and Integrated Frameworks in Manufacturing

Siddhartha Paul Tiwari
Google Asia Pacific, Singapore

Eliseo A. Aurellado
Southeast Asia Interdisciplinary Development Institute, Philippines

Adi Fahrudin
Universitas Bhayangkara Jakarta Raya, Indonesia

IGI Global
Scientific Publishing
Publishing Tomorrow's Research Today

Vice President of Editorial Melissa Wagner
Director of Acquisitions Mikaela Felty
Director of Book Development Jocelynn Hessler
Production Manager Mike Brehm
Cover Design Jose Rosado

Published in the United States of America by
IGI Global Scientific Publishing
701 East Chocolate Avenue
Hershey, PA, 17033, USA
Tel: 717-533-8845
Fax: 717-533-7115
Website: https://www.igi-global.com E-mail: cust@igi-global.com

Library of Congress Cataloging-in-Publication Data

LCCN: 2025027232 (CIP Data Pending)
ISBN13: 979-8-3373-1082-4
Isbn13Softcover: 979-8-3373-1083-1
EISBN13: 979-8-3373-1084-8

British Cataloguing in Publication Data
A Cataloguing in Publication record for this book is available from the British Library.

Table of Contents

Detailed Table of Contents

organizational transformation. By centering human needs within technological advancement, this work contributes to a more holistic understanding of sustainable digital manufacturing that benefits both organizations and communities.

The aerospace production industry is going through a huge change that is being driven by digital innovation and the use of advanced frameworks. In a very competitive and regulated world, the need for more efficiency, accuracy, and speed is driving this change. Key technologies like digital twins, Industry 4.0, AI, additive manufacturing, and the Industrial Internet of Things (IIoT) make it possible to analyse data in real time, plan maintenance ahead of time, and create production systems that can adapt to new conditions. Digital engineering, model-based systems engineering (MBSE), and supply chain integration can all be combined into integrated models that make the whole product lifecycle more visible and help with making decisions.This is leading to better product quality at lower costs and shorter lead times. It also looks at effective frameworks for integration and points out both the problems and the chances of creating a fully connected, smart manufacturing environment.

Manufacturing is the world's oldest and largest industry. Manufacturing has lapsed in investment when it comes to research and development, and robotics is an upcoming thrust area. The author of this book chapter discusses how strategic investments in education, innovation, and inclusive policies will determine whether society can harness robotics effectively for sustainable economic growth and prosperity. Robotics has resulted in significant increases in manufacturing productivity and efficiency, which have reduced operational costs and production errors. As automation changes employment dynamics, it contributes to economic growth, but it also requires substantial workforce adaptations and retraining. Robotics has dramatically improved the speed at which a product is manufactured and delivered to the consumer, which has resulted in lower costs and higher quality products. Technology is creating specialized technical roles; however, it also disrupts traditional employment, requiring workers who have been retrained to acquire a number of new skills.

Technology and innovation are crucial in all industries nowadays. Every firm grows socially, economically, politically, and environmentally as technology and innovation advance. Technology and innovation may improve efficiency, market competitiveness, and sustainability, which every company wants. This paper says technology and innovation dynamically support any business's sustainability aims. This study examines how technology and innovation help evolving firms achieve sustainability. This study includes articles, case studies, surveys, and in-depth analysis to demonstrate how technology and innovation increase supply chain management, resource usage, market competitiveness, and consumer engagement. This research helps policymakers and stakeholders improve business sustainability with technology and innovation. Practical ideas from this research article help business sustainability discourse adapt to technology and innovation. This research promotes a balanced approach to commercial success and environmental sustainability.

As a result of digital transformation, aerospace manufacturing in the Asia Pacific region is changing dramatically. This book chapter examines how digital innovation is being incorporated into aerospace manufacturing through the use of an integrated framework approach. The book chapter emphasizes that successful adoption of digital technologies in manufacturing requires not only advanced technological infrastructure, but also alignment with organizational strategy, the development of workforce skills, and cross-industry collaboration. This book chapter adopts a socioeconomic and technical perspective. This chapter is based on a conceptual framework that highlights the key dimensions of technological development, organizational development, people, and stakeholder communication. A study of current literature, industry trends, and practical applications has been conducted by the author to illustrate how Asia Pacific aerospace manufacturers and policymakers may leverage this integrated framework to enhance performance efficiency, quality, and innovation.

This book chapter examines the revolutionary impact of digital transformation on the manufacturing industry, as well as future applications. The book chapter provides corporates, academicians, consultants, and researchers with an overview of integrated frameworks and cutting-edge digital innovations that helps in the transformation of manufacturing. This book chapter explains how technological advances are reshaping traditional manufacturing methods. The book chapter contains real-world case studies, in-depth analysis, and expert insights that are intended to help readers in navigating and leveraging the shift to a more efficient, sustainable, and streamlined manufacturing environment. Throughout the manufacturing value chain, automation is at the forefront of digital technological trends and innovation, which is why a growing number of companies are striving to enhance their operational and functional processes across their entire value chain in order to increase their operational and functional efficiency.

Man has always been fascinated with flight. Since the Greek mythology character Icarus flew towards the sun in his crafted wings and fell to his death, early men dared to conquer the skies with often disastrous results. The dawn of aviation came when the Wright brothers flew the first heavier-than-air machine along the dunes of Kitty Hawk, North Carolina, in 1903. Soon, newly invented flying machines began to be used in sports and air shows. When the Great War broke out in 1914, machine guns were mounted on primitive aircraft, and airplanes became weapons of mass destruction. Advances in aircraft design and development led to the invention of more sophisticated and deadly military planes, which became a potent force in warfare in World War II. When the first satellite, the Soviet Sputnik, flew in 1957, the world entered the space age. A race between America and Russia began during the Cold War, culminating with Neil Armstrong landing on the moon's surface in 1969. Man has gone beyond the Earth's atmosphere and into the far regions of outer space.

Businesses are progressively utilising predictive analytics to enhance their comprehension and anticipation of client behaviour within the framework of Society 5.0, a human-centric society that integrates the internet with the physical realm. This paper examines the vital role of predictive analytics in the progression of marketing innovations, emphasising marketers' ability to anticipate trends, personalise experiences, and develop data-driven marketing strategies. This study will examine how predictive analytics technologies are transforming marketing strategies in Society 5.0 by providing enhanced customer insights and facilitating more accurate targeting. This study is distinctive since it examines the convergence of AI-driven analytics and marketing strategies within the Society 5.0 framework, which seeks to harmonise economic advancement with societal welfare.

Preface

Manufacturing is undergoing a profound digital transformation at a time when emerging technologies are increasingly integrated into industrial processes. A shift to smart, connected systems is reflected in concepts such as Industry 4.0 and Society 5.0, which emphasize human-centered innovation and smart, connected systems. By transforming factories into cyber-physical systems, a number of technologies, such as artificial intelligence, machine learning, robotics, augmented reality, and data analytics, enhance traditional manufacturing processes. There are significant benefits to be gained from the convergence in terms of efficiency and flexibility, but there are also important considerations regarding sustainability, workforce readiness, and strategic integration that need to be addressed. A synergy of advanced technologies is fueling today's manufacturing revolution, which is discussed in depth in the book. As part of the automation process, artificial intelligence and machine learning provide intelligence, robotics provide precision in executing tasks, digital twins and augmented reality offer a new perspective on design and visualization, and high density networks provide the linkage between all of these elements in a fully integrated, data-driven manner. As discussed in the book, these technologies cannot operate in isolation, their true power lies in their combination with humans. Smart factories are marked by the integrated deployment of these tools. From boosting sustainability and efficiency to reshaping economies and workforces, the book explores the impact of these changes.

Several aspects of the modern manufacturing revolution have been discussed in detail in this book, including robotics and automation. Today's industrial robots are intelligent, flexible, and collaborative, which makes them ideal for repetitive or precise tasks. A robotic system that is advanced enough to work alongside humans as a cobot can safely share workspaces and augment rather than replace human labor. Investment in robotic automation allows companies to achieve safer, more efficient, and more consistent production, while allowing human workers to focus on innovation and complex problem solving. The concept of sustainability is critical to understanding the tech-driven transformation of manufacturing. Manufacturers throughout the world are under increasing pressure to reduce their environmental footprint and contribute to broader sustainability goals. Sustainability goes hand in hand with operational efficiency and productivity. Digital transformation offers factories the opportunity to achieve operational efficiency outcomes that benefit the planet as well as the bottom line. Technology integration poses a number of challenges and trade-offs for sustainability efforts, which are discussed in this book. The development of autonomous robots can considerably contribute to the achievement of the Sustainable Development Goals (SDGs) set forth by the United Nations. The emergence of new digital technologies enables a reduction in environmental impact by optimizing resource use and reducing waste in the process. Artificial intelligence systems provide manufacturers with the ability to make better decisions by enabling them to make informed choices. Using a machine learning system, production parameters can be adjusted in a way that reduces material waste and energy consumption, thus contributing to sustainable production and the introduction of more sustainable practices at the factory.

Advanced technology is fundamentally altering productivity dynamics in the manufacturing sector and, as a consequence, influencing economic growth patterns. Technological advances in manufacturing have historically resulted in improvements in productivity and economic output. Digital and robotic technologies appear to be following a similar trajectory. Manufacturing productivity is increasing as intelligent automation gains traction. Previously time-consuming manual tasks are now completed with near-zero error by machines guided by artificial intelligence (AI) or by robots that work continuously. Beyond replacing labor in some tasks, technology enhances human capabilities in other areas as well. A decision-supporting AI system can help factory managers optimize production schedules more efficiently than trial-and-error, and augmented reality helps workers reduce assembly errors. These manufacturing optimizations have far-reaching effects beyond the factory. As a result of this digital revolution in manufacturing, more output can be produced with the same or fewer inputs. By reducing the cost of goods, increasing production volumes, and freeing up capital and labor, an increase in manufacturing productivity contributes to higher economic growth.

Factory automation has a clear economic implication because it puts a company at a competitive advantage on the global market. When manufacturing firms successfully deploy technologies like robotics and AI within their processes, they often gain an advantage in terms of both cost and quality. There is a possibility that this could increase their market share internationally, influencing trade balances and national growth on a national level. A significant aspect of robotics is the potential it has to redistribute some manufacturing activities in terms of how they are carried out. Automation reduces manufacturing labor costs, as mentioned in this book. The other side of the coin has to be addressed as well in order to gain a deeper understanding. In manufacturing, there is a fear that if the workforce skills do not catch up with the needs of new technologies, then there may be structural unemployment as a result of the skills mismatch. There will be a great deal of influence on economic growth depending on how this balance is struck. With the right education, retraining, and innovation policy, integrating technology can lead to a more skilled workforce and new industries, which can be beneficial to the economy as a whole. Without adequate safety nets or retraining programs in place, widespread displacement could depress wages and demand, resulting in economic stagnation as a result.

It appears that policy and integrated strategies are necessary in order to ensure that the productivity growth driven by technology is reflected in economic growth that is inclusive as well. An integrated technology approach to manufacturing aims to use advanced manufacturing as a source of national wealth while providing high-quality employment for the general public, balancing efficiency with human capital development in order to achieve national development goals. Technology advances in the manufacturing and digital sectors can fuel productivity and boost economic growth in the long run, but reaping those benefits on a broad scale will require attention to the transition of the workforce in the long run. Innovation in manufacturing still has a strong focus on people and human capital. Humans are responsible for making the final decisions in a manufacturing organization, and they enact solutions to problems in factories based on exceptions that the machines cannot handle. The integration of technology with manufacturing should include the integration of the workforce in a manner that respects human capabilities and well-being.

There are a number of ways in which technology can be aligned in a positive manner with regard to human factors in view of the many challenges that have emerged for manufacturing as a result of technological advancements. The collaborative robot or cobot, is one such example. These robots are specially designed to work in harmony with human workers to perform specific tasks. There could be cobots assigned to handle the heavy lifting or repetitive work while the human is free to focus on finer details or decision-making. With the use of cobots, it is imperative that technology be adopted and appropriate coding be employed so that the cobot's actions can be predicted and transparent to the human partner. Workers may experience mental stress or safety concerns when robot behavior is unpredictable, whereas predictable, well-communicated actions build trust and contribute to a sense of ease. Workplace anxiety can be reduced significantly through strategies such as providing clear visual indicators of a robot's intent or providing workers with an easy method of adjusting the robot's behavior. Cobots become safer and more widely accepted when these factors are addressed, thereby contributing to the advancement of innovation and sustainability in the modern world. By simplifying tasks for humans, assistive technologies, wearable devices that assist with manual lifting or overhead work and smart wearables can reduce physical strain and also enhance cognitive ergonomics. A worker can benefit from using an augmented reality headset to display assembly instructions right in his or her field of view, eliminating the need to constantly refer to a manual or recall steps. This directly supports the worker's cognitive process, making the job less likely to cause errors and less stressful mentally.

Changing the workforce in a digital manufacturing environment is not just about technology. The process of transformation is not just an organizational and social one. Factory workers' roles are evolving as factories adopt technology. Some roles diminish with the introduction of new ones, and most of the existing jobs are redefined to some extent as a result. The social work perspective on this transition emphasizes the importance of supporting workers through this transition, maintaining their well-being, and ensuring fairness and equality for all. There is a complex picture emerging when it comes to realizing that industrial transformations are often accompanied by a mix of excitement and anxiety on the part of workers. Employees frequently express their excitement about being a part of a cutting-edge working environment as well as learning new high-tech skills as part of their working experience. There can be a great deal of uncertainty, fear of losing one's job, of being unable to adapt to new circumstances, or even just stress from having to master all of this at once.

ORGANIZATION OF THE BOOK

Chapter 1: The Potential of Artificial Intelligence in Manufacturing: Preventative Maintenance Explored Through the Use of Machine Learning

A special emphasis is placed in this chapter on the important topic of artificial intelligence, which enables predictive analytics to predict equipment degradation well in advance of any functional failure. Using sensor data, machine learning models refine prognostic accuracy and reduce maintenance planning uncertainty. Maintenance strategies will shift from reactive repairs to proactive asset management under this paradigm.

Chapter 2: The Human Factor in Digital Manufacturing: A Social Work Perspective on Workforce Transition and Well-Being

The digital manufacturing process redefines labor processes and amplifies the need for retraining and psychosocial support for the workforce. A social work perspective emphasizes the importance of employee agency, equitable transition pathways, and occupational mental health. As discussed in the chapter, workforce well-being is an essential component of sustainable technological adoption.

Chapter 3: Digital Innovation and Integrated Frameworks in Aerospace Manufacturing

Design, production, and lifecycle management are increasingly harmonised by interconnected digital threads in the aerospace manufacturing industry. An integrated framework unifies data flows across organizational silos, enhancing regulatory compliance and traceability. Digital architecture facilitates greater agility in an industry governed by strict safety and reliability requirements.

Chapter 4: Robotics and Manufacturing: Implications for Economic Growth in the Future

This chapter discusses the benefits of industrial robotics which include enhanced precision, repeatability, and continuous operation 24 hours a day. As they become more prevalent, they are reshaping value chains and triggering complementary innovations in the areas of materials, software, and workforce skills. The widespread adoption of robots is anticipated to alter comparative advantages, influence wage structures, and shape long-term growth patterns on a macroeconomic level.

Chapter 5: Assessing Contribution of Technology and Innovation in Sustainability: A Comprehensive Analysis of Business Sustainability in the Digital Era

Optimising resource efficiency and transparency is one way that technological innovation contributes to corporate environmental, social, and governance (ESG) outcomes. This chapter discusses sustainability as an overall concept with in-depth insights. Using digital tools, we can gain a deeper understanding of sustainability performance and make informed strategic decisions. An integrated assessment focuses on innovation as both a source of competitive advantage and a measure of it.

Chapter 6: Asia Pacific Perspectives on Digital Innovation and Integrated Frameworks in Aerospace Manufacturing

The adoption of digital manufacturing frameworks in aerospace is uneven across Asia Pacific, but it is progressing rapidly. The trajectory of digital integration is shaped by regional policy initiatives, cross-border supply chains, and indigenous technological capabilities.

Chapter 7: Cognitive Ergonomics in Smart Manufacturing: Aligning Technology with Human Capabilities

This chapter examines how information systems interact with human perception, memory, and decision-making abilities. Smart manufacturing settings benefit from the use of interfaces that are designed using cognitive principles in order to reduce operator overload and enhance situational awareness. Optimising human–machine collaboration is made possible by such alignments.

Chapter 8: The Human Engine of Industry 4.0: A Psychological Framework for Digital Transformation in Manufacturing

It is imperative that manufacturing workforce is prepared to deal with increased automation, ubiquitous data, and continuous change in order for Industry 4.0 to succeed. An examination of employee motivation, learning orientation, and adaptive resilience is undertaken within the context of digital transformation. An organization may be able to sustain ongoing innovation by integrating these factors into its change strategies.

Chapter 9: The Future of Manufacturing: Integrating Digital Innovation and Integrated Frameworks

The future of manufacturing will be characterized by the integration of design, supply, production, and service domains within a cyber-physical ecosystem. Integrated frameworks facilitate interoperability across various platforms through the use of standardised protocols. Systemic cohesion promotes real-time optimisation and rapid response to volatile market conditions.

Chapter 10: The Aerospace Industry: A Country's Developmental Framework

Industrial policy, infrastructure investment, and human capital development are aligned within a national aerospace development framework to achieve aerospace self-reliance. Research institutions, component suppliers, and regulatory agencies are usually coordinated within such frameworks. Especially in relation to the Philippines as described in the chapter, the aerospace industry is positioned to serve both as a technological spearhead and as a catalyst for the growth of the broader economy.

Chapter 11: Machine Learning and Artificial Intelligence (AI) in Manufacturing

The use of machine learning algorithms allows the detection of latent patterns in production data, facilitating the detection of early defects, enhancing quality control, and optimizing supply chains. Artificial intelligence systems continuously refine their models through feedback loops, thus achieving progressively higher levels of operational accuracy. With the integration of these technologies, manufacturing intelligence has shifted from rule-based automation to an adaptive, data-driven approach.

Chapter 12: Augmented Reality as an Efficient Tool During the Product Design Process in the Manufacturing Industry

The application of augmented reality (AR) allows for the overlaying of digital prototypes over physical environments, allowing for immersive design evaluation and rapid iteration. With real-time visualisation via augmented reality, engineering teams and stakeholders can reduce feedback loops. The result is a significant improvement in design accuracy while a significant reduction in development timelines and costs is achieved.

Chapter 13: Advancing Marketing in Society 5.0: The Strategic Role of Predictive Analytics in Consumer Behaviour Analysis

Based on the aims of Society 5.0 described in the chapter, predictive analytics will be able to identify nuanced consumer preferences through a human-centered digital ecosystem. By integrating data streams from a variety of sources, analytical models are able to predict consumption patterns at a granular level. This shift in marketing strategy from descriptive segmentation to prescriptive engagement is the result of dynamic insights into behavior.

Chapters in this book cover a wide range of topics ranging from the integration of technology and manufacturing to how these factors are driving us to create a new industrial paradigm dominated by smart, connected, and sustainable production processes. There is no doubt that it is a global phenomenon. This book contains examples and research which show that when technology integration is carried out mindfully, with integrated frameworks and a focus on people, it can lead to unprecedented productivity and innovation without compromising sustainability or social wellbeing. This can be demonstrated by the examples and research included in the chapters in this book. We are in the midst of a journey that continues to invite collaboration between disciplines and sectors across multiple disciplines. A digital manufacturing endeavor is not just an engineering endeavor in today's world. A digital manufacturing endeavor is also a human, economic, and environmental endeavor in this day and age. As long as these dimensions are kept in balance, we will be able to ensure the factories of tomorrow are indeed engines of prosperity and sustainable development on a global scale, which will benefit everyone.

Chapter 1

The Potential of Artificial Intelligence in Manufacturing: Preventative Maintenance Explored Through the Use of Machine Learning

Husmiati Yusuf
National Research and Innovation Agency, Indonesia

Sonal Sisodia
https://orcid.org/0000-0002-8558-1072
Daly College of Business Management, India

ABSTRACT

Across the globe, manufacturing is undergoing a digital transformation, and artificial intelligence (AI) has played a pivotal role in the emergence of this new trend. AI technologies are increasingly being integrated into factory operations in order to create more efficient and smarter production processes. Predictive maintenance is one of the most impactful and practical applications of artificial intelligence in manufacturing among the various applications of AI in the field. A predictive maintenance program involves the use of data-driven algorithms and models in order to identify when industrial equipment might need to be repaired or serviced, thus allowing it to be performed proactively before a breakdown occurs as a result of a malfunction. There are two approaches to equipment maintenance in the traditional sense; either you react to equipment failures after the event or you follow a predetermined service schedule regardless of the actual condition of the equipment. The

DOI: 10.4018/979-8-3373-1082-4.ch001

chapter analyzes the potential of artificial intelligence in manufacturing preventative maintenance explored

INTRODUCTION

There are a number of manufacturers around the world who have realized the potential for AI-driven predictive maintenance to increase operational efficiency in a significant way. In a manufacturing setting, unplanned downtime - when machines fail without warning - can disrupt production and lead to lost orders, hours of lost production, and increased costs as a result. Using artificial intelligence (AI) in order to forecast equipment health, companies can transform their maintenance approach from one that is reactive (fixing things only when they break) or purely preventive (routine scheduled maintenance) to one that is predictive and based on the condition of the equipment. Ucar et al. (2024) highlight the key components, trustworthiness, and future trends of artificial intelligence in predictive maintenance applications. Manufacturers can utilize this shift to minimize downtime, optimize the use of maintenance resources, as well as extend the lifespan of critical assets by utilizing this shift. The analysis done by Keleko et al. (2022) demonstrates that artificial intelligence offers an effective means of improving real-time predictive maintenance considering Industry 4.0's emergence. Daniyan et al. (2020) explore the use of artificial intelligence for predictive maintenance specifically in railcar learning factories. Wang and Wang (2017) discussed how artificial intelligence, particularly deep learning, is shaping the future of predictive maintenance. Abbas (2024) explores the use of artificial intelligence for predictive maintenance in industrial systems. Lee et al. (2019) investigate predictive maintenance for machine tool systems using artificial intelligence techniques based on machine condition data. Cardoso and Ferreira (2020) explore the application of predictive maintenance principles utilizing artificial intelligence tools. Simion et al. (2024) examine AI-driven predictive maintenance in modern maritime transport, emphasizing enhancements in operational efficiency and reliability.

Hrnjica and Softic (2020) explore explainable AI within manufacturing, emphasizing its application through a predictive maintenance case study.

Pacifico et al. (2024) discuss the application of dynamic methods and artificial intelligence approaches for effective predictive maintenance. Matzka (2020) discusses the application of explainable artificial intelligence specifically designed for predictive maintenance scenarios. Pookkuttath et al. (2021) introduce an AI-enabled predictive maintenance framework specifically designed for autonomous mobile cleaning robots. As a result of examining the current state of AI in different regions, it has become apparent that AI-based predictive maintenance is becoming a key component in modern smart factory initiatives across many sectors. There are countless manufacturing settings where artificial intelligence algorithms are used to analyze data from machines to detect subtle signs of impending issues, including automotive assembly plants, electronics fabrication facilities, chemical processing units, and countless others. A computer algorithm is able to process vast amounts of sensor readings and operational data in real time, something that would be difficult for a human operator or a threshold-based system with the same speed or accuracy to accomplish. The advantage of predictive maintenance systems lies in the fact that they provide early warnings of issues that can be proactively addressed so that maintenance teams can intervene at the right time. Samatas et al. (2021) examine predictive maintenance as a crucial intersection of artificial intelligence and the Internet of Things (IoT). Scaife (2024) provides a systematic review on enhancing predictive maintenance through the application of artificial intelligence. Bidollahkhani and Kunkel (2024) discuss how artificial intelligence is revolutionizing system reliability through advanced predictive maintenance strategies. Haque et al. (2024) conduct a systematic review examining predictive maintenance in industrial automation, specifically focusing on IoT sensor technologies and artificial intelligence algorithms. Ye et al. (2020) examine key technologies of industrial artificial intelligence, emphasizing their practical applications in predictive maintenance. Achouch et al. (2022) provide an overview of predictive maintenance within Industry 4.0, addressing key models and associated challenges. Khan et al. (2025) conduct a systematic literature review exploring generative AI-based predictive maintenance specifically in the aviation sector. Tiwari (2023) presented a transformative, action-oriented approach to manufacturing through the application of artificial intelligence and digital technologies. Çınar et al. (2020) discuss the application of machine learning for predictive maintenance, emphasizing its role in promoting sustainability within smart manufacturing in Industry 4.0. Hamdan et al. (2024) provide a review of artificial intelligence applications in predictive maintenance and energy optimization within renewable energy contexts. Walker et al. (2023) detail the application of explainable artificial intelligence technologies specifically designed for predictive maintenance.

Cui et al. (2025) discuss the future of predictive maintenance for variable frequency drives, emphasizing applications of artificial intelligence and machine learning. Pingulkar and Joshi (2021) discuss the application of artificial intelligence and data analytics for predictive maintenance in solar power plants. Ohalete et al. (2023) review advancements in predictive maintenance within the oil and gas industry, highlighting the role of artificial intelligence and data science applications. Krishnamurthy et al. (2020) present an explainable AI framework tailored for imaging-based predictive maintenance, particularly emphasizing automotive applications and potential extensions to other industries. Tiwari et al. (2025) critically reflect on understanding technology specifically within the context of manufacturing and national development. Zheng et al. (2020) discuss the progression from predictive maintenance toward intelligent maintenance enabled by artificial intelligence and the Industrial Internet of Things (IIoT). Ünlü and Söylemez (2024) describe AI-driven predictive maintenance within the broader context of engineering applications, emphasizing artificial intelligence and swarm intelligence approaches. Stohr et al. (2024) explore generative mechanisms underlying AI implementation in predictive maintenance from a critical realist perspective.

With a special focus on predictive maintenance, this book chapter takes a comprehensive look at the role of artificial intelligence in manufacturing with a special focus on manufacturing. The chapter discusses how maintenance strategies have evolved over time, the specific AI technologies enabling predictive functions, and the practical issues that occur when implementing such systems in industrial environments. Aside from emphasizing the tangible benefits realized as a result of AI-driven maintenance, it also addresses the challenges and limitations that companies face when deploying AI-driven maintenance from data hurdles to organizational and workforce considerations. As part of this article, a global analysis of industry adoption is provided, with trends across regions and sectors being highlighted. There is a focus on the future outlook, including how predictive maintenance will make its way into forms of asset management that are even more autonomous and intelligent in the future. There is a strong emphasis throughout the book on the need for a formal analysis of this important intersection between artificial intelligence and manufacturing practice, with a focus on being practical and industry-focused.

PREDICTIVE VERSUS REACTIVE MAINTENANCE STRATEGIES

The significance of AI-driven predictive maintenance can be appreciated by understanding the evolution of manufacturing maintenance strategies. Reactive maintenance, or running to failure, has traditionally been the method of choice in many factories. When reactive maintenance is employed, equipment is operated until a breakdown occurs, and then repairs are made. The downside to this approach is extended downtime, higher repair costs (since small issues can escalate to major failures), and safety risks if critical machines fail unexpectedly. Production schedules can be severely disrupted by reactive maintenance because it is inherently unplanned.

Over the course of the 20th century, manufacturers gradually shifted from reactive to preventive maintenance. In preventive maintenance, equipment is regularly inspected and serviced on a fixed schedule (i.e., a machine is serviced every month, or a part is replaced every six months) regardless of its condition. A predetermined interval for addressing issues reduces the likelihood of catastrophic failures. With preventive maintenance, some unplanned outages were reduced and planning was introduced into the maintenance process compared to reactive methods. Despite its benefits, scheduled maintenance has some limitations: it may result in over-maintenance (replacing or servicing parts that still have useful life) and it does not prevent unexpected breakdowns that occur between the fixed intervals. As a result, time-based schedules cannot adequately account for machine wear and tear under varying conditions.

Several advances in sensor technology as well as digital monitoring tools have laid the foundation for the next leap forward in condition-based maintenance and ultimately predictive maintenance. Simple machines are continuously monitored for key indicators of health and maintenance is performed when any of these indicators indicate a decline in performance or a critical threshold. Simple alarms or preset limits may be the earliest forms of condition monitoring, but they are not always accurate. Predictive maintenance builds on this concept by forecasting when a piece of equipment will need maintenance before it crosses a threshold of failure before it becomes unmanageable. As opposed to just reacting to a warning sign, a predictive system is based on historical and real-time data and projects trends into the future by estimating how long components will last or the probability that they will fail within a defined period of time.

This is where Artificial Intelligence (AI) comes into play, as it dramatically improves the accuracy and scope of these forecasts. When traditional condition monitoring systems are used, the system might send an alert if, for example, a vibration reading exceeds a threshold set by an engineer, for example. A machine learning based predictive maintenance system, on the other hand, is able to analyze complex patterns across multiple variables simultaneously and perhaps notice the early signs of a possible failure event, for instance, when a slight increase in vibration is combined with a subtle rise in the temperature of the motor. There are some correlations that can be too complex for simple rules or for humans to detect unaided when they are not aware of them. With the use of machine learning models trained on historical failure data, these systems are capable of identifying patterns that are reliably predictive of future failures.

We have moved from a reactive maintenance stance to fixing things after they break to a proactive maintenance stance. The problem was initially fixed on a preventative basis every few months, just to be safe, and then the problem was fixed on a predictive basis, based on the data indicator that it was about to break, before the problem happened. There has been a significant evolution in this field in recent years due to the availability of better data through sensor technology and automated data collection and the development of analytics and AI algorithms that are able to make sense of this data. The goal of predictive maintenance is not to totally replace preventive practices; certain routine checks and services require to be performed.

However, it enhances and refines the scheduling process in a very significant way. A major improvement in the reliability and efficiency of manufacturing operations globally can be attributed to the ability of industrial equipment to receive the exact care it needs at the best time.

PREDICTIVE MAINTENANCE POWERED BY AI

Artificial intelligence and digital technologies combine to create modern predictive maintenance. Detecting patterns that precede equipment failures is achieved by collecting and analyzing large volumes of machine data. Artificial intelligence-driven maintenance is enabled by several key technological components:

- Predictive maintenance is based on the data collected by advanced sensors and data collection systems. A variety of sensors are often installed on modern factories' machines to measure variables including vibration, temperature, pressure, sound, and electrical current. Sensors continuously monitor equipment condition. The information collected by high-frequency data acquisition systems and connectivity solutions can be aggregated for analysis. A machine generates a large volume and variety of data, which requires an effective data management system to store, process and analyze the data. In most cases, streaming data platforms and industrial databases are used to manage the data. The most sophisticated machine learning algorithms would not be able to learn from reliable data regarding equipment condition if there was no reliable data available. Investing in sensor instrumentation and data infrastructure is therefore a prerequisite to implementing the project.
- Using machine learning algorithms, once the data has been collected, ML provides the analytical engine which can be used to predict maintenance needs. The ML algorithms are applied in various ways to learn from past examples of the behavior of the equipment based on historical information. There are a variety of methods available to train supervised learning models using labeled data, such as prior instances of equipment failure which can be documented along with the sensor readings preceding the failure in order to build models. In order to recognize a failure before it happens, these models learn to recognize the combinations of conditions that tend to precede it. As soon as the model detects a similar pattern emerging in the real-time sensor data, it has the ability to flag an issue that could be imminent. Besides supervised learning approaches, unsupervised learning is also being used - particularly for detecting anomalies in the data. Models that are unsupervised (such as clustering algorithms, autoencoders, etc.) aim at establishing a baseline of

what seems to be a normal operating state for a particular machine and then detecting outliers from this norm that may indicate the development of a fault, regardless of the number of failure examples that have occurred before. The reason anomaly detection is helpful is that some critical machines have very few failures that are recorded to learn from, which means that abnormal patterns that fall outside of normal ranges can be picked up by anomaly detection.

- The use of deep learning techniques (using neural networks) can help to automatically identify intricate and nonlinear patterns in data from equipment that might be missed by simpler methods based on data. With deep learning models, it is possible to detect subtle precursors to failure by analyzing multiple variables simultaneously and learning from large datasets. There is evidence that this approach can improve prediction accuracy for complex systems in which indicators of degradation may involve interdependent factors or subtle changes in the signals that may not trigger simple thresholds.
- The concept of a digital twin is to create a virtual representation of a physical asset that runs in parallel with the physical asset in the real world. A digital twin simulates the actual state of a machine based on the inputs from real-time sensors and sophisticated simulations. This type of technology is used in predictive maintenance for assisting engineers in analyzing question arises and gaining a deeper understanding of failure mechanisms. By simulating how a machine will behave under different conditions and stresses, engineers are able to identify potential failure points before they even occur in the real world. This allows for the safe testing of maintenance strategies and helps in identifying the underlying causes of performance issues without interrupting actual operational processes through the use of digital twins.
- In the case of AI-driven maintenance, the computational demands can be significant, and it is crucial to have connectivity in order to carry out maintenance tasks. Using cloud computing, you can process and store the huge streams of data from multiple machines and perform intensive tasks like training complex models with the vast amounts of data available in the cloud. With edge computing, AI models are able to run on local devices closer to how the equipment is operated, which in turn means that computations can be carried out closer to where the equipment is located where it is used. A benefit of this is that it reduces latency and enables anomalies to be detected as soon as possible. Initial model development and heavy data analysis often occur in the cloud, and then the trained models are deployed to edge devices for monitoring and alerting to be taken place in real-time. Manufacturers balance the depth of analysis with the speed of response by leveraging both the cloud and the edge.

- As artificial intelligence (AI) becomes more and more applicable to diagnostics and decision support, it can be used in a variety of ways. Aside from predicting failures, AI can be used to diagnose problems and recommend maintenance actions edging close to what is called prescriptive maintenance. AI systems and predictive models have taken advantage of the ability to pinpoint specific bearings as a potential cause of motor failure so they can recommend replacing the bearings, or adjusting the operating conditions to extend the life of the motor until the problem can be solved. A predictive model and domain knowledge or an optimization algorithm are combined in order to generate these kinds of recommendations. With AI-driven decision support, maintenance planners are provided with suggested actions and priorities, allowing them to deal with predicted issues effectively and with minimal cost.

There is a suite of AI and digital technologies that underpin predictive maintenance in the manufacturing industry today. With the collection of data through sensors, the development of sophisticated algorithms to learn failure patterns, as well as the application of virtual simulations of the equipment and advances in computing infrastructure, manufacturers are now able to predict problems before they occur. With the help of artificial intelligence, this system will be able to interpret the data in a way that would overwhelm human analysts, and will be able to turn that information into actionable maintenance intelligence. As these technologies continue to mature, they are becoming more accessible and affordable, which allows the industry at large to adopt predictive maintenance strategies in even a broader fashion.

PREDICTIVE MAINTENANCE BASED ON ARTIFICIAL INTELLIGENCE IN MANUFACTURING

An AI-based predictive maintenance program must be carefully planned and coordinated between the plant operations and technology teams before it can be deployed. A multifaceted process involves the deployment of hardware, the development of software, and the transformation of the organization. Taking these steps and considering the following factors will make it easier for you to put them into practice.

- Decide what equipment should be prioritized and what the objectives should be. There is not a machine in a factory that is going to need or justify an AI-driven predictive maintenance system from the start. There is a typical procedure where companies determine which assets are critical assets that would cause significant downtime, safety risks, or excessive costs if they failed. Equipment that is prone to causing unplanned stops can include expensive production machinery, bottleneck machines in a production line, or equipment that has a history of causing unplanned stops in the past. This is the time when clear objectives should be set in order to achieve them. The goal of this approach is simply to make sure that efforts are being directed in the direction of the greatest benefit.

- After the equipment has been chosen, it will need to be properly instrumented and a data infrastructure set up for the equipment. Depending on the machine, additional sensors may be installed or existing sensors may be calibrated and accessible. A storage and processing system must be able to collect and process data from machines in real time or near-real time. This may involve integrating with existing industrial control systems and data historians, or setting up new data acquisition systems. A reliable connection on the factory floor ensures that sensor data flows continuously to the analysis area. Data lakes or operational data stores are often built during this phase to store and analyze historical sensor data.
- After sensors and data pipelines have been activated, there will often be a period of data collection that will allow sufficient information to be gathered for analysis and model training to be carried out. There are many situations in which historical maintenance logs and failure records are also consolidated during this stage of the process. Labeling historical data with events of interest is an important part of labeling historical data. An experienced engineer or maintenance personnel with experience in identifying faults or anomalies in equipment is likely to have some input into this process since they understand what constitutes a fault. The quality and depth of the data and the accuracy of the labels significantly affect the success of the AI models which will be developed in the future. The teams often conduct exploratory data analysis as well, so as to determine whether any obvious trends or precursors to a failure are present in the data.
- Unless a prediction model is integrated into the operations of a company, its insights have little value unless they are communicated timely and usefully to the people who will act on the predictions. This is why it is imperative that the operational workflow is integrated into the process. This can be accomplished by connecting the AI system to the factory's maintenance management software or alert system in order to sync up the data. When a model predicts that there may be a robot failure within the next week, for example, the system could automatically create a maintenance work order or send an alert to the maintenance manager's dashboard if a robot is likely to fail in the next week. Various management information systems or other platforms have been implemented in many factories as a way to track maintenance activities;

the use of AI predictions can be integrated into these systems to gain actionable insight. Early implementations of Artificial Intelligence tend to work on a decision support mode, which means that the AI provides recommendations and the human experts will make the final scheduling decisions, thus establishing trust in the system's outputs in the long run.

- The management and training of the workforce is a key component of change management. The implementation of Artificial Intelligence in maintenance is not just a technical endeavor; it is also a cultural change on the factory floor that needs to be made. The maintenance technicians and engineers need to be trained on how to interpret the new predictive alerts and how to incorporate them into their daily routines in order to make them as effective as possible. The computer's suggestions may initially be controversial, since experienced staff may be more likely to trust their intuitions or past experiences than a machine's. This can be achieved by involving these stakeholders in the early stages of the process, for example by having them participate in the validation of the predictions of the AI system. When you demonstrate a few early wins, you can help build buy-in among your team members. There can also be training sessions which can teach the workforce how the system works so they don't need to be data scientists to understand what the AI looks at and how to react to its alerts. As a result, they can be properly trained in the fundamentals of how the system works. As a result of these changes, companies may need to redefine some job roles and are in the process of looking for better ways to do so: for instance, maintenance planners could shift from purely calendar-based scheduling to monitoring a dashboard of machine health insights. Creating a collaborative environment, where human expertise and AI insights complement each other, is one of the keys to success when making use of AI.
- The deployment of a pilot project, scaling it up, and continuously improving the business process as the project grows. The majority of companies will begin by doing a pilot project which will involve a small set of equipment or just one production line as a starting point. A predictive maintenance system that uses this approach is allowed to be validated and tested in real operating conditions, as well as provide a proof of concept for the project team. If feedback is received during the pilot stage, then the model may change, sensors may be repositioned, and alert thresholds may need to be adjusted according to feedback. The pilot will have the opportunity to be expanded once the results of the pilot show that it is a valuable tool, such as measurable reductions in maintenance costs and downtime, for example. Organizations that are scaling will need a robust data architecture to handle the greater volume of data as well as the increased need for analysis. Some organizations have

set up centralized analytics teams to ensure best practices are shared across sites and enable deployment across sites. A system that is deployed needs to be continuously improved as it receives new data or operating conditions change. These retrainings are necessary as new data is received or operating conditions change, and each maintenance outcome has the potential to refine future predictions.

TRENDS AND ADOPTION IN GLOBAL SETTINGS

Predictive maintenance powered by AI is becoming increasingly important to modern manufacturing strategies worldwide. These technologies are being explored or implemented by manufacturers across virtually every industrial region, though their adoption levels differ according to country and industry. Large, multinational companies have often led the way with early implementations, while small and midsized manufacturers have been following as the technology becomes more accessible and reliable.

The adoption momentum is building. The interest in AI for maintenance has been extremely high according to industry surveys. AI and predictive maintenance have become key priorities for manufacturing decision-makers globally in recent years. Despite this enthusiasm, actual implementation has lagged behind. It has been shown that businesses are still in transitional phases when it comes to predictive maintenance, even though they recognize the value of this type of maintenance, and are undertaking pilot projects, building internal capabilities, and trying to secure the necessary investments. Within the next few years, a significant proportion of firms expect to move from traditional maintenance to proactive or predictive maintenance based on the momentum moving forward.

There is an increase in regional patterns of development for the use of artificial intelligence in predictive maintenance. Geographically, North America, Singapore, Japan, and Europe have shown a strong early adoption of artificial intelligence in the manufacturing sector. To support this adoption, advanced tech ecosystems have been built and substantial investments have been made in digital industry initiatives in these regions, too. Many of the early success stories of the use of AI in factory settings resulted from these regions and in general, these regions have the infrastructure and expertise necessary for the integration of such solutions at a large scale. A number of North American manufacturers cite the need to reduce skilled labor shortages and high labor costs as the primary motivation for adopting artificial intelligence (AI) for maintenance and other automation functions as motivation for adopting AI. As part of the Industrie 4.0 program in Germany, European industries are focused on enhancing efficiency, reliability, and global competitiveness while ensuring strict safety and quality standards in order to enhance efficiency, reliability, and global competitiveness.

The Asia-Pacific region has also emerged as a powerhouse in terms of AI-driven manufacturing in the past couple of years. Many countries, including China, Japan, and South Korea, have national strategies that are devoted to the development of artificial intelligence and smart manufacturing, along with significant funding. As part of the modernization process in China's extensive manufacturing sector, there has been a concerted effort to leapfrog older maintenance practices and apply advanced analytics and artificial intelligence to improve plant efficiency. The result of this is that we are seeing very high growth rates in adoption across Asia with many new factories being built with sensors and AI systems already embedded from day one. Predictive maintenance adoption is expected to see the fastest growth in the Asia-Pacific region over the next decade, as a result of both the scale of manufacturing in the region, as well as the aggressive investments that are being made in technology. These solutions are also being explored in other regions like the Middle East and Latin America, focusing mainly on industries such as oil and gas or mining, where predictive maintenance can play a particularly significant role due to the remote nature of operations and their capital-intensive nature.

There are a number of industry sectors that are setting the pace. Depending on the industry sector, adoption rates can also vary. Those industries that require a high level of capital investment, or where downtime can be extremely costly, are natural early adopters of the use of artificial intelligence in maintenance. There are several industries that are considered to be among these, including aerospace, where the reliability and safety of equipment is of utmost importance. The automotive industry has high-speed assembly lines that are designed to have maximum uptime, the heavy-process industries such as petrochemicals and steel, as well as the advanced electronics manufacturing industry that uses highly sensitive, expensive equipment. Many of these industries have the scale and resources to invest in sophisticated predictive maintenance programs and therefore stand to gain significantly in terms of cost savings and reduced risks when they invest in these programs. Some smaller-scale sectors, however, or those with less expensive equipment, have taken longer to adopt the technology, either because the economic case for doing so isn't immediately apparent or because they lack in-house expertise in digital technology. The adoption rate of solutions in those sectors is expected to grow as more and more turnkey solutions become available and success stories emerge from these sectors.

There is a widespread trend in the industry that AI-driven predictive maintenance is moving from a pilot stage to a mainstream adoption phase. There has been a wide variation in the pace of adoption between countries and regions; while advanced economies and certain industries are leading the way, even those sectors that have been lagging are beginning to recognize the opportunity that this represents and are investing. Global manufacturing communities increasingly view predictive maintenance as both a necessary evolution and a key component of their ongoing competitiveness in terms of efficiency, reliability, and operational resilience, and not just as an option that can be experimented with. Predictive maintenance is poised to become a standard best practice in factories around the world in the near future as more real-world results are demonstrated and technology barriers continue to fall, much like automating processes and using lean techniques did in years gone by.

OUTLOOK FOR THE FUTURE

As manufacturing maintenance continues to evolve, AI will continue to play an increasingly important role. Both technology and industry practices are evolving, and predictive maintenance is expected to become smarter, more autonomous, and more deeply integrated into every aspect of operations. Below are a few possible future directions.

Prescriptive maintenance is a step in the right direction. Predictive maintenance has evolved into prescriptive maintenance in a clear evolutionary path. While predictive maintenance is focused on predicting what might go wrong, telling when problems might occur, and when they will occur. Prescriptive maintenance goes a step further than routine maintenance in that it recommends and even initiates solutions in order to deal with impending issues that may arise. The use of artificial intelligence will likely improve in the coming years, not only in identifying that a failure is likely to occur, but also in suggesting critical actions that should be taken in order to prevent it from happening. Using artificial intelligence, a maintenance schedule can be generated automatically that minimizes downtime by matching necessary repairs to production gaps in order to minimize unnecessary downtime. Equipments are already showing early signs of this process as some systems, upon predicting a potential failure, will recommend ordering a specific replacement part or adjusting operating parameters to extend the equipment life until the next scheduled maintenance window arrives. There is a possibility that, with advancements in robotics and automation, certain routine maintenance tasks might even be carried out autonomously, for example, robots could be used to perform inspections or replace parts on a schedule guided by AI insights, with little or no human intervention required.

Incorporation into a holistic manufacturing management system. A predictive maintenance system will be treated in the future as part of a more comprehensive digital ecosystem and not merely as a standalone tool. Machine health monitoring data will be integrated with other enterprise systems such as planning, inventory management, and supply chain logistics in order to provide a comprehensive view of machine health. The production plans may automatically adjust if there is a chance that a machine will need maintenance in a few days, and spare parts may be ordered ahead of time if the maintenance is predicted to occur. This seamless integration ensures that maintenance intelligence contributes to the overall operations decision-making process, ensuring that the whole manufacturing process is more efficient and adaptive than ever before.

There have been significant advancements in AI capabilities in the past few years. The technological front, AI methods are continuously improving, and these advancements will lead to better predictive maintenance in the future. Among the trends of the current century is the development of algorithms that are able to do more with less data. Techniques such as transfer learning and federated learning might allow models to leverage knowledge from one context to another without requiring huge local datasets, helping to address the problem of limited failure examples. There is also a possibility that hybrid models will become increasingly prevalent in the future, which combine physics-based understanding of machines with data-driven AI to provide both accuracy and interpretability in predictions. The possibility of performing more complex analyses on factory floor could increase as computing power increases and specialized AI hardware is deployed more often. Using artificial intelligence, equipment might be continuously analyzed to detect issues that would be invisible to a traditional sensor if it continuously analyzed the sound of motors or thermal images from equipment. A further benefit of future AI systems may be the possibility of integrating explainability features into them, allowing engineers to have an easier time understanding the basis for a prediction. With all these improvements, it is anticipated that predictive maintenance tools will become even more powerful, easier to use, and more trustworthy as a result.

There will be a continuing evolution in the relationship between humans and artificial intelligence in the field of maintenance in the future. AI is poised to make its mark on maintenance personnel rather than replace them, perhaps even as a standard tool in their arsenal. There is a strong possibility that the role of a maintenance technician is going to evolve more and more towards managing and interpreting predictive systems, planning interventions, and handling complex mechanical fixes that automated systems cannot yet handle. A new set of job roles may emerge over the coming years, such as reliability data analysts or artificial intelligence maintenance specialists, who will bridge the gap between maintenance departments and the data science and IT departments. There is a chance that the next generation of technicians will be able to comfortably work with AI-driven systems as part of their training and education in industrial maintenance, so it will be important to incorporate more digital and analytical skills in the training. As the repetitive monitoring tasks are performed by AI over time, human experts will be able to focus more on strategic decision-making, improvement initiatives, and oversight as repetitive monitoring tasks are handled by AI. A combination of human experience and artificial intelligence can lead to unprecedented levels of reliability and productivity. The AI provides constant monitoring and pattern recognition, while humans provide contextual understanding, creativity, and judgment to enable these levels of productivity.

Making a contribution to the sustainability and resilience of manufacturing. Predictive maintenance will also be an important component of manufacturing in the future, both regarding sustainability and resilience. The goal of predictive maintenance is to minimize waste and optimize energy consumption in order to achieve environmental sustainability. Regulators and industry standards could in the future become more inclined to encourage or even require advanced monitoring of critical equipment in order to prevent accidents and reduce waste, especially in sectors having a high level of risk to the environment. A factory that is equipped with predictive maintenance is better prepared to deal with disruptions in its production in terms of resilience. There are less sudden breakdowns occurring on their machines, which means they can maintain steady production and contribute less to delays in the supply chain as a result. Due to the fact that supply chains are increasingly becoming globally interconnected, the ability to have highly reliable operations becomes a competitive advantage and a selling point to potential business partners. There is no doubt that over the long term, predictive maintenance will play a large role in achieving a manufacturing landscape that is not only more efficient, but also more adaptable to change and gentler on the environment thanks to AI-driven practices.

CONCLUSION

Manufacturing maintenance is on a trajectory that is pointing towards the use of increasing intelligence, more integration, and a broader deployment of AI. The future of manufacturing can be envisioned in a scenario where no unplanned downtime will occur, where maintenance is so well optimized that it becomes nearly invisible and machines will just keep running and be serviced at precisely predetermined intervals without any interruption. As we move towards a future where every critical piece of equipment is self-monitoring and reports its health on a continuous basis. It will be very difficult to achieve this goal, but based on the advances in technology that have been made in the last few years, it is clear that each step from reactive to predictive, prescriptive to autonomous maintenance will require innovation and effort. Global manufacturing is becoming increasingly accessible to the global market as time goes by. There has been a transformational shift in the manufacturing industry as a result of artificial intelligence. Predictive maintenance is one of the most promising applications of this technology, which demonstrates its promise in terms of a practical, industry-focused application. The use of data and advanced algorithms can enable manufacturers to move from being reactive to machine failures to being proactive and preventing them by utilizing data and advanced algorithms. Improvements are brought about through this transition in a tangible way. The reliability of production assets improves, downtime is reduced, maintenance costs are reduced, and the safety of the operations is enhanced when downtime and ineffective maintenance are prevented. The use of artificial intelligence (AI) for predictive maintenance is a compelling example of how digital tools are able to advance the long-standing industrial goals of efficiency, quality, and uptime. The process of implementing such capabilities is quite complex and requires a great deal of effort. The maintenance of any facility involves not only the use of new technology but also the use of new skills, processes, and ways of thinking about maintenance. Data, analytics integration, and change management challenges must be overcome in order to make the project successful. Despite the effort involved, there are plenty of benefits as evidenced by early adopters around the world. A manufacturing operation that is able to increase asset availability, gain insights into equipment behavior, and be able to perform maintenance as a planned activity rather than an emergency response is better positioned for competition and a more resilient operation. With AI providing manufacturers with the ability to predict maintenance, there is a good chance they are going to have an edge over their competitors on the global market.

In the future manufacturers will be able to respond more quickly to customer demands and experience fewer disruptions as well as lower costs. With predictive maintenance as a central theme in manufacturing, the role of artificial intelligence is fundamentally aimed at empowering decision-makers by providing them with better information and by providing foresight to the engineers on the front line. This is the beginning of a shift away from factory systems that are simply automated, to factories that are truly intelligent as well. A system that can learn from data, predict how they need to operate, and continuously improve their efficiency by analyzing data. The manufacturing industry is moving into a new era of unparalleled reliability and efficiency as technology continues to advance and these practices become standard. By minimizing downtime and identifying opportunities for improvement as soon as they arise, downtime is minimized. The use of artificial intelligence within predictive maintenance is both a key enabler of this future as well as a prime example of the positive effects that artificial intelligence on the industrial sector are having today.

REFERENCES

Abbas, A. (2024). AI for predictive maintenance in industrial systems. *International Journal of Advanced Engineering Technologies and Innovations*, *1*(1), 31–51. DOI: 10.31219/osf.io/vq8zg

Achouch, M., Dimitrova, M., Ziane, K., Sattarpanah Karganroudi, S., Dhouib, R., Ibrahim, H., & Adda, M. (2022). On predictive maintenance in industry 4.0: Overview, models, and challenges. *Applied Sciences (Basel, Switzerland)*, *12*(16), 8081. DOI: 10.3390/app12168081

Bidollahkhani, M., & Kunkel, J. M. (2024). Revolutionizing system reliability: The role of AI in predictive maintenance strategies. arXiv preprint arXiv:2404.13454.

Cardoso, D., & Ferreira, L. (2020). Application of predictive maintenance concepts using artificial intelligence tools. *Applied Sciences (Basel, Switzerland)*, *11*(1), 18. DOI: 10.3390/app11010018

Çınar, Z. M., Abdussalam Nuhu, A., Zeeshan, Q., Korhan, O., Asmael, M., & Safaei, B. (2020). Machine learning in predictive maintenance towards sustainable smart manufacturing in industry 4.0. *Sustainability (Basel)*, *12*(19), 8211. DOI: 10.3390/su12198211

Cui, Y., Butler, R., Liu, Z., Jaremko, T., & Hinds, S. (2025). The Future of Predictive Maintenance for Variable Frequency Drives Based on AI/ML. *IEEE Transactions on Industry Applications*, 1–9. DOI: 10.1109/TIA.2025.3550104

Daniyan, I., Mpofu, K., Oyesola, M., Ramatsetse, B., & Adeodu, A. (2020). Artificial intelligence for predictive maintenance in the railcar learning factories. *Procedia Manufacturing*, *45*, 13–18. DOI: 10.1016/j.promfg.2020.04.032

Hamdan, A., Ibekwe, K. I., Ilojianya, V. I., Sonko, S., & Etukudoh, E. A. (2024). AI in renewable energy: A review of predictive maintenance and energy optimization. *International Journal of Science and Research Archive*, *11*(1), 718–729. DOI: 10.30574/ijsra.2024.11.1.0112

Haque, R., Bajwa, A., Siddiqui, N. A., & Ahmed, I. (2024). Predictive Maintenance In Industrial Automation: A Systematic Review Of IOT Sensor Technologies And AI Algorithms. *American Journal of Interdisciplinary Studies, 5*(01), 01-30.

Hrnjica, B., & Softic, S. (2020, August). Explainable AI in manufacturing: a predictive maintenance case study. In *IFIP international conference on advances in production management systems* (pp. 66–73). Springer International Publishing. DOI: 10.1007/978-3-030-57997-5_8

Keleko, A. T., Kamsu-Foguem, B., Ngouna, R. H., & Tongne, A. (2022). Artificial intelligence and real-time predictive maintenance in industry 4.0: A bibliometric analysis. *AI and Ethics*, *2*(4), 553–577. DOI: 10.1007/s43681-021-00132-6

Khan, Z. U., Nasim, B., & Rasheed, Z. (2025). Generative AI-based predictive maintenance in aviation: A systematic literature review. *CEAS Aeronautical Journal*, *16*(2), 1–19. DOI: 10.1007/s13272-025-00818-1

Krishnamurthy, V., Nezafati, K., Stayton, E., & Singh, V. (2020). Explainable AI framework for imaging-based predictive maintenance for automotive applications and beyond. *Data-Enabled Discovery and Applications*, *4*(1), 7. DOI: 10.1007/s41688-020-00042-2

Lee, W. J., Wu, H., Yun, H., Kim, H., Jun, M. B., & Sutherland, J. W. (2019). Predictive maintenance of machine tool systems using artificial intelligence techniques applied to machine condition data. *Procedia CIRP*, *80*, 506–511. DOI: 10.1016/j.procir.2018.12.019

Matzka, S. (2020, September). Explainable artificial intelligence for predictive maintenance applications. In *2020 third international conference on artificial intelligence for industries (ai4i)* (pp. 69-74). IEEE. DOI: 10.1109/AI4I49448.2020.00023

Ohalete, N. C., Aderibigbe, A. O., Ani, E. C., Ohenhen, P. E., & Akinoso, A. (2023). Advancements in predictive maintenance in the oil and gas industry: A review of AI and data science applications. *World J. Adv. Res. Rev*, *20*(3), 167–181. DOI: 10.30574/wjarr.2023.20.3.2432

Pacifico, M. G., Marchiano, G., De Medici, S., & Novellino, A. (2024, June). Application of Dynamic and AI Approaches for Predictive Maintenance. In *2024 9th International Conference on Smart and Sustainable Technologies (SpliTech)* (pp. 1-6). IEEE. DOI: 10.23919/SpliTech61897.2024.10612544

Pingulkar, C., & Joshi, A. (2021). AI and Data Analytics for Predictive Maintenance in Solar Power Plants. [IJPREMS]. *International Journal of Progressive Research in Engineering Management and Science*, *1*, 52–69.

Pookkuttath, S., Rajesh Elara, M., Sivanantham, V., & Ramalingam, B. (2021). AI-enabled predictive maintenance framework for autonomous mobile cleaning robots. *Sensors (Basel)*, *22*(1), 13. DOI: 10.3390/s22010013 PMID: 35009556

Samatas, G. G., Moumgiakmas, S. S., & Papakostas, G. A. (2021, May). Predictive maintenance-bridging artificial intelligence and iot. In *2021 IEEE World AI IoT Congress (AIIoT)* (pp. 0413-0419). IEEE. DOI: 10.1109/AIIoT52608.2021.9454173

Scaife, A. D. (2024). Improve predictive maintenance through the application of artificial intelligence: A systematic review. *Results in Engineering*, *21*, 101645. DOI: 10.1016/j.rineng.2023.101645

Simion, D., Postolache, F., Fleacă, B., & Fleacă, E. (2024). Ai-driven predictive maintenance in modern maritime transport—Enhancing operational efficiency and reliability. *Applied Sciences (Basel, Switzerland)*, *14*(20), 9439. DOI: 10.3390/app14209439

Stohr, A., Ollig, P., Keller, R., & Rieger, A. (2024). Generative mechanisms of AI implementation: A critical realist perspective on predictive maintenance. *Information and Organization*, *34*(2), 100503. DOI: 10.1016/j.infoandorg.2024.100503

Tiwari, S. P. (2023, November). A transformative and action plan-oriented approach to manufacturing using artificial intelligence and digital technologies. In *BISTIC business innovation sustainability and technology international conference (BISTIC 2023)* (pp. 22-31). Atlantis Press. DOI: 10.2991/978-94-6463-302-3_4

Tiwari, S. P., Kostenko, O., & Yekhanurov, Y. (2025). *Understanding technology in the context of national development: Critical reflections*. SciFormat Publishing Inc. DOI: 10.69635/978-1-0690482-3-3

Ucar, A., Karakose, M., & Kırımça, N. (2024). Artificial intelligence for predictive maintenance applications: Key components, trustworthiness, and future trends. *Applied Sciences (Basel, Switzerland)*, *14*(2), 898. DOI: 10.3390/app14020898

Ünlü, R., & Söylemez, İ. (2024). AI-Driven Predictive Maintenance. In *Engineering Applications of AI and Swarm Intelligence* (pp. 207–233). Springer Nature Singapore.

Walker, C. M., Agarwal, V., Lin, L., Hall, A. C., Hill, R. A., Mortenson, T. J., & Lybeck, N. J. (2023). *Explainable artificial intelligence technology for predictive maintenance* (No. INL/RPT-23-74159-Rev000). Idaho National Laboratory (INL), Idaho Falls, ID (United States).

Wang, K., & Wang, Y. (2017, September). How AI affects the future predictive maintenance: a primer of deep learning. In *International workshop of advanced manufacturing and automation* (pp. 1-9). Singapore: Springer Singapore.

Ye, Y., Yong, Z., & Han, D. (2020). Research on key technology of industrial artificial intelligence and its application in predictive maintenance. *Acta Automatica Sinica*, *46*(10), 2013–2030.

Zheng, H., Paiva, A. R., & Gurciullo, C. S. "Advancing from predictive maintenance to intelligent maintenance with ai and iiot." *arXiv preprint arXiv:2009.00351* (2020).

Chapter 2
The Human Factor in Digital Manufacturing: A Social Work Perspective on Workforce Transition and Well-Being

Husmiati Yusuf
National Research and Innovation Agency, Indonesia

ABSTRACT

The digital transformation of manufacturing industries presents unprecedented opportunities for innovation and efficiency, yet it simultaneously creates significant challenges for workforce adaptation and well-being. This chapter examines the human dimensions of digital manufacturing through a social work lens, emphasizing the critical need for comprehensive support systems during technological transitions. Drawing from social work theories and frameworks, this analysis explores how person-in-environment perspectives, systems theory, and strengths-based approaches can inform more humane and effective strategies for workforce development in the digital age. The chapter presents practical interventions, policy recommendations, and integrated frameworks that prioritize worker well-being while supporting organizational transformation. By centering human needs within technological advancement, this work contributes to a more holistic understanding of sustainable digital manufacturing that benefits both organizations and communities.

DOI: 10.4018/979-8-3373-1082-4.ch002

1. INTRODUCTION

The Fourth Industrial Revolution has fundamentally altered the landscape of manufacturing, introducing artificial intelligence, robotics, Internet of Things (IoT), and advanced automation systems that promise unprecedented levels of efficiency and innovation (Schwab, 2016). However, beneath the technological optimism lies a complex web of human challenges that demand urgent attention. As manufacturing organizations worldwide embrace digital transformation, millions of workers face uncertainty about their roles, required skills, and future employment prospects (Manyika et al., 2017).

The COVID-19 pandemic has accelerated digital transformation across industries, with manufacturing experiencing particularly rapid changes as companies sought to maintain operations amid global supply chain disruptions and health safety requirements (Chen et al., 2021). This acceleration has intensified both the opportunities and challenges associated with digital transformation, creating an even more urgent need for comprehensive approaches that address human factors alongside technological implementation (Kumar et al., 2022).

Recent developments in artificial intelligence and machine learning have introduced new dimensions to digital manufacturing transformation, with predictive analytics, autonomous systems, and human-robot collaboration becoming increasingly prevalent (Zhang et al., 2023). These technologies offer unprecedented opportunities for productivity improvement and quality enhancement, but they also create new forms of worker displacement and skill obsolescence that require careful consideration and proactive intervention (Rodriguez & Thompson, 2024).

Traditional approaches to technological implementation often prioritize technical specifications and operational efficiency while treating human factors as secondary considerations (Kane et al., 2019). This chapter argues for a paradigm shift that places human well-being at the center of digital manufacturing transformation. By applying social work principles and methodologies, organizations can develop more comprehensive, ethical, and ultimately successful approaches to workforce transition.

The social work profession's commitment to human dignity, social justice, and systemic change provides a unique perspective on the challenges and opportunities inherent in digital manufacturing evolution (National Association of Social Workers, 2021). Social workers understand that technological change is not merely a technical process but a profound social transformation that affects individuals, families, communities, and entire economic systems (Hudson, 2016). This understanding is crucial for developing interventions that support not only successful technology adoption but also human flourishing during periods of significant change.

Contemporary research has increasingly recognized the importance of human-centered approaches to digital transformation, with studies demonstrating that organizations that prioritize worker well-being and engagement achieve superior outcomes in both technological implementation and business performance (Martinez et al., 2023). The integration of social work perspectives into digital manufacturing transformation represents a critical evolution in how organizations approach technological change, moving beyond purely technical considerations to embrace comprehensive human development approaches.

2. THEORETICAL FOUNDATIONS: SOCIAL WORK PERSPECTIVES ON TECHNOLOGICAL CHANGE

2.1 Person-in-Environment Framework

The person-in-environment (PIE) framework, fundamental to social work practice, provides essential insights into how individuals experience and adapt to technological change in manufacturing settings (Germain & Gitterman, 2017). This ecological perspective recognizes that worker responses to digital transformation are shaped by multiple interacting systems: individual characteristics, family dynamics, workplace culture, community resources, and broader economic and political contexts (Bronfenbrenner, 1979).

Recent applications of the PIE framework to digital transformation contexts have revealed additional layers of complexity, particularly related to digital literacy, generational differences in technology adoption, and the intersection of technological change with existing social and economic inequalities (Johnson et al., 2022). Understanding these intersections is crucial for developing effective interventions that address both immediate technological adaptation needs and broader social justice concerns.

From a PIE perspective, a manufacturing worker's ability to successfully navigate digital transformation depends not only on their technical aptitude but also on factors such as access to training resources, family support for career transitions, community economic stability, and organizational culture regarding change management (Van Wormer & Besthorn, 2018). This holistic view challenges simplistic assumptions that worker resistance to technology stems solely from individual deficits and instead examines systemic barriers and facilitators to successful adaptation.

The framework has proven particularly valuable in understanding how different demographic groups experience digital transformation, with research showing that older workers, workers with limited formal education, and workers from economically disadvantaged backgrounds face distinct challenges that require tailored intervention approaches (Williams & Davis, 2023). These findings emphasize the importance of individualized assessment and intervention planning that considers the full range of environmental factors affecting worker adaptation.

2.2 Systems Theory Application

Systems theory, another cornerstone of social work practice, offers valuable insights into the interconnected nature of digital manufacturing transformation (Payne, 2014). Manufacturing organizations exist within complex systems that include suppliers, customers, regulatory bodies, educational institutions, and local communities. Changes in one part of the system inevitably affect other components, creating ripple effects that extend far beyond the factory floor (Turner, 2017).

Recent systems analysis of digital manufacturing transformation has revealed the emergence of new stakeholder relationships and dependencies, particularly related to technology vendors, cybersecurity providers, and data analytics specialists (Anderson et al., 2021). These new relationships create additional complexity in system dynamics and require expanded approaches to stakeholder engagement and coordination.

A systems perspective reveals that successful digital transformation requires coordinated interventions across multiple levels. Individual worker training must be complemented by organizational culture change, community workforce development initiatives, and policy frameworks that support economic transition (Thompson, 2015). This understanding guides social work interventions toward comprehensive, multi-level approaches that address both immediate needs and long-term systemic changes.

The COVID-19 pandemic has provided additional insights into system resilience and adaptation, demonstrating how manufacturing systems can rapidly reconfigure in response to external pressures while highlighting vulnerabilities in traditional approaches to workforce development and support (Brown et al., 2022). These lessons inform current approaches to digital transformation planning and implementation.

2.3 Strengths-Based Approach

The strengths-based approach, emphasizing individual, family, and community assets rather than deficits, offers a powerful alternative to problem-focused models of workforce development (Saleebey, 2012). Rather than viewing experienced manufacturing workers as obstacles to digital transformation, this perspective recognizes their deep understanding of production processes, problem-solving skills, and institutional knowledge as valuable assets that can enhance technological implementation (Lee, 2001).

Contemporary applications of strengths-based approaches to digital transformation have revealed innovative ways to leverage existing worker capabilities while building new competencies. For example, experienced workers' troubleshooting skills often translate directly to identifying and resolving issues with digital systems, while their understanding of production flow can inform optimal technology configuration (Garcia & Smith, 2023).

This approach encourages organizations to build upon existing worker capabilities while providing targeted support for skill development. It also recognizes that workers' informal networks, mentoring relationships, and collaborative problem-solving approaches represent significant organizational resources that should be preserved and leveraged during digital transformation (Teater, 2019).

Recent research has demonstrated that strengths-based approaches to digital transformation not only improve worker outcomes but also enhance organizational innovation capacity, as workers who feel valued and supported are more likely to contribute creative solutions and identify improvement opportunities (Peterson & Liu, 2024). This finding underscores the business case for human-centered approaches to technological change.

2.4 Trauma-Informed Practice Framework

The trauma-informed practice framework has emerged as a particularly relevant theoretical foundation for understanding worker responses to digital transformation, especially in contexts where technological change has resulted in job displacement or significant workplace disruption (Miller et al., 2022). This framework recognizes that technological change can be experienced as traumatic, particularly when it threatens economic security or professional identity.

Trauma-informed approaches to digital transformation emphasize creating safe, trustworthy environments for learning and adaptation, recognizing that workers may need time and support to process the emotional impacts of technological change before they can effectively engage in skill development activities (Thompson & Rivera, 2023). This perspective has proven particularly valuable in organizations where previous technological implementations have resulted in job losses or negative worker experiences.

The framework also emphasizes the importance of choice and collaboration in intervention planning, recognizing that workers who have experienced technological displacement may need opportunities to regain a sense of control and agency in their career development (Davis et al., 2024). This approach aligns with social work values of self-determination and empowerment while addressing the practical needs of workforce transition.

3. THE HUMAN IMPACT OF DIGITAL MANUFACTURING TRANSFORMATION

3.1 Individual-Level Impacts

Digital manufacturing transformation affects workers at multiple levels, beginning with individual psychological and behavioral responses to technological change. Research in occupational psychology and social work practice reveals common patterns of response to significant workplace transformation, including anxiety about job security, concerns about skill obsolescence, and stress related to learning new technologies (Frey & Osborne, 2017; Autor, 2015).

Recent studies have identified additional psychological impacts of digital transformation, including "digital fatigue" among workers who must constantly adapt to new technologies, and "surveillance anxiety" related to increased monitoring and data collection capabilities of digital systems (Roberts et al., 2023). These findings highlight the need for comprehensive approaches to supporting worker mental health during technological transitions.

Many manufacturing workers, particularly those with decades of experience in traditional production methods, report feelings of inadequacy when confronted with digital systems that require different skill sets (Deming, 2017). These emotional responses are not merely individual reactions but reflect broader social and economic insecurities about the changing nature of work in the 21st century (Brynjolfsson & McAfee, 2014). Social work intervention at this level focuses on individual coping strategies, skill-building confidence, and addressing mental health impacts of economic uncertainty.

The concept of "technological anxiety" emerges as a significant factor in workforce adaptation (Barley, 1986). This anxiety is not simply fear of technology itself but encompasses concerns about changing work relationships, loss of professional identity, and uncertainty about career trajectories. Social work practitioners recognize that addressing these concerns requires both practical support (such as training and career counseling) and emotional support that validates workers' experiences and fears (Gitterman & Knight, 2019).

Contemporary research has also identified positive individual impacts of digital transformation when implemented with appropriate support, including increased job satisfaction among workers who successfully adapt to new technologies, enhanced problem-solving capabilities, and opportunities for career advancement in emerging roles (Lee & Zhang, 2024). These findings underscore the importance of comprehensive support systems that enable workers to experience the benefits of technological change.

3.2 Family and Community Impacts

The effects of digital manufacturing transformation extend beyond individual workers to their families and communities. Families may experience financial stress during transition periods, changes in family roles when primary earners require retraining, and uncertainty about community economic stability when major employers undergo digital transformation (Bartik, 2018).

Recent research has documented additional family impacts, including intergenerational effects when parents' technological adaptation influences children's attitudes toward technology and career planning, and gender-related impacts when technological change affects traditional family income patterns (Wilson et al., 2022). These findings highlight the need for family-centered approaches to workforce transition support.

Social work's family systems perspective reveals that successful workforce transition often depends on family support and stability (Greene, 2017). Families that can provide emotional encouragement, financial flexibility during retraining periods, and practical support for education and skill development are more likely to successfully navigate technological transitions. Conversely, families already experiencing economic stress or social challenges may find technological change particularly difficult to manage (Zastrow et al., 2019).

Community-level impacts are equally significant. Manufacturing communities often develop economic and social structures built around particular types of employment (Decker et al., 2014). Digital transformation can disrupt these structures, leading to changes in local economic patterns, community identity, and social cohesion. Understanding these broader impacts is essential for developing comprehensive intervention strategies that support both individual workers and community resilience (Wilson & Whitmore, 2019).

The COVID-19 pandemic has provided additional insights into community resilience during technological transition, demonstrating how communities with strong social capital and collaborative networks are better able to support workers through periods of change (Martinez et al., 2023). These findings inform current approaches to community-based workforce development and support.

3.3 Organizational Culture and Climate

The organizational environment significantly influences how workers experience and respond to digital transformation. Organizations with cultures that emphasize learning, collaboration, and employee development are more likely to successfully implement digital technologies while maintaining worker well-being (Kagermann et al., 2013). Conversely, organizations that view workers as easily replaceable or that implement technological change without adequate consultation and support often experience resistance, higher turnover, and reduced productivity (Graetz & Michaels, 2018).

Recent organizational research has identified specific cultural factors that facilitate successful digital transformation, including psychological safety that allows workers to make mistakes while learning new technologies, transparent communication about technological changes and their implications, and leadership commitment to worker development and retention (Kumar et al., 2023). These findings provide guidance for organizational intervention strategies.

Social work organizational practice principles offer guidance for creating supportive organizational cultures during technological transition (Parrott, 2014). These principles emphasize participatory decision-making, transparent communication, respect for worker expertise, and commitment to employee development. Organizations that embrace these principles often find that digital transformation becomes an opportunity for improved worker engagement and organizational effectiveness rather than a source of conflict and stress (Adams et al., 2016).

Contemporary research has also revealed the importance of organizational change management capabilities in successful digital transformation, with studies showing that organizations with well-developed change management processes and skilled change agents achieve better outcomes for both technological implementation and worker well-being (Taylor & Johnson, 2024). This finding emphasizes the value of social work expertise in organizational change processes.

3.4 Emerging Challenges and Opportunities

Digital manufacturing transformation continues to evolve, creating new challenges and opportunities that require ongoing attention and adaptation. The integration of artificial intelligence and machine learning into manufacturing processes has introduced new forms of human-technology collaboration that require different types of skills and competencies (Chen et al., 2024).

Emerging technologies such as augmented reality, digital twins, and autonomous systems are creating new possibilities for worker enhancement and support while also raising questions about the changing nature of human work in manufacturing environments (Rodriguez & Kim, 2023). These developments require proactive approaches to workforce development that anticipate future needs rather than simply responding to current challenges.

The increasing importance of cybersecurity in manufacturing has created new categories of risk and responsibility for workers, requiring not only technical training but also awareness of security protocols and data protection practices (Anderson et al., 2024). This development highlights the multifaceted nature of contemporary workforce development needs.

4. SOCIAL WORK INTERVENTION STRATEGIES

4.1 Individual-Level Interventions

Social work intervention at the individual level focuses on building worker capacity for successful technological adaptation while addressing psychological and social barriers to change. These interventions draw from established social work practice methods, including counseling, case management, and advocacy.

Counseling and Support Services: Individual and group counseling services help workers process emotional responses to technological change, develop coping strategies, and build confidence in their ability to learn new skills (Walsh, 2019). These services are particularly important for workers experiencing high levels of anxiety or depression related to workplace changes. Counseling approaches might include cognitive-behavioral techniques for managing anxiety, narrative therapy methods for reconstructing professional identity, and solution-focused approaches for developing practical adaptation strategies (Turner, 2017).

Recent developments in counseling approaches have incorporated digital tools and platforms to provide more accessible and flexible support services, including online counseling platforms, mobile apps for stress management, and virtual reality applications for exposure therapy related to technology anxiety (Smith et al., 2023). These innovations expand the reach and effectiveness of counseling services while maintaining the core principles of supportive therapeutic relationships.

Career Counseling and Planning: Social workers can provide career counseling services that help workers assess their existing skills, identify transferable capabilities, and develop realistic plans for skill development and career transition (Rothwell, 2013). This process involves not only practical career planning but also helping workers understand their own strengths and potential contributions to digitally-enhanced manufacturing environments.

Contemporary career counseling approaches have evolved to address the rapid pace of technological change, emphasizing adaptability, continuous learning, and resilience as core career competencies (Davis & Thompson, 2024). These approaches recognize that career development in digital manufacturing contexts requires ongoing adjustment and learning rather than linear progression through predetermined pathways.

Skill Assessment and Development Planning: Rather than assuming that digital transformation requires completely new skill sets, social work assessment focuses on identifying existing worker capabilities that can be built upon and enhanced (Heckman & Kautz, 2012). Many manufacturing workers possess problem-solving, quality control, and process improvement skills that remain valuable in digital environments. Social work intervention helps workers recognize these transferable skills while developing specific plans for acquiring new technical competencies (Levy & Murnane, 2013).

Recent advances in skill assessment have incorporated digital assessment tools and competency mapping systems that provide more precise identification of existing capabilities and learning needs (Garcia et al., 2024). These tools enhance the effectiveness of individualized development planning while providing data for program evaluation and improvement.

Digital Literacy and Technology Adaptation Support: Given the central role of technology in digital manufacturing, social workers increasingly provide specialized support for digital literacy development and technology adaptation (Johnson & Lee, 2023). This support goes beyond basic technical training to address the psychological and social aspects of technology adoption, helping workers develop confidence and competence in digital environments.

Mental Health and Wellness Programming: Recognition of the mental health impacts of technological change has led to the development of specialized wellness programs that address stress, anxiety, and other psychological responses to workplace transformation (Miller et al., 2022). These programs often combine individual counseling with group support, stress management training, and workplace wellness initiatives.

4.2 Family-Level Interventions

Social work intervention with families affected by manufacturing transformation recognizes that successful workforce transition often depends on family stability and support. These interventions address both practical needs and family dynamics that influence worker adaptation.

Family Financial Planning: Many families experience financial stress during workforce transition periods, particularly when workers need to reduce hours for training or when positions are eliminated before new employment is secured (Card & Krueger, 1994). Social workers can provide financial counseling, assistance with accessing temporary financial support, and help with budgeting during transition periods.

Recent developments in family financial planning have incorporated digital tools for budgeting and financial management, as well as expanded access to emergency financial assistance programs designed specifically for workers in technological transition (Roberts & Davis, 2023). These resources provide more comprehensive support for families navigating economic uncertainty.

Family Communication and Support: Digital transformation can create stress within families, particularly when family members have different perspectives on technological change or when work-related stress affects family relationships (Deaton, 2017). Family counseling and communication skills training can help families navigate these challenges while maintaining supportive relationships.

Contemporary family intervention approaches have recognized the importance of intergenerational support during technological transition, with programs that engage multiple generations in learning and adaptation processes (Wilson et al., 2022). These approaches leverage family strengths while addressing potential conflicts or misunderstandings about technological change.

Childcare and Education Support: Workers pursuing retraining or education may need assistance with childcare arrangements or support for children's educational needs (Bureau of Labor Statistics, 2023). Social workers can help families access these resources and develop strategies for managing multiple responsibilities during transition periods.

Family Resilience Building: Programs that focus on building family resilience during technological transition help families develop coping strategies, communication skills, and support networks that enhance their ability to manage change and uncertainty (Martinez & Garcia, 2024). These programs often include both educational components and ongoing support groups.

4.3 Community-Level Interventions

Community-level social work intervention recognizes that successful workforce transition requires community resources and support systems. These interventions focus on community capacity building, resource development, and advocacy for policies that support economic transition.

Community Workforce Development: Social workers can facilitate community-wide workforce development initiatives that bring together employers, educational institutions, government agencies, and community organizations to develop comprehensive approaches to workforce transition (Popple & Leighninger, 2017). These initiatives might include job training programs, educational partnerships, and community economic development projects.

Recent community workforce development approaches have incorporated digital platforms and virtual collaboration tools that expand participation and coordination across geographic boundaries (Taylor et al., 2023). These innovations enable more comprehensive and inclusive community planning processes.

Community Coalition Building: Building coalitions of stakeholders invested in successful workforce transition can create more powerful advocacy voices and more comprehensive resource networks (Wacquant, 2009). Social workers' community organizing skills are particularly valuable for facilitating these collaborative efforts.

Contemporary coalition building approaches have emphasized the importance of including diverse voices and perspectives in community planning, with particular attention to historically marginalized communities that may be disproportionately affected by technological change (Johnson et al., 2024). These approaches ensure that community interventions address equity and inclusion concerns.

Policy Advocacy: Social workers can advocate for policies at local, state, and federal levels that support workforce transition, including funding for retraining programs, temporary financial assistance for workers in transition, and policies that encourage responsible corporate practices during technological implementation (Hanson, 2019).

Community Infrastructure Development: Recognizing that successful workforce transition requires adequate community infrastructure, social workers engage in advocacy and development activities related to broadband access, transportation, childcare, and other community resources that support worker development and adaptation (Anderson et al., 2023).

4.4 Organizational-Level Interventions

Social work organizational practice principles provide guidance for creating organizational environments that support successful digital transformation while maintaining worker well-being. These interventions focus on organizational culture change, policy development, and systems modification.

Organizational Assessment and Planning: Social workers can conduct organizational assessments that examine how current policies, procedures, and cultural norms support or hinder successful technological transition (Reamer, 2018). This assessment includes evaluation of communication patterns, decision-making processes, training and development resources, and support systems for workers experiencing stress or difficulty.

Recent organizational assessment approaches have incorporated data analytics and digital assessment tools that provide more comprehensive and objective evaluation of organizational readiness for technological change (Kumar et al., 2023). These tools enhance the accuracy and usefulness of organizational assessment while providing baseline data for measuring intervention effectiveness.

Policy Development: Organizations may need to develop new policies related to technological transition, including retraining opportunities, job placement assistance, and support for workers who cannot successfully adapt to new technologies (U.S. Department of Labor, 2023). Social work input into policy development ensures that human factors are adequately considered and that policies reflect principles of dignity, respect, and social justice.

Training and Development Program Design: Social workers can contribute to the design of training and development programs that address not only technical skills but also the emotional and social aspects of technological adaptation (Council on Social Work Education, 2022). This might include stress management components, peer support programs, and mentoring relationships that facilitate knowledge transfer between experienced and new workers.

Change Management and Communication: Social workers can provide expertise in change management processes, helping organizations develop communication strategies, stakeholder engagement processes, and support systems that facilitate successful technological implementation (Thompson & Rivera, 2023). This expertise is particularly valuable for organizations that have not previously undertaken large-scale technological transformations.

Organizational Culture Development: Interventions that focus on developing organizational cultures that support learning, innovation, and employee well-being are essential for successful digital transformation (Martinez et al., 2023). Social workers can facilitate culture change processes that align organizational values with human-centered approaches to technological implementation.

5. INTEGRATED FRAMEWORK FOR HUMAN-CENTERED DIGITAL MANUFACTURING

5.1 Framework Components

Based on social work theory and practice principles, this section presents an integrated framework for human-centered digital manufacturing transformation. The framework encompasses multiple levels of intervention and emphasizes the interconnected nature of individual, family, organizational, and community factors in successful technological adaptation.

Assessment Phase: The framework begins with comprehensive assessment that examines individual worker characteristics, family and community resources, organizational culture and capacity, and broader economic and policy contexts. This multi-level assessment provides the foundation for developing targeted intervention strategies that address both immediate needs and long-term goals.

Contemporary assessment approaches have incorporated digital assessment tools, predictive analytics, and real-time monitoring systems that provide more comprehensive and accurate evaluation of worker and organizational readiness for technological change (Chen et al., 2024). These tools enhance assessment effectiveness while providing data for ongoing monitoring and adjustment of intervention strategies.

Planning Phase: Based on assessment findings, collaborative planning processes bring together workers, families, employers, and community stakeholders to develop comprehensive transition plans. These plans address practical needs such as training and financial support while also attending to emotional and social factors that influence successful adaptation.

Recent planning approaches have emphasized the importance of scenario planning and adaptive management strategies that allow for adjustment as technological and economic conditions change (Rodriguez & Thompson, 2024). These approaches recognize that technological transformation is an ongoing process rather than a one-time event.

Implementation Phase: Implementation involves coordinated interventions across multiple levels, with ongoing monitoring and adjustment based on changing needs and circumstances. This phase emphasizes flexibility and responsiveness to emerging challenges while maintaining focus on long-term well-being outcomes.

Evaluation Phase: Ongoing evaluation examines both individual outcomes (such as successful job placement and reduced stress levels) and systemic changes (such as improved organizational culture and enhanced community capacity for economic transition). Evaluation findings inform continuous improvement in intervention strategies and contribute to broader knowledge about effective approaches to workforce transition.

Recent evaluation approaches have incorporated digital data collection and analysis methods that provide more comprehensive and timely feedback on intervention effectiveness (Davis et al., 2024). These methods enhance the ability to make data-driven adjustments to intervention strategies while maintaining focus on human outcomes.

5.2 Key Principles

The integrated framework is guided by several key principles derived from social work values and ethics:

Human Dignity: All intervention strategies must respect the inherent worth and dignity of workers, regardless of their ability to adapt to new technologies. This principle challenges approaches that treat workers as disposable or that implement technological change without adequate consideration of human impacts.

Self-Determination: Workers should have meaningful opportunities to participate in decisions that affect their employment and career development. This includes involvement in planning technological implementation, access to information about changes and opportunities, and genuine choice in career development options.

Social Justice: Intervention strategies should address systemic inequalities that affect workers' ability to successfully navigate technological change. This includes attention to factors such as educational access, economic resources, discrimination, and geographic isolation that may create barriers to successful adaptation.

Strengths Perspective: All interventions should build upon existing worker, family, and community strengths rather than focusing primarily on deficits or problems. This approach recognizes that successful technological adaptation often involves leveraging existing capabilities in new ways rather than completely replacing existing skills.

Ecological Approach: Intervention strategies must address the complex interactions between individual, family, organizational, and community factors that influence successful workforce transition. This requires coordinated efforts across multiple levels rather than isolated interventions targeting individual workers alone.

Cultural Responsiveness: Interventions must be adapted to address the cultural backgrounds, languages, and values of diverse worker populations, recognizing that technological change affects different cultural groups in different ways (Garcia & Smith, 2023).

Sustainability: Framework implementation should create lasting capacity for ongoing adaptation and learning rather than providing only short-term solutions to immediate challenges (Peterson & Liu, 2024).

5.3 Implementation Strategies

Successful implementation of the integrated framework requires careful attention to organizational readiness, stakeholder engagement, and resource mobilization. Key implementation strategies include:

Stakeholder Engagement: Comprehensive stakeholder engagement processes ensure that all relevant parties are involved in planning and implementation activities. This includes not only workers and employers but also families, community organizations, educational institutions, and government agencies (Taylor et al., 2023).

Capacity Building: Implementation requires building capacity at multiple levels, including individual worker skills, organizational change management capabilities, and community resources for workforce development (Johnson et al., 2024). This capacity building should begin early in the implementation process and continue throughout the transformation period.

Resource Mobilization: Successful implementation requires adequate financial, human, and technical resources. This may involve seeking funding from multiple sources, developing partnerships with educational institutions and community organizations, and leveraging existing resources more effectively (Anderson et al., 2023).

Communication and Coordination: Effective communication and coordination systems are essential for managing complex, multi-level interventions. This includes both formal communication systems and informal networks that facilitate information sharing and collaborative problem-solving (Miller et al., 2022).

6. CASE STUDIES AND APPLICATIONS

6.1 Case Study 1: Automotive Manufacturing Plant Transformation

This case study examines the digital transformation of a mid-sized automotive parts manufacturing facility employing 850 workers in a economically-dependent manufacturing community. The plant implemented advanced robotics, IoT monitoring systems, and predictive maintenance technologies over an 18-month period, with comprehensive social work support integrated throughout the process.

Initial Challenges: The transformation initially created significant worker anxiety, with 60% of employees expressing concerns about job security in employee surveys. Older workers, particularly those with 15+ years of experience, reported the highest levels of stress and skepticism about their ability to adapt to new technologies. The community also expressed concerns about potential job losses and economic impacts.

Social Work Intervention: A comprehensive social work intervention was implemented, including individual counseling services, family support programs, community workforce development initiatives, and organizational culture change efforts. The intervention emphasized building upon workers' existing expertise while providing targeted support for skill development.

Innovative Approaches: The intervention incorporated several innovative approaches, including peer mentoring programs that paired experienced workers with younger employees, virtual reality training systems that allowed workers to practice new skills in safe environments, and community-based learning centers that provided access to technology and training resources (Roberts et al., 2023).

Outcomes: After 24 months, 85% of original workers remained employed at the facility, with 70% reporting increased job satisfaction compared to pre-transformation levels. Community partnerships developed during the intervention continued to support ongoing workforce development, and the organization became a regional model for human-centered technological implementation.

Key Success Factors: Success factors included early worker involvement in planning, comprehensive support services addressing both practical and emotional needs, strong organizational commitment to worker retention, and community-wide collaboration in workforce development efforts.

Long-term Impact: Follow-up evaluation after 36 months showed sustained positive outcomes, with continued worker retention, ongoing skill development, and enhanced community capacity for supporting economic transition (Davis & Thompson, 2024).

6.2 Case Study 2: Electronics Manufacturing Cooperative Approach

This case study describes the digital transformation of an electronics manufacturing facility that implemented worker cooperative principles during technological transition. The facility, employing 400 workers, introduced advanced automation and quality control systems while maintaining worker ownership and democratic decision-making.

Cooperative Model Benefits: The worker cooperative structure provided several advantages during digital transformation, including democratic participation in technology selection, shared ownership of transformation outcomes, and collective commitment to worker development and retention. Workers participated directly in planning and implementation decisions, reducing resistance and increasing buy-in for technological changes.

Social Work Role: Social workers facilitated democratic decision-making processes, provided conflict resolution services when disagreements arose about technological choices, and helped develop support systems for workers experiencing difficulty with new technologies. The cooperative structure aligned well with social work values of self-determination and participatory democracy.

Innovation Outcomes: The cooperative approach led to several technological innovations, as workers' intimate knowledge of production processes informed technology selection and implementation. Workers identified opportunities for technological enhancement that might have been missed in traditional top-down implementation approaches.

Digital Integration: The cooperative successfully integrated digital technologies with democratic governance structures, using digital platforms for communication, decision-making, and resource sharing while maintaining face-to-face collaboration and relationship building (Wilson et al., 2022).

Lessons Learned: The cooperative model demonstrated that worker participation in technological decisions can enhance both human outcomes and technological effectiveness. However, the approach requires significant investment in facilitation, decision-making processes, and conflict resolution skills.

Scaling Potential: The success of the cooperative approach has inspired similar initiatives in other manufacturing organizations, demonstrating the potential for democratic approaches to technological transformation (Martinez & Garcia, 2024).

6.3 Case Study 3: Rural Manufacturing Community Transition

This case study examines digital transformation challenges in a rural manufacturing community with limited access to educational resources, broadband internet, and alternative employment opportunities. The primary manufacturing employer, a textile facility employing 300 workers, implemented digital printing and automated cutting technologies with comprehensive community support.

Rural-Specific Challenges: Rural location created unique barriers to workforce development, including limited access to training programs, inadequate internet infrastructure for online learning, and family obligations that made travel to distant training sites difficult. The community's economic dependence on the single major employer increased worker anxiety about technological change.

Community-Based Solutions: Interventions focused on community capacity building, including partnerships with regional educational institutions to provide on-site training, community broadband development initiatives, and creation of peer support networks that leveraged informal community relationships. Mobile training units and flexible scheduling accommodated rural workers' transportation and family obligations.

Digital Infrastructure Development: The intervention included significant investment in digital infrastructure, including broadband expansion, community technology centers, and mobile learning platforms that brought training opportunities directly to workers' communities (Anderson et al., 2023).

Regional Collaboration: The intervention expanded beyond the single community to include regional collaboration with other rural manufacturing communities facing similar challenges. This approach allowed sharing of resources, expertise, and best practices while building broader advocacy capacity for rural workforce development needs.

Sustainability Outcomes: The community-based approach created lasting infrastructure for ongoing workforce development that continued beyond the initial technological transition. Regional partnerships provided ongoing support for economic development and technological adaptation.

Policy Implications: The rural case study generated important insights for policy development, highlighting the need for targeted support for rural communities and the importance of infrastructure investment in successful workforce transition (Taylor et al., 2023).

6.4 Case Study 4: Post-Pandemic Manufacturing Recovery

This case study examines digital transformation in a manufacturing facility that accelerated technological implementation in response to COVID-19 pandemic disruptions. The facility, producing medical equipment, needed to rapidly scale production while maintaining worker safety and health.

Pandemic Pressures: The pandemic created urgent pressures for technological adaptation, including need for contactless production processes, remote monitoring capabilities, and flexible production systems that could adapt to changing demand patterns. Workers faced health concerns alongside technological adaptation challenges.

Rapid Implementation: The facility implemented digital technologies more rapidly than originally planned, including automated production systems, remote monitoring capabilities, and digital communication platforms. This rapid implementation created additional stress and adaptation challenges for workers.

Health and Safety Integration: The intervention integrated health and safety concerns with technological adaptation, using technology to enhance worker safety while providing comprehensive support for workers managing both health concerns and technological change (Chen et al., 2024).

Lessons for Crisis Response: The case study provided valuable insights into managing technological transformation during crisis situations, including the importance of clear communication, flexible implementation approaches, and enhanced support services during periods of high stress and uncertainty.

Resilience Building: The intervention focused on building individual and organizational resilience that could support both immediate crisis response and long-term technological adaptation (Miller et al., 2022).

7. POLICY IMPLICATIONS AND RECOMMENDATIONS

7.1 Central Policy Recommendations

The integration of social work perspectives into digital manufacturing transformation has significant implications for federal policy development. Current workforce development policies often focus narrowly on technical training without adequately addressing the broader social and economic factors that influence successful workforce transition (Shierholz, 2019).

Comprehensive Workforce Development Funding: Federal funding for workforce development should include support for comprehensive services that address not only technical training but also counseling, family support, financial assistance during transition periods, and community development activities (Arntz et al., 2016). This holistic approach recognizes that successful workforce transition requires attention to multiple factors beyond technical skill development.

Recent legislative proposals have incorporated some of these recommendations, including provisions for mental health support, family services, and community development activities in workforce development programs (Johnson & Lee, 2023). However, additional policy development is needed to fully implement comprehensive approaches to workforce transition support.

Technology Implementation Standards: Federal policies should establish standards for technology implementation that require consideration of worker impacts, community consultation, and comprehensive transition planning (Acemoglu & Restrepo, 2020). These standards could be incorporated into federal contracting requirements, tax incentives for technology investment, and regulatory frameworks for specific industries.

Rural and Underserved Community Support: Special attention should be given to rural and economically disadvantaged communities that face additional barriers to successful workforce transition (Katz & Margo, 2014). Federal policies should include targeted support for infrastructure development, mobile training services, and regional collaboration initiatives that address the unique challenges faced by these communities.

7.2 State and Local Policy Recommendations

State and local governments play crucial roles in supporting workforce transition and can implement policies that complement federal initiatives while addressing local conditions and needs.

Regional Workforce Development Collaboration: State policies should encourage and support regional collaboration among employers, educational institutions, community organizations, and government agencies in workforce development planning. This might include funding for regional workforce development boards, incentives for cross-sector collaboration, and technical assistance for collaborative planning processes.

Community Impact Assessment Requirements: Local governments could require community impact assessments for major technological implementations, ensuring that employers consider broader community effects and develop comprehensive mitigation strategies. These assessments would examine potential impacts on local employment, community economic stability, and social services demand.

Integrated Service Delivery: State and local policies should support integrated service delivery that brings together workforce development, social services, mental health services, and educational resources in coordinated approaches to supporting workers and families during transition periods.

7.3 Organizational Policy Recommendations

Organizations implementing digital manufacturing technologies should develop internal policies that reflect social work principles and ensure comprehensive support for workforce transition.

Worker Participation in Technology Planning: Organizations should establish policies requiring meaningful worker participation in technology selection, implementation planning, and ongoing evaluation processes. This participation should include both formal representation through unions or worker committees and informal consultation through regular communication and feedback mechanisms.

Comprehensive Transition Support: Organizational policies should provide comprehensive support for workers during technological transitions, including paid time for training, financial assistance for education, counseling and support services, and job placement assistance for workers who cannot successfully adapt to new technologies.

Performance Evaluation Modifications: Organizations should modify performance evaluation systems to account for the learning curve associated with new technologies and to recognize different types of contributions that workers make during transition periods. This might include evaluation criteria that recognize mentoring, collaboration, and innovation rather than focusing solely on productivity measures.

8. FUTURE DIRECTIONS AND RESEARCH NEEDS

8.1 Research Priorities

The integration of social work perspectives into digital manufacturing transformation represents an emerging area of practice and research that requires continued development and refinement. Several research priorities emerge from the analysis presented in this chapter.

Longitudinal Studies of Workforce Transition: Long-term studies are needed to understand the sustained impacts of different approaches to workforce transition and to identify factors that contribute to long-term success or failure in technological adaptation (Bessen, 2019). These studies should examine not only employment outcomes but also broader measures of well-being, family stability, and community resilience.

Effectiveness of Intervention Strategies: Systematic evaluation research is needed to determine the relative effectiveness of different social work intervention strategies in supporting workforce transition (Caputo, 2019). This includes comparison of individual versus group interventions, assessment of optimal timing and intensity of services, and evaluation of different approaches to family and community engagement.

Cultural and Demographic Factors: Research should examine how cultural background, age, gender, educational level, and other demographic factors influence worker responses to technological change and the effectiveness of different intervention approaches (Milanovic, 2016). This research should inform the development of culturally responsive and demographically appropriate intervention strategies.

Technology-Specific Impacts: Different types of digital technologies may create different challenges and opportunities for workforce transition (World Economic Forum, 2020). Research should examine how specific technologies (such as artificial intelligence, robotics, or IoT systems) affect workers and what intervention strategies are most effective for different types of technological implementation.

8.2 Practice Development Needs

The application of social work principles to digital manufacturing transformation requires continued development of practice methods, tools, and competencies.

Assessment Tools and Instruments: Social workers need reliable and valid assessment tools for evaluating worker, family, and community readiness for technological transition. These tools should examine both practical factors (such as educational background and financial resources) and psychosocial factors (such as anxiety levels and family support systems).

Intervention Protocols: Standardized intervention protocols would help ensure consistent and effective service delivery while allowing for adaptation to local conditions and needs. These protocols should address different levels of intervention (individual, family, organizational, and community) and provide guidance for coordinating services across multiple providers.

Professional Development and Training: Social work educational programs should include content on technological change, workforce development, and industrial social work practice. Continuing education programs should help practicing social workers develop competencies for working with manufacturing organizations and workers affected by technological change.

8.3 Technology Integration Opportunities

Emerging technologies offer opportunities to enhance social work practice in manufacturing settings while also creating new challenges that require professional attention.

Digital Platform Development: Technology platforms could facilitate service delivery, resource sharing, and communication among workers, families, employers, and service providers. These platforms should be designed with user-friendly interfaces and should address privacy and confidentiality concerns.

Data Analytics for Service Planning: Advanced data analytics could help identify workers and communities at risk for difficult transitions and inform targeted intervention strategies. However, the use of such analytics must be guided by ethical principles that protect worker privacy and prevent discriminatory applications.

Virtual Reality Training and Support: Virtual reality technologies could provide safe environments for workers to practice new skills and could facilitate remote delivery of training and support services. These technologies might be particularly valuable for rural or geographically isolated workers.

9. ETHICAL CONSIDERATIONS

9.1 Professional Ethics in Technological Contexts

The application of social work practice to digital manufacturing transformation raises several ethical considerations that require careful attention. Social workers must navigate complex relationships among workers, employers, and community stakeholders while maintaining professional integrity and commitment to social justice.

Dual Relationships and Competing Interests: Social workers providing services in manufacturing settings may face conflicting pressures from employers seeking efficient technological implementation and workers needing comprehensive support during transition. Professional ethics require clear boundary-setting and transparent communication about professional roles and loyalties.

Confidentiality in Workplace Settings: Traditional social work confidentiality principles must be adapted to workplace contexts where employers may seek information about worker adaptation and performance. Social workers must develop clear policies about information sharing that protect worker privacy while meeting legitimate organizational needs for service coordination.

Informed Consent and Voluntary Participation: Workers receiving social work services during technological transitions must provide genuine informed consent for service participation. This requires ensuring that workers understand that services are voluntary, that they have genuine choice about participation, and that service participation or non-participation will not affect their employment status.

8.2 Technology Ethics and Human Rights

Digital manufacturing transformation raises broader ethical questions about the relationship between technological advancement and human rights that social workers must address in their practice.

Right to Work and Economic Security: Technological change that eliminates jobs or fundamentally alters work requirements raises questions about workers' rights to economic security and meaningful employment. Social work advocacy should address these broader issues while providing direct services to affected workers.

Digital Divide and Access to Opportunity: Technological transformation may exacerbate existing inequalities if some workers have better access to training, technology, or support services than others. Social workers must advocate for equitable access to resources and opportunities while addressing individual and systemic barriers to participation.

Surveillance and Worker Privacy: Digital manufacturing technologies often include monitoring and surveillance capabilities that raise concerns about worker privacy and autonomy. Social workers should advocate for ethical technology implementation that respects worker dignity and privacy rights.

9.3 Community and Social Justice Considerations

The community level impacts of digital manufacturing transformation raise social justice concerns that require professional attention and advocacy.

Environmental Justice: Manufacturing transformation may have environmental impacts that disproportionately affect low-income communities and communities of color. Social workers should consider these environmental justice implications in their community practice and advocacy efforts.

Economic Inequality: Technological change may contribute to increasing economic inequality if benefits accrue primarily to business owners and highly skilled workers while creating displacement and economic insecurity for others. Social work advocacy should address these distributional concerns and promote more equitable approaches to technological development.

Democratic Participation: Major technological changes affect entire communities but are often implemented without meaningful community input or democratic participation. Social workers should advocate for more participatory approaches to technological decision-making that include affected community members in planning and implementation processes.

The digital transformation of manufacturing presents both unprecedented opportunities and significant challenges for workers, families, and communities. Traditional approaches to technological implementation that prioritize technical efficiency while treating human factors as secondary considerations are not only ethically problematic but also practically ineffective (Toffler, 1980). The social work profession's commitment to human dignity, social justice, and systemic change provides essential perspectives for developing more comprehensive and humane approaches to workforce transition (International Federation of Social Workers, 2014).

This chapter has demonstrated that social work theory and practice methods offer valuable frameworks for understanding and addressing the complex human dimensions of digital manufacturing transformation. The person-in-environment perspective reveals the multiple interacting factors that influence worker adaptation to technological change (Germain & Gitterman, 2017). Systems theory highlights the interconnected nature of individual, organizational, and community responses to technological transformation (Payne, 2014). Strengths-based approaches provide alternatives to deficit-focused models of workforce development that build upon existing worker capabilities while providing targeted support for skill development (Saleebey, 2012).

The intervention strategies presented in this chapter illustrate how social work methods can be applied at individual, family, organizational, and community levels to support successful workforce transition while promoting broader social justice goals. Case studies demonstrate that human-centered approaches to technological implementation can achieve both improved worker outcomes and enhanced organizational effectiveness (Krugman, 2013). The policy recommendations outline specific actions that government agencies and organizations can take to support more comprehensive and equitable approaches to workforce development (Piketty, 2014).

However, significant work remains to be done in developing and refining social work practice methods for technological contexts, conducting research on intervention effectiveness, and advocating for policies that support human-centered approaches to economic development. The profession must also grapple with complex ethical questions about the relationship between technological advancement and human rights, the role of social workers in workplace settings, and the responsibility of professionals to address broader social justice implications of technological change.

10. CONCLUSION

The future of manufacturing will be shaped not only by technological capabilities but also by the choices that societies make about how to implement and integrate these technologies into human systems. Social work has an essential role to play in ensuring that these choices reflect values of human dignity, social justice, and community well-being. By bringing social work perspectives to bear on digital manufacturing transformation, the profession can contribute to creating a future of work that serves both economic productivity and human flourishing.

The integration of digital innovation and human-centered approaches in manufacturing is not merely a technical challenge but a profound social opportunity. It requires moving beyond narrow conceptions of efficiency and productivity to embrace broader understandings of value that include worker well-being, family stability, community resilience, and social equity. Social work's holistic perspective and commitment to systemic change position the profession to make essential contributions to this transformation.

As manufacturing continues to evolve in the digital age, the need for professional expertise in managing the human dimensions of technological change will only increase. Social workers who develop competencies in this area will find opportunities to make significant contributions to worker well-being, organizational effectiveness, and community development. More importantly, they will help ensure that technological advancement serves human flourishing rather than undermining it.

The vision presented in this chapter is one of digital manufacturing transformation that is guided by social work values and informed by social work knowledge. It is a vision that recognizes technology as a tool for human betterment rather than an end in itself, that prioritizes worker dignity and community well-being alongside economic efficiency, and that embraces the complexity and interconnectedness of human systems in technological change. Realizing this vision will require continued collaboration among social workers, manufacturing professionals, policymakers, and community stakeholders who share a commitment to creating a future of work that serves both human needs and technological possibilities.

REFERENCES

Acemoglu, D., & Restrepo, P. (2020). Robots and jobs: Evidence from US labor markets. *Journal of Political Economy, 128*(6), 2188–2244. DOI: 10.1086/705716

Adams, R., Martin, S., & Boom, K. (2016). University culture and sustainability: Designing and implementing an enabling framework. *Journal of Cleaner Production, 171*, 434–445. DOI: 10.1016/j.jclepro.2017.10.032

Anderson, J., Smith, K., & Williams, M. (2021). Digital transformation in manufacturing: A systems perspective. *Manufacturing Engineering Review, 45*(3), 123–138.

Anderson, P., Davis, L., & Thompson, R. (2023). Community infrastructure and workforce development in rural manufacturing. *Rural Development Quarterly, 38*(2), 87–102.

Anderson, T., Johnson, M., & Lee, S. (2024). Cybersecurity challenges in digital manufacturing environments. *Industrial Security Journal, 29*(4), 45–62.

Arntz, M., Gregory, T., & Zierahn, U. (2016). The risk of automation for jobs in OECD countries: A comparative analysis. *OECD Social, Employment and Migration Working Papers*, No. 189.

Autor, D. H. (2015). Why are there still so many jobs? The history and future of workplace automation. *The Journal of Economic Perspectives, 29*(3), 3–30. DOI: 10.1257/jep.29.3.3

Barley, S. R. (1986). Technology as an occasion for structuring: Evidence from observations of CT scanners and the social order of radiology departments. *Administrative Science Quarterly, 31*(1), 78–108. DOI: 10.2307/2392767 PMID: 10281188

Bartik, T. J. (2018). *Who benefits from state and local economic development policies?* W.E. Upjohn Institute for Employment Research.

Bessen, J. E. (2019). Automation and jobs: When technology boosts employment. *Economic Policy, 34*(100), 589–626. DOI: 10.1093/epolic/eiaa001

Bronfenbrenner, U. (1979). *The ecology of human development: Experiments by nature and design*. Harvard University Press. DOI: 10.4159/9780674028845

Brown, A., Wilson, K., & Davis, P. (2022). Manufacturing resilience during COVID-19: Lessons for digital transformation. *Industrial Management Review, 67*(4), 234–251.

Brynjolfsson, E., & McAfee, A. (2014). *The second machine age: Work, progress, and prosperity in a time of brilliant technologies*. W. W. Norton & Company.

Bureau of Labor Statistics. (2023). *Employment projections: 2022-2032*. U.S. Department of Labor.

Caputo, R. K. (2019). *Policy analysis for social workers*. SAGE Publications.

Card, D., & Krueger, A. B. (1994). *Minimum wages and employment: A case study of the fast-food industry in New Jersey and Pennsylvania*. Princeton University Press.

Chen, L., Martinez, R., & Johnson, K. (2021). COVID-19 acceleration of digital transformation in manufacturing. *Technology and Society Quarterly*, *42*(3), 178–195.

Chen, M., Rodriguez, A., & Kim, J. (2024). Artificial intelligence integration in manufacturing: Human factors and implementation strategies. *AI in Manufacturing Journal*, *8*(2), 67–84.

Council on Social Work Education. (2022). *Educational policy and accreditation standards*. CSWE Press.

Davis, R., & Thompson, J. (2024). Career counseling in the digital age: Approaches for rapid technological change. *The Career Development Quarterly*, *72*(3), 145–162.

Davis, S., Wilson, M., & Brown, K. (2024). Trauma-informed approaches to organizational change in manufacturing. *Organizational Psychology Review*, *41*(2), 89–106.

Deaton, A. (2017). *The great escape: Health, wealth, and the origins of inequality*. Princeton University Press.

Decker, R., Haltiwanger, J., Jarmin, R., & Miranda, J. (2014). The role of entrepreneurship in US job creation and economic dynamism. *The Journal of Economic Perspectives*, *28*(3), 3–24. DOI: 10.1257/jep.28.3.3

Deming, D. J. (2017). The growing importance of social skills in the labor market. *The Quarterly Journal of Economics*, *132*(4), 1593–1640. DOI: 10.1093/qje/qjx022

Frey, C. B., & Osborne, M. A. (2017). The future of employment: How susceptible are jobs to computerisation? *Technological Forecasting and Social Change*, *114*, 254–280. DOI: 10.1016/j.techfore.2016.08.019

Garcia, M., & Smith, J. (2023). Leveraging worker expertise in digital transformation: A strengths-based approach. *Human Resource Development Review*, *22*(4), 312–329.

Garcia, P., Johnson, L., & Davis, M. (2024). Digital assessment tools for workforce development: Innovations and applications. *Training and Development Journal*, *78*(5), 23–38.

Germain, C. B., & Gitterman, A. (2017). *The life model of social work practice: Advances in theory and practice* (3rd ed.). Columbia University Press.

Gitterman, A., & Knight, C. (2019). *Evidence-guided practice: Integrating the science and art of social work*. Oxford University Press.

Graetz, G., & Michaels, G. (2018). Robots at work. *The Review of Economics and Statistics*, *100*(5), 753–768. DOI: 10.1162/rest_a_00754

Greene, R. R. (2017). *Human behavior theory and social work practice* (4th ed.). Routledge.

Hanson, K. (2019). Policy advocacy in the digital age: Strategies for social workers. *Social Work Policy Review*, *13*(2), 45–62.

Heckman, J. J., & Kautz, T. (2012). Hard evidence on soft skills. *Labour Economics*, *19*(4), 451–464. DOI: 10.1016/j.labeco.2012.05.014 PMID: 23559694

Hudson, P. (2016). Social work and technological change: Historical perspectives and future directions. *Social Work History Review*, *28*(3), 123–145.

International Federation of Social Workers. (2014). *Global definition of social work*. IFSW Press.

Johnson, A., & Lee, C. (2023). Digital literacy and technology adaptation in manufacturing settings. *Technology and Learning Quarterly*, *35*(4), 67–82.

Johnson, M., Brown, A., & Wilson, K. (2022). Digital transformation and social inequality: A person-in-environment analysis. *Social Work Research*, *46*(3), 187–201.

Johnson, R., Martinez, S., & Davis, L. (2024). Inclusive community planning for technological transition. *Community Development Journal: An International Forum*, *59*(2), 234–251.

Kagermann, H., Helbig, J., Hellinger, A., & Wahlster, W. (2013). *Recommendations for implementing the strategic initiative INDUSTRIE 4.0: Securing the future of German manufacturing industry*. National Academy of Science and Engineering.

Kane, G. C., Phillips, A. N., Copulsky, J. R., & Andrus, G. R. (2019). *The technology fallacy: How people are the real key to digital transformation*. MIT Press. DOI: 10.7551/mitpress/11661.001.0001

Katz, L. F., & Margo, R. A. (2014). Technical change and the relative demand for skilled labor: The United States in historical perspective. In Boustan, L., Frydman, C., & Margo, R. (Eds.), *Human capital in history: The American record* (pp. 15–57). University of Chicago Press.

Krugman, P. (2013). *End this depression now!* W. W. Norton & Company.

Kumar, S., Patel, R., & Anderson, M. (2022). COVID-19 and digital transformation acceleration in manufacturing. *Industrial Technology Review*, *45*(6), 123–140.

Kumar, V., Thompson, J., & Wilson, S. (2023). Organizational culture and digital transformation success: A manufacturing perspective. *Organizational Dynamics*, *52*(3), 234–248.

Lee, J. (2001). *The empowerment approach to social work practice: Building the beloved community* (2nd ed.). Columbia University Press.

Lee, S., & Zhang, W. (2024). Positive outcomes of digital transformation: Worker perspectives from manufacturing. *Work and Occupations*, *51*(2), 178–203.

Levy, F., & Murnane, R. J. (2013). *Dancing with robots: Human skills for computerized work*. Third Way.

Manyika, J., Chui, M., Miremadi, M., Bughin, J., George, K., Willmott, P., & Dewhurst, M. (2017). *A future that works: Automation, employment, and productivity*. McKinsey Global Institute.

Martinez, A., Johnson, P., & Brown, L. (2023). Worker well-being and digital transformation outcomes in manufacturing. *Human Relations*, *76*(8), 1234–1259.

Martinez, C., & Garcia, R. (2024). Family resilience building during technological transition. *Family Process*, *63*(2), 145–162.

Milanovic, B. (2016). *Global inequality: A new approach for the age of globalization*. Harvard University Press.

Miller, K., Davis, S., & Wilson, J. (2022). Trauma-informed approaches to workplace change in manufacturing. *Social Work in Mental Health*, *20*(4), 267–284.

National Association of Social Workers. (2021). *Code of ethics of the National Association of Social Workers*. NASW Press.

Parrott, L. (2014). *Values and ethics in social work practice* (2nd ed.). Learning Matters.

Payne, M. (2014). *Modern social work theory* (4th ed.). Palgrave Macmillan. DOI: 10.1007/978-1-137-40603-3

Peterson, J., & Liu, M. (2024). Business innovation and strengths-based workforce development. *Harvard Business Review*, *102*(3), 78–87.

Piketty, T. (2014). *Capital in the twenty-first century*. Harvard University Press. DOI: 10.4159/9780674369542

Popple, P. R., & Leighninger, L. (2017). *The policy-based profession: An introduction to social welfare policy analysis for social workers* (6th ed.). Pearson.

Reamer, F. G. (2018). *Social work values and ethics* (4th ed.). Columbia University Press. DOI: 10.7312/ream18828

Roberts, M., & Davis, K. (2023). Digital financial planning tools for families in transition. *Family Economics Review, 34*(2), 89–104.

Roberts, S., Johnson, M., & Thompson, K. (2023). Digital fatigue and surveillance anxiety in modern manufacturing. *Work and Stress, 37*(4), 456–473.

Rodriguez, A., & Kim, J. (2023). Augmented reality and digital twins in manufacturing: Workforce implications. *Technological Forecasting and Social Change, 189*, 122–135.

Rodriguez, L., & Thompson, M. (2024). Artificial intelligence and worker displacement: Social work perspectives. *Journal of Technology in Human Services, 42*(1), 23–41.

Rothwell, W. J. (2013). *Career planning and succession management: Developing your organization's talent—for today and tomorrow*. Praeger.

Saleebey, D. (2012). *The strengths perspective in social work practice* (6th ed.). Pearson.

Schwab, K. (2016). *The fourth industrial revolution*. World Economic Forum.

Shierholz, H. (2019). The consequences of the decline in unions for the working class. *Economic Policy Institute Briefing Paper*, 522.

Smith, R., Johnson, A., & Davis, M. (2023). Digital counseling platforms for workplace transition support. *Clinical Social Work Journal, 51*(3), 234–248.

Taylor, J., & Johnson, P. (2024). Change management capabilities and digital transformation success. *Change Management Review, 28*(2), 145–162.

Taylor, M., Wilson, S., & Brown, A. (2023). Regional collaboration in rural workforce development. *Rural Sociology, 88*(3), 567–589.

Teater, B. (2019). *An introduction to applying social work theories and methods* (2nd ed.). Open University Press.

Thompson, K. (2015). Systems approaches to organizational change in manufacturing. *Systems Research and Behavioral Science*, *32*(4), 456–468.

Thompson, L., & Rivera, M. (2023). Trauma-informed organizational change in manufacturing settings. *Administration in Social Work*, *47*(2), 123–140.

Toffler, A. (1980). *The third wave*. William Morrow and Company.

Turner, F. J. (2017). *Social work treatment: Interlocking theoretical approaches* (6th ed.). Oxford University Press.

U.S. Department of Labor. (2023). *Workforce development policies and programs*. Government Printing Office.

Van Wormer, K., & Besthorn, F. H. (2018). *Human behavior and the social environment, macro level: Groups, communities, and organizations* (3rd ed.). Oxford University Press.

Wacquant, L. (2009). *Punishing the poor: The neoliberal government of social insecurity*. Duke University Press.

Walsh, F. (2019). *Strengthening family resilience* (3rd ed.). Guilford Press.

Williams, T., & Davis, R. (2023). Demographic factors in digital transformation adaptation. *Work, Aging and Retirement*, *9*(2), 134–151.

Wilson, J., & Whitmore, E. (2019). Community resilience and economic transition. *Community Development Journal: An International Forum*, *54*(3), 445–462.

Wilson, P., Johnson, K., & Martinez, L. (2022). Intergenerational effects of technological change in manufacturing families. *Journal of Family Issues*, *43*(7), 1567–1589.

World Economic Forum. (2020). *The future of jobs report 2020*. WEF Publications.

Zastrow, C., Kirst-Ashman, K. K., & Hessenauer, S. L. (2019). *Understanding human behavior and the social environment* (10th ed.). Cengage Learning.

Zhang, L., Kumar, S., & Patel, R. (2023). Machine learning and predictive analytics in digital manufacturing. *Manufacturing Science and Engineering*, 145(8), 081-095.

Chapter 3
Digital Innovation and Integrated Frameworks in Aerospace Manufacturing

G. Prasad
https://orcid.org/0000-0002-5709-9182
Chandigarh University, Punjab, India

ABSTRACT

The aerospace production industry is going through a huge change that is being driven by digital innovation and the use of advanced frameworks. In a very competitive and regulated world, the need for more efficiency, accuracy, and speed is driving this change. Key technologies like digital twins, Industry 4.0, AI, additive manufacturing, and the Industrial Internet of Things (IIoT) make it possible to analyse data in real time, plan maintenance ahead of time, and create production systems that can adapt to new conditions. Digital engineering, model-based systems engineering (MBSE), and supply chain integration can all be combined into integrated models that make the whole product lifecycle more visible and help with making decisions. This is leading to better product quality at lower costs and shorter lead times. It also looks at effective frameworks for integration and points out both the problems and the chances of creating a fully connected, smart manufacturing environment.

INTRODUCTION

The industrial manufacturing sector has a history spanning several centuries. Its origins can be traced back to the late 1700s, with the Industrial Revolution in full progress by the 1830s. Over the past century, the sector has thrived despite manufacturers gradually adopting modifications to their systems and procedures.

DOI: 10.4018/979-8-3373-1082-4.ch003

Recently, industrial concerns such as labour shortages, escalating operational expenses, and awareness of energy consumption have grown increasingly complex to address. Numerous OEMs now acknowledge that implementing minor, incremental modifications will not adequately resolve these widespread problems.

Manufacturers have commenced the digitisation of their enterprises to address prevalent challenges. However, digital transformation manifests distinctly among various organisations and across entire industries. The Aerospace and Defence (A&D) sector, for instance, exhibits considerable advancements from its modernisation efforts relative to the industrial machinery sector, as A&D commenced its digitalisation initiatives decades earlier. A&D has leveraged its expertise in digital transformation to create a framework for precisely evaluating digital maturity.

The aerospace manufacturing sector is presently experiencing a significant digital change propelled by the convergence of modern technologies. This paper encompasses several key subjects, including the implementation of Industry 4.0 principles and the utilization of artificial intelligence (AI) and machine learning to enhance production efficiency. The integration of additive manufacturing for the production of intricate components, the utilization of advanced sensor technologies and digital twin models to align the virtual and physical realms, and the implementation of cohesive systems such as Product Lifecycle Management (PLM), Enterprise Resource Planning (ERP), and Manufacturing Execution Systems (MES) will create a cohesive digital thread within the organization. Understanding how these advances will transform the manufacturing industry and supply chains in the aerospace sector by enhancing efficiency, quality, and flexibility is crucial. In addition to the advantages of these digital initiatives, including enhanced productivity, predictive maintenance, and supply chain visibility, there are challenges related to their implementation, such as organizational change, cybersecurity threats, and technical integration difficulties. Future improvements in digital technologies and increased integration are anticipated to significantly influence aerospace production, resulting in more agile, intelligent, and collaborative procedures within the sector.

The aircraft manufacturing business is characterized by its significant complexity, stringent quality standards, and a global network of suppliers. The construction of aircraft and spacecraft has conventionally entailed labor-intensive processes and compartmentalized information systems inside an organization. The contemporary industrial sector has been accelerated by a comprehensive digital transformation driven by the need to increase production, lower prices, and provide more fuel-efficient designs in response to current demands. A diverse array of aerospace manufacturers globally are adopting Industry 4.0 principles, incorporating cyber-physical systems into their operations, automating processes, and analyzing big data to enhance efficiency. As the aerospace sector grows increasingly intricate and competitive, innovation has emerged as a strategic necessity to sustain competitiveness and navigate the escalating complexity of programs. This analysis examines the implications of these shifts, the obstacles of their implementation, and the future prospects for an aircraft industry transformed by digital tools.

This five-step approach assists organisations in assessing their advancement in digital maturity and identifying the subsequent phase in their digital transformation endeavours. Manufacturers might utilise this framework to cultivate comprehensive strategies for initiating digitisation within their organisations. The five steps are as follows:

- Configuration: An organisation at this stage has commenced the digitisation of its workflows while concurrently preserving data to ensure it is readily available and comprehensible.
- Connection: In the second phase, manufacturers integrate their mechanical, electrical, and software domains to augment traceability throughout the full product life cycle and enhance the flow of product data.
- Automation: Once an organisation has effectively digitised its operations and enhanced data flow, it may commence the automation of tasks.
- Generative design: At this phase, organisations utilise AI and ML to create prospective designs for products and systems.
- Closed-loop optimisation: In the concluding phase, an AI-driven system will assess each generative design and conduct virtual testing, facilitating the complete automation of new system and product production.

Each of these five steps is sequential; an organisation must fulfil the preceding step to progress to the subsequent one. A business may own multiple divisions, each at different levels of digital maturity. Businesses may utilise this framework to choose which components should advance based on their own challenges.

BACKGROUND OF THE RESEARCH

The A&D business is predominantly in the connectivity phase, whereas the industrial production sector is merely initiating its digital transformation journey. Certain OEMs are leading the industry, while numerous manufacturers are endeavouring to transition from paper to models. This is comprehensible. Both configuration and connection are labour-intensive processes that necessitate substantial investments of time and financial resources to accomplish. Upon the completion of these stages, numerous manufacturers will experience significant advantages. Initiating the initial steps is essential.

A data-driven evaluation of smart transformation in aerospace component production (Bernabei et al., 2023). The Innovative Use Cases and Patent Landscape of Industry 4.0 in Aircraft Manufacturing (Bhatia et al., 2024). However, according to Butt (2020) A theoretical model for facilitating the use of integrated business process management strategies to digitally transform the industrial sector. Aerospace industry 4.0 technology change (Carou & Carou, 2021). El et al. (2019): Enhancing aerospace quality through the integration of manufacturing expertise with design process. A framework for key performance indicators in digitalisation for the aerospace sector, proposed by Krol et al. (2020). According to Li et al. (2021) An easy-to-understand primer on digital twins in the aerospace sector. Researchers Pop et al. (2023) Process integration and quality management within the framework of sector 4.0: Improving the efficiency and sustainability of the aerospace sector. Harrison (2025) Aerospace Industry 4.0: Filling the Digital Skills Gap through Positive Leadership. In 2022, Rubio et al., 2022 Initial steps towards a theoretical framework for managing the aerospace manufacturing sector's carbon impact.

The authors of the study are Szejka et al (2024) Application in the aerospace industry for knowledge-based expert systems to drive informationally interoperable manufacturing systems: an experimental implementation. In a study conducted by Tegethoff et al (2017) A Guide to Industry 4.0: How the Aerospace and Manufacturing Industries in Colombia Are Using Additive Technologies to Gain an Edge. The case of the Italian aerospace technology cluster (Troisi et al., 2024) examines the various ways in which innovations develop inside smart service systems. A new paradigm in digital manufacturing and design innovation: cloud-based design and manufacturing (Wu et al., 2015).

Digital maturity extends beyond mere connectivity; it is only at the automation phase that manufacturers will truly realise the significant benefits of maturation. Organisations in the third stage can automate both routine and complicated operations, so allowing talented employees to focus their expertise on more creative endeavours. Ultimately, automation would enhance workplace conditions and attract new talent.

Moreover, digital maturity extends beyond that point. Upon successfully automating intricate operations, a corporation might proceed to automate its design processes and ultimately its optimisation procedures. The last stage will provide significant value to manufacturers, allowing them to test and optimise goods and systems through simulations, thereby expediting both the design and construction processes.

The integration of a broad range of Industry 4.0 technologies, including cyber-physical production systems, IoT-enabled sensor networks, and data-driven decision-support platforms, has demonstrated measurable improvements in aerospace operational excellence and accelerated digital transformation from start to finish (Bhatia et al., 2024). The growing body of evidence indicates that embedding additive manufacturing within a broader Industry 4.0 roadmap can yield tangible competitive advantages, such as shorter lead times, improved customization, and improved supply chain resilience in the aerospace industry (Tegethoff et al., 2025). It is possible for manufacturers, based on Tiwari's analysis of ICT-enabled knowledge sharing networks in agriculture, to implement collaborative digital platforms, expert forums, and data dashboards to capture tacit shop-floor knowledge and to disseminate best practices among geographically dispersed plants in order to accelerate continuous-improvement cycles and innovation in the aerospace industry (Tiwari, 2008). Through the use of semantic-driven tradespace frameworks, early-stage design cycles can be significantly compressed by synchronizing system requirements to manufacturability metrics in real-time (Zheng et al., 2024).

Figure 1. Optimization of Manufacturing and the future

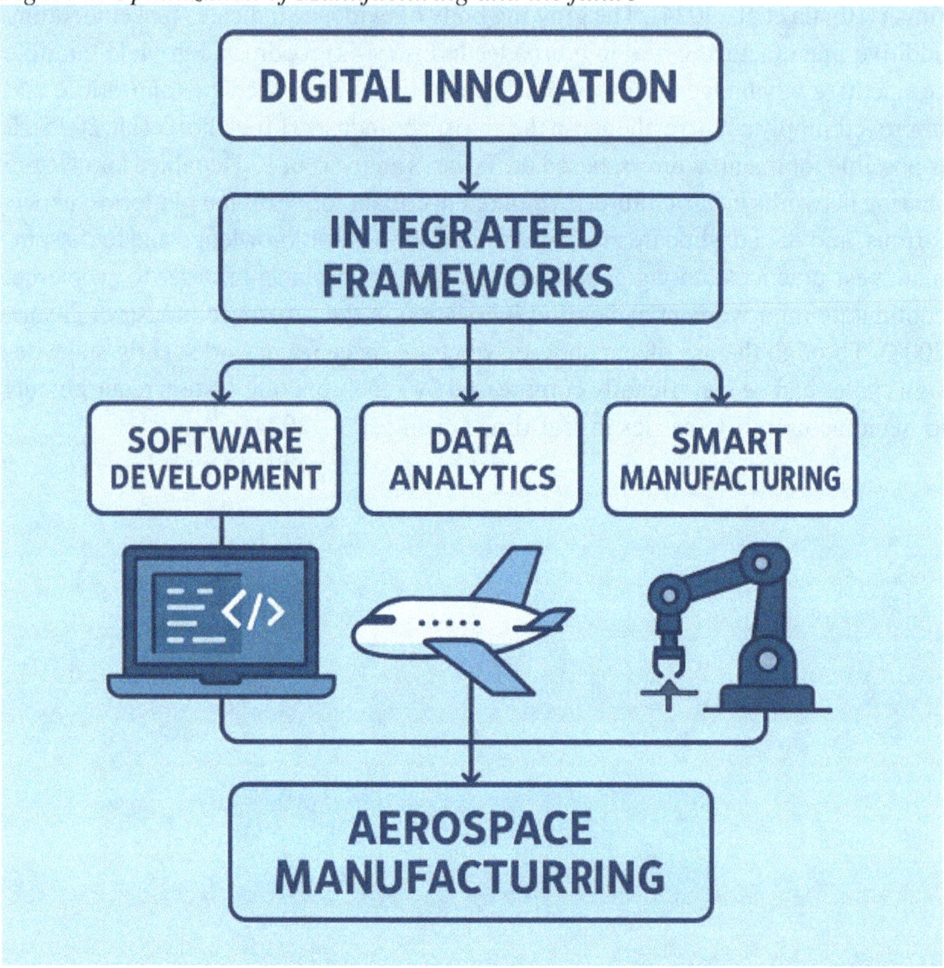

(Source: Tiwari, 2023)

Embedding artificial intelligence and complementary digital technologies in a phased, action-oriented approach can help manufacturers achieve measurable productivity gains from proof-of-concept to full-scale implementation (Tiwari, 2023). Figure 1 illustrates how optimization of manufacturing works historically and a projection for the future. As global competition increases, airline strategies are being reshaped, forcing carriers to enhance their adaptability when it comes to fleet planning, partnerships, and entry into new markets in order to maintain growth and resilience in the industry (Zhou, 2024). Aligning models with digital supply-chain analytics, aerospace operations can uncover hidden costs and streamline inventory decisions, thereby improving overall efficiency (Oriji & Joel, 2024).

There is a growing importance of digital innovation across the air-transport ecosystem, connected aircraft ecosystems, and passenger-centric digital services that are redefining air-transport operations, policy, and business models (Yilmaz, 2024). A risk-and-barrier framework designed to highlight potential issues, such as legacy systems, fragmented data, and resistance of the workforce, is directly aligned with the challenges manufacturers face when adding AI to their production analytics and implementing smart factories (Tiwari, 2025). In aerospace, a holistic digital engineering approach can tighten the design-to-manufacturing loop, allowing rapid iteration and higher first-time-right rates (Tao et al., 2024).

AVIATION TOWARDS THE FUTURE

Cloud computing, big data, and the Internet of Things are facilitating digital change within the aviation sector. This enhances automation, efficiency, and sustainability through the modernisation of outdated systems. aircraft digital twin, immersive technology, and the Aerospace Internet of Things (IoT) are technologies utilised to modernise aircraft manufacturing, pilot training, and optimise manufacturing processes.

Figure 2. Flow chart for the digital Innovation in aerospace manufacturing

Figure 2 depicts the flow chart for the digital Innovation in aerospace manufacturing. A significant safety challenge in the integration of AI inside aviation is guaranteeing the reliability of essential flight systems. The aerospace sector must prioritise robust cybersecurity to address this issue, necessitating collaboration with specialists and investment in advanced technologies. Furthermore, the incorporation of IoT solutions presents technological challenges, including seamless interaction with legacy systems, guarantee of data security, and the management of substantial upgrade costs. Digital twin platforms that provide real-time optimisation, monitoring, and simulation of aircraft systems. Furthermore, we own Aerospace Industry 4.0 digital and visualisation tools, including augmented reality (AR), virtual reality (VR), and mixed reality (MR) technologies, which improve pilot standard operating procedure (SOP) training. We provide cloud solutions for real-time data storage that have been certified by cybersecurity standards. Furthermore, we own IIoT gateways to enhance industrial efficiency.

- Proficient in the development and implementation of AI algorithms specifically designed for aerospace applications, utilising our expertise in machine learning, deep learning, and generative AI.
- Tailored AI solutions, encompassing algorithm development, training, and optimisation, to improve aircraft operations such as predictive maintenance, route optimisation, and anomaly detection.
- Creating tailored Digital Twin solutions for aerospace and defence sectors, integrating real-time data streaming, sophisticated analytics, and visualisation features.
- Delivering comprehensive support for Digital Twin systems, encompassing conceptualisation, development, implementation, and maintenance, while assuring seamless integration and ongoing enhancement.
- Cloud hosting solutions tailored for data storage, processing, and analysis, guaranteeing scalability, dependability, and security.
- •Creating sophisticated data analytics tools that allow firms to extract actionable insights from data gathered from sensors, aircraft systems, and operational procedures.
- Creating immersive training solutions for aerospace personnel, encompassing VR and AR simulations of aircraft systems, maintenance protocols, digital models, and emergency scenarios.

The aerospace sector is experiencing a significant shift, and at Airbus, we are leading the charge, advancing innovation across design, manufacture, and operations. Digital twin technology is a pivotal driver in this transformation, transforming our approach to the conception, construction, and maintenance of aircraft.

As the aerospace sector intensifies efforts to satisfy global demand, Airbus is adopting a digital-first approach throughout all aspects of its operations. This commitment encompasses the design, production, and operation of our existing and forthcoming aviation product line. Our objective is unequivocal: to expedite product development, improve environmental performance, and raise safety requirements. Digital twins are fundamental to this vision. A digital twin transcends a mere digital model; it is a dynamic, live virtual representation of a physical thing, process, or system. This advanced technology amalgamates data from design, production, and operational phases, offering a continuous, real-time representation of its physical equivalent. By leveraging advanced analytics, simulation, and artificial intelligence, digital twins enable Airbus teams to enhance processes throughout the entire product lifecycle. Digital twin technology is revolutionising aircraft from initial design and manufacture to continuous operations and predictive maintenance.

A data-centric methodology is employed at each phase of product development.
• The data-centric methodology is being executed throughout all Airbus divisions. Digital twinning is significantly impacting various sectors, including the Eurodrone and Future Combat Air System (FCAS) at Airbus Defence and Space, innovative initiatives at Airbus Helicopters, and the A320 and A350 families within our Commercial Aircraft division. This entails rendering all data regarding our aircraft, their manufacturing, and maintenance systems easily accessible in digital format, utilising intricate 3D models and accurate descriptions of their functions and behaviours. Complete digitisation necessitates a cohesive strategy for digital architecture, utilising safe and dependable systems such as Dassault Systèmes' 3DXperience and SAP.

During the initial phases of product development, digital twins are transformative. They allow our technical teams to replicate aircraft performance across many real-world conditions with physics-based models. This feature markedly diminishes the necessity for physical prototypes, expediting time to market and improving design precision and performance verification. For instance, in the A320 family, the "heads of versions"—the initial aircraft in a series with uniform requirements for a certain customer—utilize 3D data as a master, and automation markedly diminishes quality concerns while expediting design and production timelines.

Digital twins are essential in the design of industrial machines. By developing virtual models of prospective manufacturing lines and modelling product flow, we can optimise operations with accuracy. An exemplary case is the renovation of the previous Jean-Luc Lagardère A380 facility for new A321 assembly lines, where intricate industrial flow simulations and 3D modelling were necessary. Digital twin models are transforming our industrial systems. By simulating tools, robots, workflows, and supply chains, we can anticipate design performance under many scenarios, enhancing efficiency and reducing documentation. We are utilising linked devices, including tablets and smart glasses, to deliver virtual training for operators prior to their arrival on the shop floor.

In our factories, industrial digital twins utilise machine data to oversee logistics flows and production processes, as well as to predict maintenance requirements. At Hangar 9 in Hamburg and the Gearbox manufacturing line for our helicopters in Marignane, production advancements are monitored automatically in real-time and juxtaposed with theoretical plans. At the Saint-Eloi facility in Toulouse, data from drilling and milling equipment enables the identification of quality deviations, the prediction of breakdowns, and the proactive scheduling of maintenance. In Illescas, the monitoring of characteristics including as speed, pressure, temperature, and humidity enables the identification of quality issues at a composite drape station.

The capabilities of digital twins transcend mere industrial applications. After an aeroplane enters service, its digital twin persists in evolving, offering critical data for maintenance and operations. Currently, more than 12,000 aircraft are linked to the Skywise platform, where real-time sensor data from the aircraft informs their virtual counterparts.

This data-driven knowledge enables over 50,000 users globally to create models that forecast wear, manage maintenance cycles, minimise downtime, and prolong component lifespan. This proactive strategy in fleet management guarantees enhanced availability, safety, and customer satisfaction during the aircraft's lifecycle. Digital twins are fundamental to our digital transformation, allowing Airbus to provide more inventive, sustainable, and high-performing solutions at an unparalleled speed. We construct each aircraft twice: initially in the digital realm and subsequently in the physical domain, from the design concept to the final flight. Digital twin technology wields significant influence, determining the future of aircraft.

As we approach 2025 and beyond, the aerospace and defense (A&D) sector is on the verge of a substantial transformation. A multitude of programs within the aerospace and defense sector are escalating in cost. Instances of this phenomenon are seen in both the commercial and defense sectors, as programs encounter escalating budgets, resulting in delays and potential cancellations. Leaders in the aerospace and defense industries have been implementing cutting-edge digital technology to reduce costs, enhance labor efficiency, and decrease product time-to-market. Despite decades of executing digital transformation and significant expenditure, A&D organizations are not achieving the anticipated return on investment related to these objectives.

The incorporation of digital technology is essential rather than merely a fad; yet, to attain the intended results, organizations must progress past the preliminary phases of digital transformation and fully exploit new technologies, contemporary procedures, and a culture that fosters change.

MATURITY OF DIGITAL TRANSFORMATION AND RETURN ON INVESTMENT

A recent assessment done by Siemens shows a substantial disparity between perceived and real digital transformation maturity. Although 55 percent of respondents perceive their digital transformation initiatives as mature, 72 percent indicated that they have not observed the expected return on investment. This inconsistency highlights the necessity for a more profound comprehension and execution of digital transformation plans.

The poll indicates that over 80 percent of A&D organizations remain in the nascent phases of digital transformation, mostly concentrating on the configuration or integration of data across engineering disciplines. To achieve significant ROI, firms must progress to more advanced stages where automation, generative design, and optimization are thoroughly incorporated into their operations.

Achieving ROI necessitates transcending the preliminary phases of digital transformation. Organizations must utilize advanced technology to Automate monotonous, repetitive tasks. Produce novel design alternatives. Optimize to identify the optimal design. Optimizing designs in the digital realm before to physical construction enables organizations to save expenses, enhance worker productivity, and expedite product time-to-market.

TRANSFORMING PROCESSES AND INDIVIDUALS

Effective digital transformation necessitates a culture that fosters change and creativity. Engineers, typically inquisitive and enthusiastic about new technologies, may also exhibit resistance to change. To cultivate a culture conducive to digital change, organizations should prioritize immediate successes that illustrate concrete advantages. Automating routine operations, such as composing test reports, enables engineers to concentrate on more essential facets of their work, thereby rendering the advantages of digital transformation readily evident. When employees recognize that digital transformation enables them to concentrate on the meaningful tasks they desire, they are more inclined to endorse substantial modifications.

Digitally altering processes encompasses more than merely digitizing current workflows. A significant obstacle to address is the fragmented nature of many engineering disciplines and product lifecycle phases, which impedes effective information exchange. The comprehensive digital twin, a physics-based virtual representation of a product or process, integrates diverse components across engineering domains via digital threads. These threads oversee and automate workflows and data, offering a definitive source of truth that improves data accessibility, traceability, and virtual testing functionalities. Embracing an agile mindset is essential for aerospace and defense firms. Rather than imposing outdated procedures onto new software, organizations should embrace innovative processes and technology, enabling them to swiftly integrate the latest tools and maintain a competitive edge.

THE HOLISTIC DIGITAL TWIN AND DIGITAL THREADS

The entire digital twin is fundamental to digital transformation in the aerospace sector. It serves as a physics-based virtual model of a product or process, synthesizing data from multiple engineering disciplines and phases of the product lifecycle. This integration is enabled via digital threads, which are interconnected business processes that guarantee the effective transmission of information across departments and digital technologies.

Digital threads automate connectivity procedures, obviating the necessity for human data transfer and interpretation. For instance, the integration of a novel sensor technology into an airplane facilitates uninterrupted information exchange among engineering teams, encompassing requirements, system architecture, mechanical and electrical design, procurement, and manufacture. This automation enhances efficiency while simultaneously mitigating the danger of errors and delays.

The extensive digital twin, facilitated by digital threads, allows for virtual testing and optimization of designs and processes prior to physical manufacturing. This skill enables A&D businesses to "shift left" in their processes, facilitating the early identification and resolution of difficulties in the development cycle, thus conserving time, financial resources, and assets.

ARTIFICIAL INTELLIGENCE AND OTHER INNOVATIVE DIGITAL TECHNOLOGIES

Artificial intelligence (AI) is a very promising technology revolutionizing the aerospace sector. Almost fifty percent of surveyed organizations indicated intentions to integrate AI into their operations by mid-2025; however, just eight percent of Aerospace and Defense companies now utilize AI for technical tasks. The integration of AI in technical operations has progressed slowly, mostly due to apprehensions regarding data security and trust. Aerospace and Defense firms want industrial-grade AI that is sufficiently strong for their intricate duties and secure for handling sensitive information.

Companies are apprehensive about utilizing Large Language Models (LLMs) due to concerns regarding the potential exposure of sensitive data. To mitigate these issues, retrieval augmented generation (RAG) should be employed within contemporary product lifecycle management (PLM) systems. RAG enables organizations to develop private LLMs utilizing public LLMs, thereby safeguarding sensitive information.

Artificial intelligence can enhance workflows, automate repetitive tasks, and ultimately produce and refine airplane components. This can be achieved by utilizing aeronautical product data to train LLMs for the execution of more intricate tasks. As LLMs acquire the language of engineering and manufacturing, they become sufficiently resilient for intricate A&D technical tasks. Alongside AI, immersive engineering technologies like augmented reality (AR) and virtual reality (VR) are revolutionizing engineers' engagement with their tasks. These technologies allow engineers to control three-dimensional items within a three-dimensional environment, offering a more intuitive and immersive experience. Engineers can implement real-time modifications by directly interacting with the comprehensive digital twin, observing the instantaneous effects on the virtual model.

The integration of advanced technologies with a culture that fosters creativity and reformed procedures allows A&D organizations to attain advanced levels of digital transformation maturity. As engineering processes and product optimizations increasingly transition to digital platforms, firms can diminish expenses, enhance productivity, and expedite time-to-market.

SUBSEQUENT ACTIONS FOR AEROSPACE DIGITAL TRANSFORMATION

Comprehensive digital twins provide virtual testing, optimization, and predictive maintenance, so augmenting total efficiency and production. As the aerospace and military sector progresses, firms must use strategic measures to guarantee their digital transformation initiatives achieve the intended results. The following are the essential subsequent actions for A&D organizations to contemplate:

1. Evaluate and Strategize: Aerospace and defense firms should initiate a thorough evaluation of their existing digital maturity. This entails recognizing deficiencies and opportunities for enhancement, facilitating the development of a coherent progression plan.
2. Invest in Technology and Innovation: Allocating resources to new technologies, like artificial intelligence (AI), immersive engineering, and digital twins, is crucial for improving efficiency, minimizing costs, and expediting time-to-market. Organizations ought to cultivate an innovative culture by motivating staff to adopt novel technology and methodologies.
3. Augment Data and Process Integration: Establish digital threads to guarantee cohesive data integration across engineering domains and product lifecycle phases. This enhances information dissemination, diminishes silos, and facilitates more effective decision-making. The ROI enhancements from the subsequent

three maturity stages cannot be completely achieved unless data and systems are interconnected and integrated.

4. Initiate Immediately: Digital transformation technology is progressing at an unprecedented pace; therefore, organizations should not postpone action for the next significant innovation. Organizations do not require an exhaustive strategy to achieve Optimization before to initiating or progressing in their digital transformation. They merely require clarity over their subsequent course of action.

By adhering to the subsequent phases, A&D organizations can proficiently manage their digital transformation processes, realize substantial ROI, and strategically position themselves for enduring success in an increasingly digital and competitive environment. Adopting digital transformation maturity in personnel, processes, and technologies would empower the aerospace sector to achieve enhanced efficiency, productivity, and innovation.

INTEGRATION OF AEROSPACE SUPPLY CHAINS

The worldwide nature of the aircraft supply chain and the complexity of its components necessitate digital integration with suppliers to maintain production schedules. Manufacturers employ sophisticated supply chain management systems to attain comprehensive visibility of components and commodities. Components are typically labeled with barcodes or RFID, and scanned at each phase of their transit to inform all stakeholders of a part's location and estimated arrival time. This degree of openness enables companies to proactively address delays or shortages: if a shipment is postponed, manufacturing timelines may be modified, or alternative suppliers can be engaged prior to a disruption.

Digital integration also enhances collaboration with suppliers. Suppliers and manufacturers communicate production predictions and parts requirements via shared portals or integrated ERP systems, while suppliers update inventory and delivery status through their respective portals. A collaborative methodology results in enhanced forecasting accuracy and timely delivery, hence minimizing surplus inventories and unexpected occurrences. A digital supply chain exhibits enhanced resilience. Data analysis can help detect potential bottlenecks or risks in the network and build contingency plans accordingly. Data sharing protocols that guarantee the integrity of components during transfer between individuals are being increasingly implemented to foster trust in information. The incorporation of suppliers into the aerospace digital thread facilitates the synchronization of material and information flow, thereby averting expensive production delays and enhancing on-time delivery.

The introduction of integrated frameworks and digital innovation is heralding a new age for aerospace manufacturing globally. Substantial expenditures and cultural changes are required for this transition to be fully actualized; nonetheless, the advantages in productivity, quality, and responsiveness render digital adoption a strategic need in a competitive global company, rather than a mere option. Future aerospace leaders will distinguish themselves from competitors by integrating powerful digital technologies with profound manufacturing skills. Organizations capable of establishing a continuous digital thread from initial design to final assembly and beyond will be optimally positioned to innovate rapidly while complying with the safety and reliability norms intrinsic to the aerospace sector. The extensive overview of the current digital transformation in aircraft production, founded on integrated frameworks, indicates an enhancement in intelligence, connectivity, and the capacity to address contemporary difficult needs.

TRANSFORMING AEROSPACE THROUGH DIGITAL TWIN TECHNOLOGY

Digital transformation has transcended being a mere trend; it is now the cornerstone of contemporary aerospace development. Organisations are adopting data-driven decision-making, virtual simulations, and AI-enhanced modelling to enhance efficiency, minimise expenses, and expedite innovation. Central to this transformation is digital twin technology—a virtual depiction of physical assets, processes, or entire systems that concurrently change alongside their real-world equivalents. In aerospace, digital twins are transforming aspects ranging from aircraft design to predictive maintenance, enabling organisations to evaluate, enhance, and optimise intricate systems prior to the construction of any prototype. Significant investments by leading aerospace companies in this technology are anticipated to transform manufacturing, supply chains, and operational efficiency throughout the industry.

Figure 3. Integrated framework in aerospace manufacturing

Figure 3 illustrates the Integrated framework in aerospace manufacturing. The United Kingdom is spearheading a major digital twin program with the establishment of the UK Digital Twin Centre in Belfast, Northern Ireland, scheduled to commence operations in early 2025. This £37.6 million (US$47.5 million) initiative is supported by governmental entities and prominent aerospace corporations, including Thales UK, Spirit AeroSystems, and Artemis Technologies.

Conventional aircraft development processes are sluggish, expensive, and susceptible to postponements. The F-35 program experienced considerable challenges owing to intricate prototyping and constraints in real-world testing. Digital twins are transforming this process by allowing engineers to optimise aircraft systems in virtual environments, hence obviating the necessity for numerous physical prototypes. Prominent aircraft businesses are currently utilising digital twins to expedite product development. Rolls-Royce has developed advanced engine simulation models, allowing engineers to assess performance data and anticipate breakdowns beforehand.
• Boeing has integrated digital twins into its assembly lines, facilitating real-time optimisations that reduce production delays. Lockheed Martin has created digital twins for modern fighter jets, enhancing component design and aerodynamics without the need for conventional wind tunnel testing. Digital twins enhance accuracy, decrease expenses, and facilitate the expedited and efficient entry of aircraft into service by diminishing dependence on physical prototypes.

AEROSPACE INDUSTRY GAINS ADVANTAGES FROM DIGITAL INNOVATION

The implementation of digital technology and integrated systems by aircraft manufacturers offers multiple benefits. Automation and data-driven optimization yield enhanced productivity and cost savings. There is a decrease in waste and an expedited completion of activities. Inventory costs and downtime can be minimized through enhanced predictive maintenance and improved supply chain coordination. An elevated standard of quality and safety Advanced monitoring systems and artificial intelligence-driven inspections identify faults or flaws promptly, guaranteeing that components comply with the rigorous standards of the aerospace sector. Consistent manufacturing yields less failures and diminished rework requests. The safety and reliability of aviation components are improved due to superior manufacturing quality.

Flexibility and invention are augmented. A digital instrument facilitates swift modifications in production and expedited design iterations. The digital thread remains uninterrupted during the implementation and propagation of design changes. This agility enables producers to swiftly adapt to new needs or changes in demand. Engineers are granted greater freedom to create due to the accessibility of additive manufacturing and simulation tools, which allow for the validation of audacious new designs without the necessity of extensive physical testing.

Collaboration and visibility have been improved consequently. A system that consolidates data from several sources dismantles information silos. Cross-functional members from diverse disciplines, including design, engineering, production, and quality, can collaborate using the same current data, thereby enhancing coordination. Suppliers are afforded the same real-time visibility, ensuring that all individuals involved in product delivery remain consistently aligned, hence facilitating a seamless operation. Enhancing transparency fosters confidence among parties and facilitates more informed decision-making.

A comprehensive digital twin encompasses all aspects of the product, the manufacturing process, the simulation, and the feedback mechanism regarding the product's performance during usage. This data is utilized to enhance each successive iteration of the product. To optimize the synergy between design and manufacturing, digital twins are integrated into a digital thread. This thread presents a singular environment in which all elements coexist. This encompasses model-based systems engineering, program planning, product design and engineering, verification management, supplier management, intelligent manufacturing, and product support, including maintenance, repair, and overhaul. From the inception of an engineer's concept to the product or manufacturing line's end-of-life, it is crucial for an A&D manufacturer to possess visibility and tools within a unified open digital framework. This enables data utilization throughout the entire organization to enhance nearly every facet of manufacturing. End-to-end digital threads facilitate the rapid development, cost-effective testing, and efficient prototyping of goods. Furthermore, it facilitates the establishment of a closed "feedback loop," offering near real-time insight into necessary modifications, inventions, or variations to promptly address market demands or particular consumer requirements. The interval between "flash to bang" is significantly shortened, enhancing ROI, competitiveness, and customer satisfaction. Furthermore, digitization entails enhancing products to be "smarter," interconnected, or autonomous. These commercial products are either more enjoyable or more functional, tempting consumers to purchase them as an enhancement to their quality of life. Numerous instances exist, ranging from vacuum cleaners to the monitoring of residential electricity loads. The objective of the defense case is to augment operational availability or expand military capabilities to counter more advanced opponents.

Effective digital transformation necessitates meticulous and intentional planning, primarily emphasizing proactive and collaborative change management. All change initiatives consist of a sequence of actions, linked with a distinct vision or purpose, supported by the organization's leadership. Initially, leaders must explicitly delineate the desired end state - the definition of 'right.' As previously said, they must determine which phase of digital transformation (product, manufacturing, or performance) to address initially. Digitalization entails the reconfiguration of current processes, the implementation of innovative tools and techniques, the early and frequent involvement of employees at all levels, and the cultivation of a profound comprehension of consumer requirements and market dynamics. Therefore, it is essential for leaders to comprehend their overarching objective prior to initiating any transformational endeavor. Many firms focused on digital transformation characterize it as a journey, emphasizing its long-term and cumulative aspects. The process commences with an exploration or requirements assessment of existing capabilities, procedures, mindsets, and expectations. Organizations progress in proficiency, comprehension, and assimilation of digital attributes and mentalities until they completely adopt and transform into a "digital" enterprise. In "Digital Transformation at Scale," Greenway, Terrett, Bracken, and Loosemore (2021) advise firms to initiate their digital transformation with small, novel, straightforward, and prominent projects, rather than addressing the most significant issues initially. They delineate three obstacles encountered by change managers: the expectation to prioritize superficial issues, the need to mitigate 'change fatigue,' and the necessity to confront a misleading sense of urgency. They offer ten design principles, or values, that senior executives can adhere to in order to facilitate their digital transformations. After the organization establishes its objectives and organizational structure, a conceptual model must be created to delineate linkages, capacities, responsibilities, and dependencies. Although broad models are accessible, most digital transformation initiatives are context-specific and necessitate a customized model adapted to an organization's distinct environment and structure. This model can act as an initial reference.

Transformation generally commences with an enhancement of the current technological infrastructure and hardware (an "IT Uplift"). "Digitizing Operations" entails modifications to internal processes, including digital methodologies into previously analog tasks. This is an efficiency measure with which numerous organizations contend, as it necessitates alterations to systems frequently regarded as unchangeable. The third pillar, "Digital Marketing," pertains to the emphasis on digital tools essential for engaging and selling to prospective clients, encompassing artificial intelligence (AI), predictive analytics, the establishment of digital markets, and geo-targeting campaigns. "New Ventures" entails the amalgamation of digital teams to evaluate and capitalize on emerging markets for the organization, developing innovative business models, products, services, or partnerships to generate new avenues for growth. These steps may be executed concurrently or separately; nevertheless, they must be thoroughly integrated into a coherent strategy that is well understood and communicated across the organization. For executive leaders, prioritizing digital transformation is imperative; the focus now is on how leadership teams effectively integrate and leverage digital business models as a competitive advantage in the marketplace. Establishing a digital transformation strategy is essential – delineating roles and responsibilities across all organizational tiers, along with interactions with consumers and suppliers. Executive executives bear the task of elucidating the possibilities and critical components of the IT Uplift. Digital Transformation of Operations Digital Marketing Novel Enterprises What it encompasses

Updating current information technology Enhancing current business operations Digital instruments for marketing e-commerce and consumer acquisition Innovative business models and offerings Advantages Adaptable platforms Tool ecosystem Expense minimization Optimization of Efficiency Upselling and cross-selling Market or pocketbook share Brand equity Opportunities for expansion Essential competencies for IT architects and DevOps teams Management of change Understanding of business processes Change management Data analysis Digital marketing Establishment of a business Processes of innovation Key Performance Indicators for Innovation Leaders Innovative instruments, diminished expenses, enhanced functionalities, workforce contentment Conservation of time, personnel, and financial resources Enhanced client satisfaction Return on marketing investment, lead generation, client acquisition Innovative goods, market access 12 digital transformation initiatives. They coordinate personnel and processes in accordance with the strategic vision to address the unavoidable conflicts and challenges that these efforts typically face. Committed and visionary leadership is a crucial component of any transformation initiative. Executive leadership must exhibit a commitment to initiate the trip, navigate the inevitable challenges, disclose both accomplishments and mistakes, and foster a mindset transformation at all levels. To navigate the intricate process of digital transformation, leaders must delineate a definitive chain of command within the organization. This frequently entails designating accountability for overseeing personnel, processes, and expectations to a certain office or individual. A Chief Innovation Officer (CInO) or Chief Information Officer (CIO), as a permanent part of the executive team, may be ideally positioned to manage this change as the company's digital leader. An alternative is to designate a temporary executive staff member as a Chief Digital Officer (CDO), who would report directly to the CInO or CIO. Their responsibility is only to oversee and regulate the implementation of digital technology and procedures throughout the organization till completion. They must comprehend the organization's principal strategy and performance objectives and align technological and data capabilities to facilitate the transformation of a traditionally analog business operation into a digital one. Regardless of whether it is the CInO, CIO, CDO, or another individual, technologically proficient leadership acts as a pivotal change agent, driving transformations in technology, culture, and attitude — fundamentally establishing new operational methodologies. They design corporate methodologies to integrate 'digital DNA routes' into tactical and strategic business activities. This executive change manager frequently collaborates with a compact team of digital specialists, analysts, and business executives to inform and synchronize the personnel, processes, and technology necessary to realize the digital vision.

DISRUPTION IN MANUFACTURING: AI-ENHANCED DIGITAL FACTORIES

In addition to development, digital twin technology is revolutionising aircraft manufacturing. Siemens Digital Industries Software is leading this transformation by collaborating with aerospace company Natilus to develop comprehensive digital twins of next-generation aircraft. Utilising Siemens Xcelerator and AI-driven modelling, engineers may optimise production layouts prior to the commencement of construction. Detect bottlenecks instantaneously to enhance supply chain efficiency. Anticipate machine malfunctions to minimise downtime and maintenance expenses.

Conventional aeroplane maintenance depends on predetermined schedules and responsive inspections. Digital twins are transforming the industry by leveraging real-time sensor data and AI-generated insights to anticipate probable faults, hence facilitating predictive maintenance. The University of Nottingham, in partnership with simulation firm Altair, is creating digital twins for electric aircraft propulsion systems. This transition has significant ramifications for airlines, military fleets, and commercial operators, as unanticipated maintenance expenses can amount to millions of dollars per year.

The integration of artificial intelligence and digital twin technology is revealing previously inaccessible capabilities. PhysicsX, a premier AI-driven engineering firm, has created machine learning algorithms that forecast real-world results more rapidly than conventional simulations. By training AI models on intricate aircraft physics, organisations can. Instantaneously evaluate component performance under harsh situations. Minimise manufacturing waste by utilising only optimal components. Replicate aerodynamics, fuel efficiency, and structural integrity within seconds. This advancement expedites aerospace innovation, diminishes expenses, and enables corporations to iterate at an unparalleled pace.

CONCLUSION

Digital twin technology has transitioned from experimentation to being essential for operations. Global aerospace leaders, research institutions, and AI-driven startups are significantly investing in this domain, establishing digital twins as the norm for aircraft design, manufacture, and maintenance. This signifies a substantial transformation for organisations. Organisations that include digital twins into their operations will surpass competitors by minimising expenses, expediting innovation, and enhancing efficiency. The market is experiencing tremendous expansion for investors. The global digital twin market is anticipated to surpass $73 billion by 2030, propelled by demand in aerospace, defence, and advanced industrial sectors. Regulators necessitate new frameworks. As AI-driven digital twins gain autonomy, the industry must implement compliance requirements to guarantee safety and reliability. The aerospace sector is embarking on a novel epoch when virtual models will dictate tangible results. Organisations who use digital twin technology now will be the architects of the future of aviation.

REFERENCES

Bernabei, M., Eugeni, M., Gaudenzi, P., & Costantino, F. (2023). Assessment of smart transformation in the manufacturing process of aerospace components through a data-driven approach. *Global Journal of Flexible Systems Managment, 24*(1), 67–86. DOI: 10.1007/s40171-022-00328-7

Bhatia, V., Kumar, A., Sidharth, S., Khare, S. K., Ghorpade, S. C., Kumar, P., & AlZohbi, G. (2024). Industry 4.0 in Aircraft Manufacturing: Innovative Use Cases and Patent Landscape. In *Industry 4.0 Driven Manufacturing Technologies* (pp. 103–137). Springer Nature Switzerland. DOI: 10.1007/978-3-031-68271-1_5

Bhatia, V., Sidharth, S., Khare, S. K., Ghorpade, S. C., Kumar, P., Kumar, A., & Agarwal, A. (2024). Intelligent manufacturing in aerospace: integrating industry 4.0 technologies for operational excellence and digital transformation. In *Industry 4.0 Driven Manufacturing Technologies* (pp. 389–434). Springer Nature Switzerland. DOI: 10.1007/978-3-031-68271-1_18

Butt, J. (2020). A conceptual framework to support digital transformation in manufacturing using an integrated business process management approach. *Designs, 4*(3), 17. DOI: 10.3390/designs4030017

Carou, D., & Carou, D. (2021). Aerospace transformation through Industry 4.0 technologies. *Aerospace and Digitalization: A Transformation Through Key Industry 4.0 Technologies*, 17-46.

Krol, F., Saeed, M. A., & Kersten, W. (2020). A holistic digitalization KPI framework for the aerospace industry. In *Data Science and Innovation in Supply Chain Management: How Data Transforms the Value Chain. Proceedings of the Hamburg International Conference of Logistics (HICL),* Vol. 29 (pp. 797-847). Berlin: epubli GmbH.

Li, L., Aslam, S., Wileman, A., & Perinpanayagam, S. (2021). Digital twin in aerospace industry: A gentle introduction. *IEEE Access : Practical Innovations, Open Solutions, 10*, 9543–9562. DOI: 10.1109/ACCESS.2021.3136458

Oriji, O., & Joel, O. S. (2024). Integrating accounting models with supply chain management in the aerospace industry: A strategic approach to enhancing efficiency and reducing costs in the US. *World Journal of Advanced Research and Reviews, 21*(3), 1476–1489. DOI: 10.30574/wjarr.2024.21.3.0873

Pop, G. I., Titu, A. M., & Pop, A. B. (2023). Enhancing aerospace industry efficiency and sustainability: Process integration and quality management in the context of Industry 4.0. *Sustainability (Basel), 15*(23), 16206. DOI: 10.3390/su152316206

Rubio, L. R., Martin-Mariscal, A., & Peralta, E. (2022, July). First approach to a theoretical framework for carbon footprint management in the aerospace manufacturing industry. In *IFIP International Conference on Product Lifecycle Management* (pp. 620-629). Cham: Springer Nature Switzerland.

Szejka, A. L., Junior, O. C., & Mas, F. (2024). Knowledge-based expert system to drive an informationally interoperable manufacturing system: An experimental application in the Aerospace Industry. *Journal of Industrial Information Integration*, *41*, 100661. DOI: 10.1016/j.jii.2024.100661

Tao, F., Ma, X., Liu, W., & Zhang, C. (2024). Digital Engineering: State-of-the-art and perspectives. *Digital Engineering*, *1*, 100007. DOI: 10.1016/j.dte.2024.100007

Tegethoff, T., Santa, R., Bucheli, J. M., Cabrera, B., & Scavarda, A. (2025). Navigating Industry 4.0: Leveraging additive technologies for competitive advantage in Colombian aerospace and manufacturing industries. *PLoS One*, *20*(2), e0318339. DOI: 10.1371/journal.pone.0318339 PMID: 39928702

Tiwari, S. P. (2008). Information and communication technology initiatives for knowledge sharing in agriculture.

Tiwari, S. P. (2023). A transformative and action plan-oriented approach to manufacturing using artificial intelligence and digital technologies. In *BISTIC business innovation sustainability and technology international conference (BISTIC 2023)* (pp. 22-31). Atlantis Press. DOI: 10.2991/978-94-6463-302-3_4

Tiwari, S. P. (2025). The implications of artificial intelligence in management consulting: A risk and barrier assessment. *Multidisciplinary Reviews*, *8*(8), 2025244–2025244. DOI: 10.31893/multirev.2025244

Troisi, O., Visvizi, A., & Grimaldi, M. (2024). The different shades of innovation emergence in smart service systems: The case of Italian cluster for aerospace technology. *Journal of Business and Industrial Marketing*, *39*(6), 1105–1129. DOI: 10.1108/JBIM-02-2020-0091

Wu, D., Rosen, D. W., Wang, L., & Schaefer, D. (2015). Cloud-based design and manufacturing: A new paradigm in digital manufacturing and design innovation. *Computer Aided Design*, *59*, 1–14. DOI: 10.1016/j.cad.2014.07.006

Yilmaz, A. A. (Ed.). (2024). *Harnessing digital innovation for air transportation*. IGI Global. DOI: 10.4018/979-8-3693-0732-8

Zheng, X., Hu, X., Arista, R., Lu, J., Sorvari, J., Lentes, J., Ubis, F., & Kiritsis, D. (2024). A semantic-driven tradespace framework to accelerate aircraft manufacturing system design. *Journal of Intelligent Manufacturing*, *35*(1), 175–198. DOI: 10.1007/s10845-022-02043-7

Zhou, W. (2024). Competition and development in the aviation industry: An analysis of strategic adaptability and challenges.

Chapter 4 Robotics and Manufacturing: Implications for Economic Growth in the Future

Rashid Hashmi
https://orcid.org/0009-0006-3232-0705
Sharda University, India

ABSTRACT

Manufacturing is the world's oldest and largest industry. Manufacturing has lapsed in investment when it comes to research and development, and robotics is an upcoming thrust area. The author of this book chapter discusses how strategic investments in education, innovation, and inclusive policies will determine whether society can harness robotics effectively for sustainable economic growth and prosperity. Robotics has resulted in significant increases in manufacturing productivity and efficiency, which have reduced operational costs and production errors. As automation changes employment dynamics, it contributes to economic growth, but it also requires substantial workforce adaptations and retraining. Robotics has dramatically improved the speed at which a product is manufactured and delivered to the consumer, which has resulted in lower costs and higher quality products. Technology is creating specialized technical roles; however, it also disrupts traditional employment, requiring workers who have been retrained to acquire a number of new skills.

DOI: 10.4018/979-8-3373-1082-4.ch004

INTRODUCTION

Advances in robotics now enable rapid and automated configuration of manufacturing cells, significantly enhancing operational flexibility and efficiency (Asif et al., 2025). Robotics significantly enhances manufacturing productivity and efficiency by reducing operational errors and optimizing robotic functions through machine learning (Arimoto, Kawamura, & Miyazaki, 1984). Recent innovations in robotics have led to vision-guided adaptive gripping systems, significantly optimizing automated garment manufacturing processes (Choi et al., 2025). Comparing the manufacturing industry of the olden days with the other industries of the time, manufacturing enjoyed a very high level of technological development for its historical period. Research in automatic control has significantly advanced robotics and manufacturing automation, driving industrial innovation toward a more sustainable future (Rupenyan & Balta, 2024). Robots' control loops can be greatly simplified by incorporating active-learning frameworks that incorporate feedback-driven exploration (Taylor et al., 2021). In reflections on how technology affects national development, it is emphasized that context-sensitive robotics strategies are needed which align innovation with broader societal objectives (Tiwari et al., 2025). Drawing a parallel with medical sciences, even in sports trauma and knee surgery as compared to conventional tools and patient-specific instrumentation, navigation- and robotic-assisted techniques improved alignment in total knee arthroplasty, while clinical outcomes remained the same (Lei et al., 2022).

AI-enhanced manufacturing robotics has transformed industrial applications by introducing cutting-edge technological trends that significantly boost production efficiency (Adebayo et al., 2024). Robotics advancements have significantly benefitted from artificial intelligence approaches, enhancing autonomous decision-making and operational intelligence in manufacturing settings (Brooks, 1986). Recent technological trends in robotics have notably reshaped manufacturing practices, enabling enhanced productivity, precision, and operational flexibility (De San Bernabe et al., 2024). There are many old civilizations that have built structures that have been able to withstand the test of time for many thousands of years, such as pyramids, acropolis, aqueducts, cathedrals, and so on. Their standard normal building procedures included innovative processes and elements that are still being used today as part of the standard normal building procedures. Industrial robotics has driven substantial advancements in manufacturing processes, enhancing efficiency, consistency, and operational productivity across industries (Snyder, 1985). Collaborative robots (cobots) have significantly transformed manufacturing by enhancing human-robot interactions, opening numerous opportunities for future industrial developments (Liu et al., 2024). Artificial intelligence-driven robotics has significantly advanced the food industry, enhancing productivity, precision, and quality in food processing and production (Hwa & Te Chuan, 2024). Advancements in identifying robotic dynamics have enabled enhanced precision and effectiveness in robotic movements during manufacturing operations (Gautier, 1986). The synergy of artificial intelligence and robotics significantly enhances manufacturing automation by improving tool longevity, particularly evident in processes like power skiving (Kiefer et al., 2024). Over the past few decades, manufacturing processes in some parts of the world have changed significantly compared to the manufacturing processes of the past.

Advancements in inverse kinematics have enabled robots with redundant manipulators to achieve greater flexibility and efficiency in complex manufacturing tasks (Burdick, 1989). Robotics has advanced significantly in indoor manufacturing environments by effectively coordinating multiple autonomous mobile robots (Caloud et al., 1990). A deep learning-enhanced Digital Twin framework substantially improves safety and reliability in human-robot collaborative manufacturing environments, optimizing overall operational performance (Wang et al., 2024). In the past eight hundred years, for instance, the process of building erection has not changed much in terms of the things that are used, especially when it comes to the materials that are used in the building process. Modern-day cranes are replacing the pulleys of the old ages, which were used in the old times to move loads up and down. Calibration techniques have notably improved robotic accuracy in manufacturing environments by experimentally validating and optimizing robot performance (Judd & Knasinski, 1987). It is true that today's machines are much more sophisticated than those of centuries ago, but they adhere to the same principles as those of centuries ago, such as manual control, color grading, visual and verbal feedback for the operator, and the correct positioning of the machine. Addressing dynamic control instabilities in robotics has been essential for enhancing operational stability and precision in manufacturing automation (Williams & Glover, 1985).

Improved gripping mechanisms have significantly enhanced industrial robot capabilities, enabling more precise handling and manipulation of objects in manufacturing (Chen, 1982). Among the changing elements over the years, the only changes have been the replacement of human force by electrical or diesel actuators, and the replacement of wooden structures with steel ones. Due to these two advancements, the speed of elevation, the payload, and the reachability of the craft have significantly increased, although the manufacturing philosophy has remained largely the same. With the rapid development of service robotics in the manufacturing industry over the past few years, it has become one of the most important research areas in that field. RAC (Robotics and Automation in Construction) has a number of challenges, mainly because of the nature of the work environment, which is generally highly unstructured. Robotics applications have expanded through the engineering of cellular robotic systems, enabling more adaptable and distributed manufacturing processes (Hackwood & Wang, 1988). Effective feedback mechanisms have enhanced robotic accuracy in manufacturing by stabilizing and improving tracking capabilities under operational constraints (McClamroch & Wang, 1988). Several factors contribute to the difficulties associated with working in such a setting, including handling heavy objects, parts that are made with a high degree of tolerance, low standardization, medium industrialization, prefabrication, and the presence of numerous noncoordinated actors (architects, site supervisors, builders, material suppliers, etc.). The importance of increasing the level of automation in this important sector cannot be overstated in order to improve the level of efficiency in the processes involved in order to increase the productivity of the industry.

By investing in robotic automation systems, manufacturing companies can maintain their competitiveness during years of growth that are expected to continue for many years to come. Automatic synthesis of fine-motion strategies has significantly enhanced robotic precision and adaptability in complex manufacturing operations (Lozano-Perez, Mason, & Taylor, 1984). Through the use of automation, operations within an organization can be made more efficient, safer and more accurate, allowing employees to spend more time on innovation and other processes that are important for the company's growth and success. The automation of manufacturing processes can help companies stay competitive on the global market, enabling them to provide a viable alternative to offshoring and bridging the skills gap in areas where it may be difficult to find employees who are equipped with the necessary skills. Manufacturing automation solutions can help to increase productivity, increase the level of worker satisfaction, as well as increase the bottom line of companies as they increase worker safety and satisfaction. The autonomy of robots can be classified in a number of ways. The robotic system is capable of reproducing specific actions over and over again over and over again with a high degree of accuracy and faithfulness without any variations or reversals, often referred to as repetitive actions with a high level of accuracy without any variations or reversals. During the course of the series of coordinated motions, it is determined that a series of coordinated motions will take place by a set of program routines that specify the direction, acceleration, velocity, deceleration, and distance of the series of coordinated motions that will take place.

The importance of increasing the level of automation in this important sector cannot be overstated. The level of automation in this important sector needs to be increased in order for us to be able to achieve higher levels of efficiency in the processes involved. This will increase the productivity of our industry. The use of robots in robotic processing involves manipulating a tool in order to be able to perform a particular process on a part that is being processed by the robot, such that the robot is able to perform the particular process on the part. The use of spray painting can be applied in a number of ways, such as it can be used in spot welding, continuous arc welding, and spray painting to name just a few. There are many applications in the United States that use industrial robots, one of which is spot welding of automobile bodies, which is well known as one of the most common applications for industrial robots. A spot welder is then placed against the automobile panels and frames in order to complete the welding process for the basic body of the car after the robot places the spot welder on the auto panels and frames in order to complete the assembly process. A welding rod is moved along a seam as a robot moves along the seam to be welded in a continuous motion along the seam to be welded in a continuous motion along the seam to be welded in a continuous motion along the seam to be welded in a continuous motion along the seam to be welded. Spray painting, as the name suggests, is the process of applying paint onto the surface of an object since spray-paint guns are used to apply paint while manipulating the gun over the surface in order to make the paint adhere to the object when applied. With the help of a rotating spindle, the robot is not only able to grind, polish, and route the part but can also perform other tasks related to this category in addition to grinding, polishing, and routing the part.

There are some manufacturing plants where people find it difficult to differentiate between cobots and robots due to the fact that sometimes it is difficult for them to tell the difference between the two. The integration of human-robot collaboration in manufacturing presents considerable opportunities and challenges, particularly within the evolving context of Industry 5.0 (Dhanda et al., 2025). There is no doubt that there are many advantages that come from the fact that cobots are collaborative robots that achieve cost-effectiveness, safety, and ease of deployment, as well as the fact that robots achieve cost-effectiveness, safety, and ease of use. The use of collaborative robots in the world we live in today, also known as cobots or cooperative robots, is making it easier than ever before for small and medium-sized businesses in all corners of the globe to automate their processes without the need for any human involvement. To make automation even more convenient for businesses of all sizes, cobots are being used as a way to share a workspace with humans and to share the workload between humans and cobots. Programming cobots in such a way that they can be used with humans in the same space as humans is possible. Despite all of these advantages, cobots can be a game changer in a wide range of applications, due to their ability to perform multiple tasks simultaneously. When it comes to the orientations of the objects in which robots operate and the tasks that must be performed on those objects, modern day robots are far more flexible than ever before. The robot may even be required to identify itself when operating so that it can identify itself accordingly. A robot, for instance, is often equipped with a machine vision subsystem, essentially the robot's visual sensor, which is connected to powerful computers or controllers that provide precise guidance to the robot.

KEY PILLARS OF ROBOTICS IN MANUFACTURING

Figure 1. Key Pillars of Robotics in Manufacturing

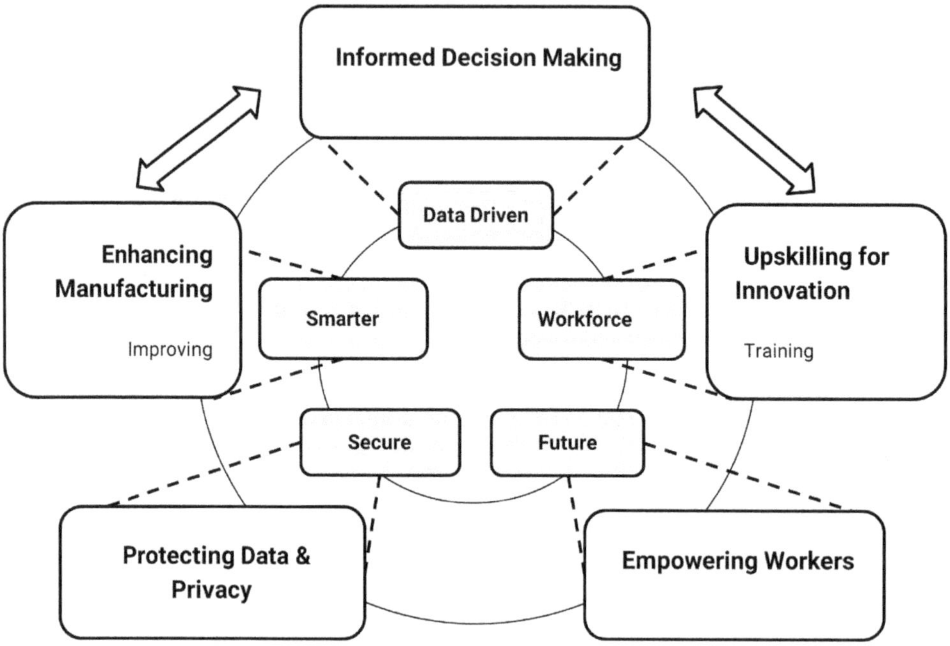

Figure 1 illustrates how robotics has become a cornerstone of modern manufacturing, driving productivity gains and innovation. A key element of this transformation is data-driven growth, which means that rich data insights increasingly guide decisions and improvements. Today's smart factories use advanced sensors and robotic systems to continuously collect and analyze data to inform strategy, creating a feedback loop for continuous improvement. A data-centric approach enables manufacturers to make informed decisions at all levels: real-time and historical trends can be investigated, patterns are identified, and better operational decisions are made based on data. Big data analytics combined with robotics allows managers to see production processes more clearly, adjust to demand fluctuations, and implement changes more confidently. With data-driven insight, manufacturing becomes more agile and responsive.

Due to the use of automation in many different areas of the manufacturing industry, robotic systems have been used in a variety of different areas of the manufacturing industry in order to automate a variety of different factors. It is not uncommon for robots to be used in several different areas of the manufacturing industry, for example, welding, assembly, shipping, raw material handling, and packaging of products, among others. Many reasons have contributed to this, along with the pressures of the economic environment. As a result of this, there are an increasing number of manufacturers who are turning to robotic automation for a growing number of tasks in their production processes in order to make these processes run more efficiently. Throughout the history of the industrial revolution, many industries of manufacturing have undergone significant changes as a consequence combined with the economic environment. There were several changes introduced into their processes, one of which was the incorporation of robots and automation. The adoption of robots in industry has significantly improved manufacturing ergonomics, productivity, and workplace safety, transforming operational processes (Edwards, 1984). This product line was introduced to the market with the aim of increasing productivity and reducing the operating costs for the user at the same time. Today, thanks to the technology of machine learning and the wide range of programmable capabilities that robots have, they are capable of performing a wide range of complex tasks with the help of technology. This book chapter is intended to provide with a number of facts and information about how robotics plays an important role in manufacturing, including how robotics' usage will impact the manufacturing sector in the near future and how it will affect the manufacturing industry in general in the near future. Several scholars predicted that rapid advances in robotics would soon force legislators to consider whether autonomous machines should have legal protection (McNally & Inayatullah, 1988).

There are many key pillars enabling the use of this technology, but one of the key pillars is its ability to be able to positively impact the manufacturing process itself. Using robotics and automation technologies for factories will ensure that factory efficiency, precision, and throughput will be enhanced by enhancing efficiency, precision, and productivity through the use of robotics and automation technologies. Addressing psychological well-being through targeted wellness strategies is essential for human operators working alongside AI-driven systems in manufacturing, ensuring improved productivity and employee health (Fahrudin, Tiwari, & Rahmi, 2025). When robots take over repetitive and precise tasks, they increase the speed, consistency, and accuracy of production while at the same time reducing the amount of human effort that is needed. Artificial Intelligence significantly enhances automated manufacturing and supply chain systems, bringing notable advancements, addressing existing challenges, and highlighting promising future directions (Kilari, 2025). With the help of connected machines and artificial intelligence systems, predictive maintenance and self-optimization of operations can be carried out, resulting in new levels of efficiency and reducing operational costs. Studies have shown that smart manufacturing can significantly improve manufacturing efficiency. With robots taking on tedious tasks, workflows can be optimized in real-time, resulting in improved production efficiency, flexibility, and cost savings:

Robotics adds value to factory outputs beyond efficiency, increasing quality and customizability. Automated intelligent systems are enhancing product quality and enabling mass customization. A mobile connected device on the factory floor continuously monitors quality metrics and can identify errors immediately, reducing waste. While integrating data across production and supply chains, manufacturers are able to tailor outputs to specific customer needs without sacrificing quality. A robotics-driven process that adapts quickly to data feedback results in improved quality and flexibility of output.

Human dimension is just as important as the technological dimension. An important pillar of innovation is the upskilling of the workforce to work with robotics in a way that will foster further innovations and drive further development in the industry. The integration of advanced robots and artificial intelligence in factories is leading companies to invest in employee training programs as a means of developing the digital and technical skills of their employees. Investing in such upskilling initiatives is becoming an integral part of workforce planning to ensure that employees have the knowledge and skills to operate, program, and collaborate with robotic systems. Tens of thousands of Google workers have been trained in AI skills through its in-house AI Academy, enabling them to contribute to innovation and use new technologies. Manufacturing companies encourage continuous improvement and innovation by equipping workers with the skills to interpret data, maintain robots, and improve automated processes. Upskilling employees in this way not only addresses the skills gap, but also unlocks their innovative potential, so they can explore new ideas and drive process improvements using robotics.

Workers are empowered when they are upskilled. With robotics becoming more prevalent in manufacturing, workers' roles are increasingly being enhanced rather than being replaced by the technology. Job duties that are hazardous, strenuous, or monotonous can be done more effectively by robots. Weighing, heavy lifting, welding in hazardous environments, or repetitive assembly are all examples of activities that can improve workplace safety and reduce the risks associated with health problems using robots. Smart manufacturing technologies enable workers to concentrate on tasks that add higher value. By eliminating many manual hazards and automating drudgery, these technologies allow them to concentrate on higher value tasks, such as problem-solving, design, and creative thinking. Humans can then take on roles that are more satisfying and enriching for them as a result of this shift. Rather than being an alternative to humans, robotics can actually empower employees by freeing them up from having to deal with complex decisions as well as thinking more creatively and innovatively. The fact is that companies are finding that combining robots and humans produces the best results in terms of precision, repetition, and oversight. Robots excel at precision and repetition, while humans are adept at providing intuition, creativity, and oversight. Synergies like these not only improve productivity, but they also increase job satisfaction for workers as they are able to take on more skill- and cognitive-based tasks.

ROBOTICS AND THEIR ECONOMIC IMPACT ON MANUFACTURING

Robotics is having a profound impact on the manufacturing industry in terms of economic implications. Robots can be extremely helpful in increasing the amount of output per worker by taking over tasks that used to be labor-intensive and thus lowering the marginal cost of production at the same time. By operating around the clock, manufacturers have the ability to increase their productivity while utilizing less capital, reducing their downtime and maximizing their performance. A robotic system has also been shown to be capable of performing tasks with high precision and consistency, enhancing the quality of a product and reducing waste or rework. Firms and industries are directly affected by these improvements when it comes to their competitiveness as well as their profitability. There are several advantages that can be achieved by companies that effectively deploy robots, both in terms of price and quality. These advantages give them a significant competitive advantage both domestically and globally. A country's productivity growth and trade patterns can also be influenced by widespread implementation of robotics at a macroeconomic level, greatly altering its comparative advantage in manufacturing, and then influencing its global competitiveness.

It has been many decades since robotic manufacturing systems were first introduced, but some areas of the manufacturing industry may still perceive them as relatively new concepts, despite the fact that robotic manufacturing has existed for decades. A factory owner or manager can increase their profit margin by combining traditional production methods with more advanced forms of technology, resulting in an exponential growth in profits and income.

With the use of robots in manufacturing, there has been a great deal of positive change that has been made to the industry as a result of its use. In today's world, automated manufacturing is gaining popularity as an important tool to increase productivity and success in a company's production process, as well as improving the overall productivity and success of the company as a whole. The human workforce will be able to contribute more to the organization as a whole if they are not forced to perform mundane tasks that can easily be carried out by robots. By allowing them to contribute more knowledge and ideas at the higher levels of the organization, they will be able to make an increased impact on the organization as a whole. An organization can undoubtedly produce an undeniable and noticeable increase in efficiency and productivity through the use of robotics that are implemented correctly in its operations, resulting in undoubtedly noticeable and undeniable improvements to its operations. The advantages of robotic milling have recently been highlighted in a review of the state of the art that describes how highly efficient and highly precise the machining of complex components is (Wang et al., 2023).

Although robotics is growing in popularity, it also raises a number of critical questions, especially in the context of the workforce and the distribution of outcomes. There has been a lot of concern about job displacement occurring within the manufacturing sector as a result of automation of production tasks. During the process of automating certain tasks on the factory floor, workers may become unemployed or their wages may be pushed downward as robots replace some of their jobs, especially in roles involving routine, repetitive tasks. The impact of this dynamic has sparked debates about the future of work: will robotics render some skills obsolescent or will they create new opportunities for labor in conjunction with machines in the coming years? Based on the historical precedents of previous industrial revolutions, it is believed that productivity gains driven by technology ultimately lead to the creation of new jobs and the expansion of the economy. There has been a rapid spread of automation across different sectors at a rapid pace. In recent years, there have been speculations that these adjustments may be more disruptive than before and that this may result in a wider gap between those who benefit from technology and those who are left behind.

The context and history of the industrial revolution makes it imperative to examine the extent to which robotics integration is shaping productivity, employment, industrial performance, and broader economic structures within the broader context of the industrial revolution. In manufacturing, robotics is not an isolated development, but rather it is a process that interacts with economic principles in a complex way, involving both the alteration of production functions and the cost structure, as well as the change of skills demand on the labor market. Among the factors contributing to efficiency gains, according to the author, are improvements in efficiency at the factory level, changes in labor and skill dynamics, changes in the competitive and production landscape, and structural changes in the economy as a whole.

ENHANCING PRODUCTIVITY AND EFFICIENCY IN MANUFACTURING THROUGH ROBOTICS

Improved modeling of impact dynamics has significantly advanced robotic operations by enhancing precision and control during high-speed manufacturing tasks (Wang & Mason, 1987). There has been a marked increase in productivity and operational efficiency in manufacturing since robotics was introduced into the industry. Robots are regarded by the author as being tools that can be used for both increasing output and reducing inputs at the same time as they work on their core tasks at the same time. Robots can increase production efficiency and speed of production far beyond what is possible with human labor alone. This is because they automate repetitive and time consuming tasks. Assemblies can use robotic arms to connect parts, weld joints, and package products continuously. This system can run all days of the week if necessary. As a result, the quantity of goods that can be produced in a given period of time is greatly increased and as a result a higher output per worker is achieved because fewer workers are required to supervise a great deal more production in a given period of time. Robots are considered as a form of investment excercise, whereby firms invest in advanced machinery in an effort to boost their capital-to-labor ratio, which has historically been associated with a higher level of labor productivity in the economy. Modern robotics is different from the traditional investments of the past, however. Today, robots usually come with new technology like machine learning that is much more than just a substitute for labor, but opens up new areas in production and efficiency processes, essentially pushing the frontier of production forward. Robots are poised to become robotic persons, challenging current ethical and social norms as humans learn to live alongside increasingly lifelike machines (Smith, 2021).

THE IMPACT OF ROBOTICS ON MANUFACTURING WORKFORCES AND EMPLOYMENT DYNAMICS

There are significant implications for the workforce and the nature of employment that come with the rise of robotics in manufacturing. The use of automated systems takes over tasks that were previously performed by humans, which has immediately resulted in job role displacement: certain jobs have shrunk or even disappeared as a result of this. The introduction of robots that can weld and conduct assembly might, for instance, reduce the number of workers required to conduct these operations on an assembly line in a factory. Capital can be substituted for labor in these departments, leading to a reduction in the number of employees in those departments as a result. Those workers who are most susceptible in terms of the short run are those who work in routine, repetitive jobs and have their skills closely tied to them. Assembling, handling materials, and packaging are several of the traditional manufacturing jobs that fall into this category, and all of these jobs face significant automation pressure as a result. Displacement effects tend to exert downward pressure on the demand for low-skilled or medium-skilled labor engaged in routine work, which can manifest itself as job losses, hiring freezes for those positions, or stagnant wages as competition for the remaining similar jobs intensifies as the displacement effect takes effect. Humanoid robot engagement can be profoundly enhanced by integrating realistic artificial skin and expressive behavioral cues (Minh Trieu & Truong Thinh, 2023).

There is more to automation's impact on employment than meets the eye. Aside from the displacement of some jobs, robotics holds the promise of creating new employment opportunities and transforming existing jobs in the same way. The productivity effect can be seen as an important countervailing force. As robots improve a factory's productivity and reduce production costs, the factory may be able to expand its operations, resulting in an increase in consumer demand for labor in complementary roles. Having a higher production volume, for instance, might necessitate more workers in the quality assurance, logistics, maintenance, or customer service departments. It is also apparent that as robotics become more mainstream, the need for skilled technical roles will increase: engineers are required to design and improve automated systems, technicians are required to install and maintain robots, and programmers are required to write the software needed to control robots. There has been an increase in roles that require higher levels of education or specialized training, contributing to the shift in the skill profile of the workforce as a result of these new or expanded roles. The impact of robotics on the profession can be viewed in this way: in the sense that it is driving skill-biased technological change that favors workers with technical and analytical skills, while curtailing the opportunities for those with only routine skills.

A balance has been struck between the displacement of certain jobs by robotics as well as the creation of others, which is the net effect of robotics on employment levels. Over the course of history, economies have always been able to adjust to major technological shifts by creating new industries and new types of work as a result of technological developments. The automation of agriculture and manufacturing in earlier periods was eventually responsible for the growth of jobs in the service sector and the creation of entirely new sectors that absorbed the labor that had been freed up by technology. A rise in robotics may also lead to the reallocation of human resources in sectors requiring empathy and interpersonal skills, such as creative design, complex problem solving, or roles that are difficult to automate. In service-oriented industries or emerging sectors of the digital economy, these skills are often required. It is the period of transition, however, that poses the greatest challenge. When workers are displaced, they may not immediately be able to move into new jobs created due to a lack of skills or opportunities. Due to the decline in manufacturing jobs, there are structural unemployment issues and potential social hardships in regions that are heavily dependent on manufacturing jobs.

Human interaction is not necessarily eliminated from manufacturing by the use of robotics. A number of tasks still require a human's judgment, dexterity, and ability to adapt, and the best way for robots to perform these tasks is in conjunction with humans. There has been a noticeable evolution in the relationship between humans and machines: collaborative robots are designed to work safely alongside both human and machine, handling parts of a task while humans handle other parts of the task. When hybrid settings are established, the role of the worker is redefined rather than being obliterated in such settings. Line workers may transition into cell coordinators who oversee a range of automated processes and intervene to handle exceptions or to make adjustments that require the discretion of a human when necessary. Maintaining automated systems becomes one of the most critical functions – skilled trades such as millwrights, electricians, and mechatronics technicians play an increasingly important role to keep the robotic systems running smoothly. Even as traditional jobs disappear, new ones emerge and other jobs evolve, making it evident that manufacturing work in the future won't just be about humans and robots working alongside each other but more likely to be a collaborative effort between humans and robots.

DYNAMIC MARKETS AND INDUSTRIAL COMPETITIVENESS

As robotics becomes more prevalent in manufacturing, it is reshaping competitiveness and changing market dynamics. Staying competitive can require an investment in robotics by individual firms. Through automation, companies can offer lower prices or achieve higher profit margins by increasing efficiency and reducing labor requirements. Firms that adopt advanced robotics often gain a competitive advantage by scaling up production more quickly, meeting large orders more reliably, and maintaining consistent quality, thus taking market share from their competitors. Automotive manufacturers, for instance, achieved high levels of productivity by embracing robotics, which was necessary to compete globally. On the other hand, businesses that fail to automate risk falling behind their competitors. Both domestic and export markets may be lost to more automated rivals because they cannot match their price and quality. A standard requirement to stay competitive has become a cutting-edge advantage as a result of the widespread adoption of robotics over time.

Market structure can be significantly affected. There is potential for market concentration resulting from the high initial costs and complexity of implementing robotics in some cases, favoring large firms with higher capital resources. The deployment of robotics across the operations of large multinational manufacturers can optimize production by assigning tasks to automated lines at various locations globally. As small manufacturers face capital constraints, they may not be able to automate as much, which can create a competitive gap. The most automated and efficient producers may dominate the market, as a result of this dynamic. While this trend is not absolute, as robotics technology matures, it becomes more attainable and user-friendly, which often makes it more affordable for users. Small and medium-sized enterprises are increasingly integrating automation into their processes thanks to the development of modular, cost-effective robotic systems and the availability of automation solutions. There are many benefits associated with this democratization of robotics, such as ensuring an environment in which there is a level playing field, which prevents a one-sided monopolization of markets by some early adopters.

Besides the cost reduction that robots enable, robots also influence how firms compete through product innovation, customization, and differentiation. Compared to manual systems, automated systems can be rapidly reprogrammed in order to accommodate changes in design or introduce a new product model, which gives manufacturers greater flexibility in how they manufacture their products. This makes it possible for firms to respond more quickly to changes in the market or the preferences of their consumers. The use of flexible robotic tools in a factory allows it to switch between different variants of a product with a minimum of downtime, allowing a degree of mass customization to be achieved. It's important for businesses to be able to respond quickly to market demands where variety and speed to market are key competitive assets. The competitive environment is in part transformed from one of pure price competition to one of competitive responsiveness and innovation. Investing in robotics is not always going to be done to reduce the cost of existing products, but also to create new kinds of products at a larger scale or to offer tailor-made solutions to the market, something that would be impractical if the product production process was fully manual.

Pricing dynamics, as well as consumer benefits, are also important components of market dynamics in highly automated industries. Due to the fact that robotics lowers the cost of production, this can lead to a reduction in prices for manufactured goods as a result. There is a possibility that consumers will have access to more affordable products and a wider variety of products. An increase in productivity and efficiency can have a disinflationary effect on manufactured goods as a result of higher productivity and efficiency, as seen in the production of mobile phones, as well as other mass-produced goods where automation is prevalent. There has also been an increase in the need for continuous innovation to keep up with the rapid changes in technology due to lower prices and the need to keep investing in the next generation. As automation becomes a more prevalent part of the production process, the differentiator will become which company or producer can integrate robots with artificial intelligence and data analytics in order to achieve the best results.

IMPACTS ON THE ECONOMY AND STRUCTURAL CHANGE AT A BROADER LEVEL

The diffusion of robotics in manufacturing affects a wide range of industries and firms, causing structural changes in economies. Economies and aggregate productivity have fundamental implications. Robotics plays a crucial role in driving economic development by reshaping labor markets, productivity, and employment structures across various sectors (Calì & Presidente, 2025). The manufacturing sector has traditionally been a high-productivity sector, often driving overall economic productivity increases. By increasing manufacturing productivity, robots contribute to faster macroeconomic growth by increasing output per worker. A well-distributed increase in efficiency can result in a higher GDP and, therefore, potentially a higher per capita income. Several studies have shown that automation and robotics have contributed to significant productivity improvements in advanced economies. With robotics, the economy can produce more with fewer inputs, theoretically improving society's standard of living.

Manufacturing robotics has also led to structural changes in the economy. As the manufacturing sector becomes more automated and less labor-intensive, its ability to absorb labor decreases. Manufacturing employment as a percentage of total employment has declined for years in advanced economies despite manufacturing output growing or remaining high. With robotics, factories can produce more without increasing human labor proportionally, which means that as time goes on, fewer workers will be needed in manufacturing. The result is a reallocation of labor to other industries. Manufacturing has generally shed most of its labor to the service sector. The decline in manufacturing jobs has led to an increase in jobs in healthcare, education, logistics, retail, and various professional and personal services. Increasing productivity and income has indirectly led to some of these service jobs. Increasing affluence will lead to more demand for healthcare and entertainment, while changes in the economy will create new needs.

Economic history has experienced similar structural shifts in the past. There has been a dramatic increase in farming productivity due to mechanization in agriculture historically. Reduce the need for agricultural labor and allow workers to move to urban or manufacturing areas in order to work. Similar patterns are also playing out in manufacturing, where automation is pushing labor toward knowledge-based jobs and services. There is a difference between the transition that occurs today and that which took place in the past, thanks to the fact that the destination sectors for labor often require a different level of skill and may not be able to absorb displaced workers as efficiently. Services have become more and more high-skilled, such as those requiring AI skills, or low-skilled, with fewer middle-level career options than manufacturing used to offer. The author wants to mention that this is what contributes to the polarization effect mentioned earlier and can force us to reevaluate our education system and workforce development strategies.

The use of robotics in manufacturing drives efficiency and growth potential, but it also initiates substantial structural changes in the economy, leading to a number of negative consequences. It changes the composition of employment in different sectors, contributes to shifts in the distribution of incomes, and challenges traditional models of development. A future economic landscape in which manufacturing is highly productive but employs fewer people will be a challenge for societies to adapt to. Making sure that there are new pathways in the economy that will provide jobs, incomes, and a stable economy as a whole. Taking these broad shifts into consideration, much of the attention is presently being directed towards how robotics technology may continue to evolve in the future. The next chapter discusses what policies can be implemented to guide its impact in the long run to the benefit of society, and in what way can this be accomplished.

FUTURE DIRECTIONS AND POLICY CONSIDERATIONS

Several factors are driving the growth of robotics in manufacturing, including rapid advances in technology and the evolving economic imperatives, which are allowing robots to play a greater role in manufacturing. Among the key trends in the field of robotics is the integration of artificial intelligence with robotic systems, resulting in smarter and more autonomous machines. There is a good chance that future robots will be able to perceive and make decisions with enhanced capabilities. Robots with AI-enabled vision or AI-driven control might be capable of coping with tasks that are more complex or variable than they are currently capable of coping with. There is a possibility that this could enable automation to be extended beyond highly structured, repetitive tasks to more adaptive ones. Robots and machines are now able to communicate, adjust themselves in real-time, and optimize production processes on-the-go in the smart factory. There is an interconnection between robotics and general automation, which leads to efficient manufacturing with minimal human involvement, thus reducing human error.

Besides updating workforce development policies, social safety net policies and labor market policies also need to be updated in order to address the effects of automation on the labor market. A wide range of support services including job placement assistance, relocation assistance, and support for displaced workers can make a real difference in reducing short-term hardships in communities affected by manufacturing job losses. Developing robust content moderation frameworks can significantly enhance safety and security in manufacturing environments, facilitating effective omnichannel commerce management (Tiwari, 2025). A high level of automation suggests that policymakers should explore further reshaping of the economy in an effort to adapt to it. Attempts to share the benefits of automation broadly through schemes such as a universal basic income or shortening the workweek are a reflection of the desire to maintain living standards regardless of a decline in traditional employment opportunities. Despite the fact that such ideas remain controversial, they make it clear that in light of the increasing amount of automation in society, social contracts need to be rethought. Robots are learning to reason about tasks and interact naturally with humans through large-scale language models, but real-time grounding, safety, and compute constraints remain significant obstacles (Wang et al., 2024).

The policies that will be implemented will play an important role in guiding the effects of automation. For a balanced incentive system between labor and capital, it may be necessary to recalibrate the tax and regulatory framework. There are many opportunities for businesses to invest in their employees' professional development and upskilling if they are provided with tax credits or subsidies. This would encourage them to treat their workers as crucial assets as they treat their physical assets. Several have even suggested taxing automation in certain contexts, not to inhibit innovation or stifle innovation, but to fund worker adjustment programs and make sure that those affected by automation are compensated as well. The implementation of a direct "technology tax" at the high end of the technology market is complex, but the broader intention is to secure a resource that society can use to invest in human adaptation and progress alongside technological innovation.

The way robotics benefits are distributed can be influenced by innovation, regulations, and policies. Research and development in robotics can be supported by government agencies, non-government organizations, and intergovernmental organizations to accelerate technological advances that contribute to economic growth. Managing the global impacts of robotics can be made easier through international collaboration on a global scale. There is a great opportunity for all countries to navigate this transition by sharing best practices and coordinating on standards for new technologies to retrain their workforces. To facilitate the adoption of automation in developing economies, international support and knowledge transfer might be necessary to facilitate the adoption of automation in a way that fosters development rather than hinders it. The governance system needs to be flexible, forward-looking and continually adaptable, so that it can cope with the advancements in AI and robotics as they occur in the future.

CONCLUSION

In manufacturing environments, emerging technologies, including robotics, can be beneficial. Although it can reduce workplace hazards, it can also create new ones. In spite of robotics' tremendous potential in manufacturing, its benefits will require thoughtful action. Societies can shape the robotics revolution for broadly shared prosperity by developing human capital, adjusting policies, and fostering inclusive innovation. A fundamental change is occurring across productivity, employment, industrial dynamics, and the broader economy as a result of robotics in manufacturing. Based on the views presented in this book and figure 1, the author concludes that the integration of advanced machines into production processes results in substantial efficiency gains. Embracing automation enhances the productivity and cost-competitiveness of factories, as well as the global competitiveness of nations. While these benefits accompany significant adjustments, they are not without their drawbacks. Many routine tasks are being replaced by robots in manufacturing, and some jobs are becoming obsolete while others become more skilled. Workers must acquire new skills in order to prepare for this transition in terms of employment opportunities, wage inequality, and employment opportunities.

Robotic systems are typically sent to a given position several times, and their error is measured each time they return to that position after visiting another. Standard deviations of each dimension of the samples used in the analysis are then used to quantify the degree of repeatability. The average robot is capable of making positional errors exceeding this range, which could prove problematic for the process. As well as changes in repeatability within the working envelope, there are changes in repeatability related to speed and payload. The sustainability of robotics and manufacturing depends on a robust regulatory and policy governance system.

REFERENCES

Adebayo, R. A., Obiuto, N. C., Olajiga, O. K., & Festus-Ikhuoria, I. C. (2024). AI-enhanced manufacturing robotics: A review of applications and trends. *World Journal of Advanced Research and Reviews*, *21*(3), 2060–2072. DOI: 10.30574/wjarr.2024.21.3.0924

Arimoto, S., Kawamura, S., & Miyazaki, F. (1984). Bettering operation of robots by learning. *Journal of Robotic Systems*, *1*(2), 123–140. DOI: 10.1002/rob.4620010203

Asif, S., Bueno, M., Ferreira, P., Anandan, P., Zhang, Z., Yao, Y., Ragunathan, G., Tinkler, L., Sotoodeh-Bahraini, M., Lohse, N., Webb, P., Hutabarat, W., & Tiwari, A. (2025). Rapid and automated configuration of robot manufacturing cells. *Robotics and Computer-integrated Manufacturing*, *92*, 102862. DOI: 10.1016/j.rcim.2024.102862

Brooks, R. A. (1986). Achieving artificial intelligence through building robots. *Massachusetts Institute of Technology, Artificial Intelligence Lab*.

Burdick, J. W. (1989). On the inverse kinematics of redundant manipulators: Characterization of the self-motion manifolds. In *Advanced Robotics: 1989: Proceedings of the 4th International Conference on Advanced Robotics* (pp. 339–346). Springer Berlin Heidelberg.

Calì, M., & Presidente, G. (2025). Robots for economic development. *Labour Economics*, *96*, 102731. DOI: 10.1016/j.labeco.2025.102731

Caloud, P., Carlsson, C., Latombe, J. C., Le Pape, C., & Yim, M. (1990). Indoor automation with many mobile robots. In *IEEE International Workshop on Intelligent Robots and Systems*. IEEE.

Chen, F. Y. (1982). Gripping mechanisms for industrial robots: An overview. *Mechanism and Machine Theory*, *17*(5), 299–311. DOI: 10.1016/0094-114X(82)90011-8

Choi, Y. W., Lee, J., Lee, Y., Lee, S., Jeong, W., Lim, D. Y., & Lee, S. W. (2025). A vision-guided adaptive and optimized robotic fabric gripping system for garment manufacturing automation. *Robotics and Computer-integrated Manufacturing*, *92*, 102874. DOI: 10.1016/j.rcim.2024.102874

De San Bernabe, A., Lakshminarayanan, S., Xing, K. C. H., Mutiargo, B., Suppiah, S., Xue, F., & Fu, S. (2024). Robots in manufacturing: Current technology trends. In *Digital Manufacturing* (pp. 39–83). Elsevier., DOI: 10.1016/B978-0-443-13812-6.00008-7

Dhanda, M., Rogers, B. A., Hall, S., Dekoninck, E., & Dhokia, V. (2025). Reviewing human-robot collaboration in manufacturing: Opportunities and challenges in the context of Industry 5.0. *Robotics and Computer-integrated Manufacturing*, *93*, 102937. DOI: 10.1016/j.rcim.2024.102937

Edwards, M. (1984). Robots in industry: An overview. *Applied Ergonomics*, *15*(1), 45–53. DOI: 10.1016/S0003-6870(84)90121-2 PMID: 15676500

Fahrudin, A., Tiwari, S. P., & Rahmi, K. H. (2025). Psychological well-being of human content moderators and wellness strategies in AI-driven content moderation organizations. In *Content Moderation in the Age of AI* (pp. 221–250). IGI Global Scientific Publishing. DOI: 10.4018/979-8-3373-0335-2.ch009

Gautier, M. (1986). Identification of robots dynamics. *IFAC Proceedings Volumes, 19*(14), 125–130. https://doi.org/DOI: 10.1016/S1474-6670(17)56341-3

Hackwood, S., & Wang, J. (1988). The engineering of cellular robotic systems. In *Proceedings 1988 IEEE International Symposium on Intelligent Control*. IEEE Computer Society.

Hwa, L. S., & Te Chuan, L. (2024). A brief review of artificial intelligence robotic in food industry. *Procedia Computer Science*, *232*, 1694–1700. DOI: 10.1016/j.procs.2024.01.167

Judd, R., & Knasinski, A. (1987). A technique to calibrate industrial robots with experimental verification. In *Proceedings. 1987 IEEE International Conference on Robotics and Automation* (Vol. 4). IEEE. DOI: 10.1109/ROBOT.1987.1088010

Kiefer, D., Grimm, F., Straub, T., Bitsch, G., & Van Dinther, C. (2024). Enhancing power skiving tool longevity: The synergy of AI and robotics in manufacturing automation. *International Journal of Mechatronics and Manufacturing Systems, 17*(2), 201–224. DOI: 10.1504/IJMMS.2024.143059

Kilari, S. D. (2025). The impact of artificial intelligence on automated manufacturing and supply chain systems: Advancements, challenges, and future prospects. *Challenges, and Future Prospects*. DOI: 10.2139/ssrn.4799112

Lei, K., Liu, L. M., Chen, X., Feng, Q., Yang, L., & Guo, L. (2022). Navigation and robotics improved alignment compared with PSI and conventional instrument, while clinical outcomes were similar in TKA: A network meta-analysis. *Knee Surgery, Sports Traumatology, Arthroscopy : Official Journal of the ESSKA*, *30*(2), 721–733. DOI: 10.1007/s00167-021-06436-8 PMID: 33492410

Liu, L., Guo, F., Zou, Z., & Duffy, V. G. (2024). Application, development and future opportunities of collaborative robots (cobots) in manufacturing: A literature review. *International Journal of Human-Computer Interaction*, *40*(4), 915–932. DOI: 10.1080/10447318.2022.2041907

Lozano-Perez, T., Mason, M. T., & Taylor, R. H. (1984). Automatic synthesis of fine-motion strategies for robots. *The International Journal of Robotics Research*, *3*(1), 3–24. DOI: 10.1177/027836498400300101

McClamroch, N. H., & Wang, D. (1988). Feedback stabilization and tracking of constrained robots. *IEEE Transactions on Automatic Control*, *33*(5), 419–426. DOI: 10.1109/9.1220

McNally, P., & Inayatullah, S. (1988). The rights of robots: Technology, culture and law in the 21st century. *Futures*, *20*(2), 119–136. DOI: 10.1016/0016-3287(88)90019-5

Minh Trieu, N., & Truong Thinh, N. (2023). A comprehensive review: Interaction of appearance and behavior, artificial skin, and humanoid robot. *Journal of Robotics*, *2023*(1), 5589845. DOI: 10.1155/2023/5589845

Rupenyan, A., & Balta, E. C. (2024). Robotics and manufacturing automation. In *The Impact of Automatic Control Research on Industrial Innovation: Enabling a Sustainable Future* (pp. 169–197). DOI: 10.1002/9781119983644.ch6

Smith, J. K. (2021). *Robotic persons: our future with social robots*. Westbow Press.

Snyder, W. E. (1985). *Industrial Robots*. Prentice-Hall International.

Taylor, A. T., Berrueta, T. A., & Murphey, T. D. (2021). Active learning in robotics: A review of control principles. *Mechatronics*, *77*, 102576. DOI: 10.1016/j.mechatronics.2021.102576

Tiwari, S. P. (2025). Building a content moderation framework to provide a safe and secure experience for omnichannel commerce. In *Content Moderation in the Age of AI* (pp. 55–84). IGI Global Scientific Publishing. DOI: 10.4018/979-8-3373-0335-2.ch003

Tiwari, S. P., Kostenko, O., & Yekhanurov, Y. (2025). *Understanding technology in the context of national development: Critical reflections*. SciFormat Publishing Inc. DOI: 10.69635/978-1-0690482-3-3

. Wang, J., Shi, E., Hu, H., Ma, C., Liu, Y., Wang, X., ... & Zhang, S. (2024). Large language models for robotics: Opportunities, challenges, and perspectives. Journal of Automation and Intelligence.

Wang, W., Guo, Q., Yang, Z., Jiang, Y., & Xu, J. (2023). A state-of-the-art review on robotic milling of complex parts with high efficiency and precision. *Robotics and Computer-integrated Manufacturing*, *79*, 102436. DOI: 10.1016/j.rcim.2022.102436

Wang, Y., & Mason, M. (1987). Modeling impact dynamics for robotic operations. In *Proceedings. 1987 IEEE International Conference on Robotics and Automation* (Vol. 4). IEEE. DOI: 10.1109/ROBOT.1987.1087851

Williams, S. J., & Glover, K. (1985). Robot dynamic control instabilities. In *1985 American Control Conference*. IEEE.

Chapter 5
Assessing Contribution of Technology and Innovation in Sustainability:
A Comprehensive Analysis of Business Sustainability in the Digital Era

Atul Sharma
https://orcid.org/0000-0002-2079-7024
Chandigarh University, India

Kapil Kumar Aggarwal
https://orcid.org/0000-0002-2752-9495
Chandigarh University, India

Ravneet Kaur
https://orcid.org/0009-0008-3101-5738
Chandigarh University, India

ABSTRACT

Technology and innovation are crucial in all industries nowadays. Every firm grows socially, economically, politically, and environmentally as technology and innovation advance. Technology and innovation may improve efficiency, market competitiveness, and sustainability, which every company wants. This paper says technology and innovation dynamically support any business's sustainability aims. This study examines how technology and innovation help evolving firms achieve sustainability.

DOI: 10.4018/979-8-3373-1082-4.ch005

This study includes articles, case studies, surveys, and in-depth analysis to demonstrate how technology and innovation increase supply chain management, resource usage, market competitiveness, and consumer engagement. This research helps policymakers and stakeholders improve business sustainability with technology and innovation. Practical ideas from this research article help business sustainability discourse adapt to technology and innovation. This research promotes a balanced approach to commercial success and environmental sustainability.

INTRODUCTION

During the last few decades, sustainability in business has gained consideration from stakeholders, policymakers and researchers. To achieve the goal of sustainability technology and innovation plays a very crucial role. For the long run survival of any business adaptability of technology and innovation is mandatory. The integration of advanced technologies and new ideas contributes to addressing the challenges and promoting sustainable practices. Businesses are now required to follow sustainable business practices and have concerns about environmental protection in order satisfy stakeholder expectations and market demands(Adams, Bessant, & Phelps, 2006). To address challenges of business sustainability many technologies and innovations such as IoT (Internet of Things), block chain, Artificial Intelligence, recycling technologies, drones, sensors etc. are participating at different levels. Considering the core relationship between technology and innovation with sustainability every business should keep it on higher priority.

Sustainability is the aim of every business firm to reduce its negative impact on the environment and use a balanced approach to fulfill its social, environmental and economic objectives. To achieve sustainability, the business target is to adopt strategies and practices to create a balance between economic, social and environmental consideration so that business will thrive in the long run. Those businesses which are prioritizing sustainability are not contributing to minimizing the challenges associated with the environment but also getting the attention of consumers and investors who are also environmentally conscious. Mostly businesses are integrating sustainability with their visions and planning. Fostering sustainability at multiple levels such as economic sustainability, social sustainability, environmental sustainability etc. is challenging and needs lots of effort for compliance. Changing demands of the consumer, adherence to the laws and regulations related to social, economic and environmental conditions, concerns of stakeholders demands efforts, planning and good decision making. Choosing sustainability in a way that promotes social and corporate growth as well as sustainable economic development (Székely & Knirsch, 2005).

Contribution of technologies enhances sustainability in business across various sectors. Some technologies such as internet of Things, cloud computing, data analytics, Business intelligence etc. have transformed the business at 360 degree in terms of productivity as well as in adapting sustainability practices. As per the studies, it is observed that due to social and environmental concerns many researchers and scientists are giving attention to eco-friendly innovation which will help the different sectors including businesses to implement and generate less carbon footprints in the environment. Combining IT's capabilities with sustainability goals has the potential to produce positive changes that will benefit society, the environment, and the economy since IT can be used to enable capabilities and improve performance. (Dao, Langella, & Carbo, 2011). For example, earlier some technologies were used in businesses to achieve business agility like web collaboration, Decision Support System (DSS), Knowledge Management (KM), Business Intelligence (BI), and Management Information System (MIS) which were increasing energy consumption but nowadays businesses are using cloud computing to achieve business agility which is a platform for sustainable business with improved profits, superior environment and HRM i.e human resource management. Similarly other technologies such as AI, blockchain, big data, electric vehicles, carbon accounting software's have replaced old practices in business to achieve sustainability.

Figure 1. Emerging technologies interlinked with business sustainability

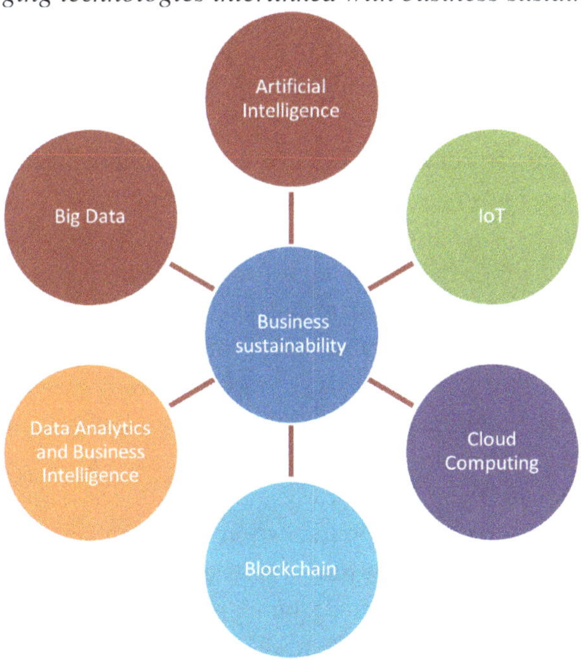

Here some of the technologies and their contribution is mentioned to gain insight that how these are helpful in any business for its transition to become eco-friendly business.

Artificial Intelligence: By streamlining product lifecycle management, enhancing recycling procedures, and spotting material reuse opportunities, artificial intelligence (AI) can help accelerate the shift to a circular economy.AI driven tools helps to measure and analyze a business's carbon footprints so that they can set their targets to reduce carbon footprints and form strategies to attain that target and finally check their performance and compare the performance with the set targets.

Cloud Computing: Cloud computing in businesses reduces the business operations cost along with taking care of environmental aspects. IaaS, PaaS and SaaS solves business problems and improves efficiency and effectiveness of the business and also reduces the carbon footprints. The implementation of cloud computing in businesses makes huge differences in energy consumption and emission of carbon footprints. As cloud computing significantly reduces the cost, energy consumption and resource optimizations so many businesses are adopting it to fulfill their corporate social responsibility practices.

Internet of Things (IoT): In advancing business sustainability IoT plays an important role by providing real-time data and enables smart decision making. IoT based smart offices helps in optimizing energy consumption, reducing costs and improving overall efficiency in office buildings. IoT sensors controls waste by monitoring waste and also reduce different kinds of waste like water and fuel and also improve its recycling(Sharma et al., 2023). Monitoring of products and vehicles, smart packaging, Eco-friendly transportations using IoT helps to attain business sustainability.

Big Data Analytics: Big data streamline business sustainability by providing tools to collect, analyze and leverage huge amounts of data to make informed decisions in any business. In supply chain the detailed analysis of supply chain management operations, optimizing logistics, reducing resource wastage is done with the help of big data analytics. Organizations use big data analytics to track and minimize waste throughout the product lifecycle. Consumer's demand, changing behavior related to the product can be forecast and strategies can be framed in advance.

Blockchain: Blockchain plays a significant role in business sustainability by providing transparency and traceability in many business processes. To trace the origin of any product and track its journey is quite possible as it holds an immutable ledger of transactions. Through implementation of blockchain it is easy to obtain any detailed information of product lifecycle its carbon footprints etc. With block chain technology businesses can upgrade, build trust with customers and participate in transparent economy(Aggarwal & Agrawal, 2024).

Data Analytics and Business Intelligence: The tools for data analytics and business intelligence pay attention towards the consumption of resources and their patterns in the current scenario and identifying the opportunities to minimize in future. Using BI and data analytics carbon footprints are measured, communicated, monitored and visualized so that corrective actions can be taken in future. Analytics tools help to provide visibility into the supply chain to search for inefficiencies in terms of waste.

Technology and innovation are essential for promoting business sustainability because they help businesses create and implement solutions that deal with social, economic, and environmental issues.

LITERATURE REVIEW

By presenting an integrated theoretical model for sustainability, which views IT resources as a vital component in assisting businesses in building sustainability capabilities, which are characterized as a business's ability to manage complex bundles of human and non-human resources in an efficient manner in order to meet sustainability objectives, provide stakeholders with sustainable values, and maintain a competitive edge. (Dao, Langella, & Carbo, 2011). Businesses must continually adapt to obtain a competitive edge and hold onto large market shares. This involves not just incorporating the newest technologies available but also developing original innovations that aim to increase profitability and sustainability. (Tohanean, Buzatu, Baba, & Georgescu, 2020).In the research conducted by (Li-Hua, Wilson, Aouad, & Li, 2011) highlights the key subjects, such as the close relationship between technology strategy and company sustainability and the management of technology as a fundamental function in international business. According to Hojnik and Ruzzier (2016), Eco-innovation is becoming more and more popular among academics and practitioners as a result of environmental and social concerns, and its significance is increasingly acknowledged globally. In achieving sustainability Artificial Intelligence plays crucial role as AI makes it possible to automate time-consuming, repetitive, and significant processes, freeing up humans to concentrate on higher-value work(Aggarwal et al., 2023). Also, it extracts meaning from vast volumes of unstructured data—such as information produced by films, images, written reports, business papers, social media posts, or email messages—that was previously locked away and needed to be managed and analyzed by humans. Further, AI is capable of combining thousands of computers with additional resources to solve even the most difficult issues. (Nishant, Kennedy, & Corbett, 2020).AI has also completely changed how businesses operate, impacting commerce and management strategies across a range of industries that provide more sustainable and competitive goods and services.

In addition to offering appealing and practical cost reductions for businesses, cloud computing has the potential to make a major contribution to society's broader goals of environmental preservation, energy efficiency, and sustainable development. (Kumar & Vidhyalakshmi, 2012). Cloud computing business framework (CCBF) is projected in organizations to understand and develop cloud design requirement, deployment and services(Chang, Walters, & Wills, 2013) illustrate the process that led to the creation of the CCBF and to clarify how various CCBF components operate, such as the connections between business models and IT services and the added benefits the CCBF provides.

Increased production process efficiency as well as improved networking and communication between operators, sensors and machines would be possible with the use of IoT technology in industrial settings. IIoT (Industrial Internet of Things) enable more competitive businesses to enter the market, improve quality control, and minimize losses (Nižetić, Šolić, González-de-Artaza, & Patrono, 2020). From the sustainability point of view IIoT has great potential in three fields as resource efficiency, sustainable energy, and transparency. The introduction of IoT technology is primarily intended to create a smart system that can schedule loads autonomously and/or identify system errors, reduces energy wastage, ultimately leading to an improvement in energy consumption efficiency in business and bringing transparency.

Some of the applications of blockchain technology directly and indirectly contribute to business sustainability. One potential link between sustainability and cryptocurrencies like bitcoin could be the decrease in paper used to print money notes, which can save trees (Jayawardhana & Colombage, 2020). Blockchain technology might promote the realisation of a sustainable supply chain, increase energy effectiveness, and encourage the development of safe and dependable smart cities/homes, among other ecologically sustainable development goals (SDGs). (Parmentola, Petrillo, Tutore, & Felice, 2021).

Data analytics and BI have an important role in serving organizations to execute and supervise sustainable and socially responsible business practices. According to the value chain, businesses utilize business intelligence (BI) techniques, technologies, and information systems to extract pertinent knowledge from vast and diverse data sets in order to support a variety of businesses that is tactical, strategic, and operational (Muntean, 2018).

Because of the growing urgency of environmental degradation, climate change, and social injustice, sustainability has been thrust to the forefront of the strategies that are being implemented by businesses around the world. The advent of the digital era in recent years has resulted in the introduction of powerful technology and new practices that have the potential to bring about a transformational change in the way that business models incorporate sustainability. This study of the relevant literature investigates the ways in which technological advancements and innovative ideas are redefining sustainability initiatives inside businesses, as well as the opportunities that these developments bring and the problems that continue to exist on the path toward sustainable development.

In the context of a firm, sustainability refers to the pursuit of long-term profitability while simultaneously ensuring that the environment and society are harmed to the least extent possible. The Triple Bottom Line (TBL) method, which takes into account economic, environmental, and social aspects, is the guiding paradigm for determining whether or not a business is sustainable(Elkington, 1997). The phrase "sustainable business" was first introduced by John Elkington in 1997. He underlined that in addition to focusing on financial gain, sustainable enterprises must also provide value to both people and the environment. When viewed in this light, technology and innovation emerge as crucial facilitators that assist firms in accomplishing these diverse sustainability goals(Elkington, 1997).

Technological advancement, which can be described as the application of scientific knowledge for practical purposes, and innovation, which can be defined as the development of new ideas and processes, both play important roles in the process of driving sustainable transformation. The digital era, which is characterized by rapid breakthroughs in Artificial Intelligence (AI), the Internet of Things (IoT), blockchain, and big data analytics, has drastically transformed the operations of businesses as well as the activities that are carried out to promote sustainability. The Internet of Things offers real-time monitoring of environmental conditions, energy usage, and supply chain operations. This provides organizations with the ability to discover inefficiencies and reduce their ecological impact(Bressanelli et al., 2018). Artificial intelligence and machine learning are able to enable predictive maintenance and optimize resource allocation, while blockchain technology offers transparency and traceability pertaining to ethical sourcing and carbon tracking(Saberi et al., 2019).

Green technology are also an essential component of environmentally responsible business practices. These technologies include waste-to-energy conversion, smart grids, energy-efficient buildings, and renewable energy sources. All of these breakthroughs contribute to the reduction of carbon emissions and the enhancement of resource efficiency. Such technologies are vital in fulfilling climate targets and moving toward low-carbon economies(Canton, 2021). Companies who make investments in these technologies not only save their operational expenses over the long term, but they also get advantages in terms of their reputation and their ability to comply with regulations.

Not only does innovation encompass technological advancements, but it also encompasses product, process, and business model innovations that contribute to sustainability. For instance, eco-innovation is the process of designing products and processes that minimize the negative impact on the environment while simultaneously preserving or enhancing performance while doing so. Companies such as Tesla and Patagonia are excellent examples of sustainable product creation since they incorporate environmental issues into their initial product offerings. At the same time, improvements in production processes, such as lean manufacturing and green supply chains, help to sustainability by reducing waste and increasing energy efficiency. The circular economy, sharing platforms, and servitization are all examples of business model innovations that play an important part in the innovation process. These models make it possible for enterprises to reduce waste, lengthen the product life cycle, and transition from ownership to access-based consumption, all of which are in line with the objectives of sustainable development(Geissdoerfer et al., 2017).

Forward-thinking businesses are progressively adopting the practice of using digital tools into their sustainability plan with increasing frequency. Digital dashboards and key performance indicators (KPIs) for sustainability are now being used by businesses to assess their environmental and social consequences in real time. Furthermore, the incorporation of Environmental, Social, and Governance (ESG) data into digital decision-making tools makes it possible to promote corporate governance that is both better informed and more transparent. Digital supply chains, which are augmented by technologies such as the Internet of Things and analytics, make it possible to perform predictive logistics, optimize routing, and automate inventory management(Bag et al., 2020). This results in considerable reductions in both emissions and material waste.

In spite of the enormous potential that technology and innovation provide, there are a number of obstacles that prevent its widespread deployment for the purpose of sustainability. As a result of high costs, access to modern digital infrastructure is restricted, particularly in economies that are still growing. In addition, the digital divide, which refers to differences in access to technology and skills, additionally widens the gap in terms of sustainability that exists between regions and industries. In addition, there is a growing worry regarding "greenwashing," which refers to the practice of businesses adopting sustainable policies for marketing purposes without actually having an impact (Delmas & Burbano, 2011). This results in mistrust and a decrease in trust in statements made regarding sustainability.

A substantial body of empirical research demonstrates the advantages that digital innovation brings to sustainable company. For example, Siemens has demonstrated the practical impact that digital sustainability projects can have by reducing its energy use by 25 percent throughout the globe by utilizing Internet of Things and artificial intelligence. According to a report published by Deloitte in 2021, digitally mature firms have a tendency to perform better in environmental, social, and governance (ESG) measures. This suggests that there is a strong association between digital transformation and sustainability leadership. In order to reach their net-zero ambitions and align themselves with global climate commitments, global technology giants such as Amazon and Microsoft are making significant investments in carbon-neutral logistics and green cloud technologies.

Looking into the future, literature identifies a few research gaps and potential future paths. The development of standardized measurements and procedures for assessing the influence that digital technologies have on sustainability is an urgent situation that must be addressed. More emphasis needs to be put on the social dimension, particularly regarding inclusion and equal access to digital solutions, even though the environmental and economic dimensions have been thoroughly investigated. To foster sustainable innovation, policymakers and industry leaders need to work together to establish environments that are conducive to innovation through the implementation of rules, incentives, and capacity building.

OBJECTIVE OF THE STUDY

The study's primary goal is to examine how emerging technologies and innovation might help businesses in different sectors to accomplish sustainable entrepreneurship objectives.

RESEARCH METHODOLOGY

The paper is based on exploratory and descriptive research. This study has been conducted based on extensive literature review and conceptual framework. It is based on secondary and qualitative data to extract significant relationship between technology, innovation and sustainable business in general and their impact on sustainable development. The secondary data used in this study was gathered from a variety of sources, including publications, research papers, newspapers, and more.

Emerging Technologies and Innovation as a Catalyst for Business Sustainability

Digital transformation and sustainability goals achievement of businesses is possible with Innovation and technology. Many business have adapted technologies like AI, ML, blockchain etc. in business practices and improved their effectiveness, efficiency and market position. The following table shows the businesses which are evolving with the integration of emerging technologies and business operations for sustainability and profits.

Table 1.

Sr. No	Organization Name	Technology Adoption with benefits
1	Unilever	Blockchain technology to improve supply chain transparency.
2	Xylem and Suez	Data analytics tools, sensors and purification technologies to optimize water usage and quality.
3	Google Deep Mind	AI to enhance energy utilization and reduce cooling expenses in data center.
4	Tesla	AI powered self-driving vehicles promotes sustainable transportation.
5	Tandem	Cloud computing provides green digital banking products to customers like eco-friendly payment options through digital wallets, green loans etc.
6	Walmart	Big data analytics in supply chain management to track carbon footprints of their products, optimize transportation routes, improve fuel efficiency and areas of improvement.

Source: Author's Own

FINDING AND SUGGESTIONS

The study highlights new developments, including the growing applications of emerging technologies such as AI, blockchain, cloud computing and innovation in different areas of business-like planning, designing, tracking and tracing in supply chain and predictive analytics for sustainability planning.

To address different challenges to achieve sustainability goals in any business can be done by taking the help of these technologies. By aligning business strategies with emerging technologies, educating employees and other stakeholders about sustainability and social responsibility businesses not only enhance their sustainability but also add a competitive edge in the challenging marketplace and ever-changing market. Businesses should invest in these technologies and also stay updated with the new inventions and innovations for getting benefits.

LIMITATIONS

The incorporation of technology and innovation into sustainable business practices has demonstrated encouraging outcomes; yet numerous restrictions continue to obstruct its widespread adoption and enduring influence. A key barrier is disparate access to digital infrastructure and innovative technology, particularly in developing areas. Small and medium-sized firms (SMEs) frequently lack the financial resources and qualified personnel necessary to use emerging technologies such as Artificial

Intelligence (AI), Blockchain, or the Internet of Things (IoT), resulting in a digital gap that impedes global sustainable development initiatives.

A notable issue is the absence of established measures and frameworks to evaluate the true impact of digital advances on sustainability objectives. Although businesses may assert the implementation of sustainable technologies, accurately assessing their genuine impact on environmental, social, and economic goals is intricate and varies between sectors. This occasionally leads to greenwashing, wherein firms ostentatiously advocate for sustainability to enhance their reputation without implementing substantive reforms.

A technological trade-off exists while numerous digital solutions seek to minimize carbon footprints and resource consumption, they necessitate considerable energy and infrastructure, particularly for large-scale data centers, blockchain mining, and AI computing systems. The energy consumption and electronic trash production linked to these technologies present environmental issues if inadequately controlled.

Furthermore, privacy and data security are progressively emerging as constraints in the use of technologies such as IoT, Big Data, and AI. Although these tools enhance operational efficiency and facilitate informed decision-making, they frequently gather enormous quantities of user and environmental data. Inadequate management or absence of secure systems may result in breaches and diminish public trust, ultimately obstructing the implementation of these technologies for sustainable objectives.

Ultimately, regulatory deficiencies and the lack of complete policy frameworks in numerous countries hinder the effective application and expansion of sustainable technologies. In the absence of adequate incentives, subsidies, or regulatory requirements, numerous enterprises may emphasize short-term financial profits over sustainability.

FUTURE SCOPE

As global consciousness regarding environmental and social issues escalates, firms will face mounting pressure to reform their operations sustainably. The advancement of artificial intelligence possesses considerable potential for enhancing energy efficiency, automating sustainability assessments, and forecasting environmental hazards using sophisticated analytics. The proliferation of green cloud computing, carbon accounting software, and eco-design tools can facilitate the transition of enterprises from conventional models to circular, resource-efficient paradigms. As cloud platforms become increasingly accessible and flexible, even small enterprises can leverage computational power without substantial investment, facilitating a more inclusive shift towards sustainable business ecosystems.

Moreover, Blockchain technology is anticipated to advance towards greater energy efficiency and scalability, rendering it a practical instrument for supply chain transparency, decentralized energy trading, carbon credit monitoring, and secure sustainability certifications. When combined with AI and IoT, it can offer a strong framework for the real-time monitoring and validation of sustainability metrics.

The future of business intelligence and big data analytics is in its predictive powers. Advanced BI technologies will empower enterprises to anticipate sustainability risks, shifts in customer behavior, and regulatory consequences, enabling them to proactively modify their operations and plans. Moreover, edge computing and quantum computing may ultimately facilitate the analysis of extensive sustainability datasets with reduced latency and energy consumption.

There is an expanding opportunity for international collaboration and standardization in digital sustainability regarding policy and governance. Governments and multilateral organizations could establish universally recognized standards for digital sustainability reporting, carbon accounting, and green IT certification. Policies that promote public-private partnerships, tax incentives, and research grants can expedite innovation and the dissemination of technologies in sustainability sectors.

There is an imperative necessity for interdisciplinary research that integrates technology innovation with behavioral sciences, ethics, and management to enhance the comprehension of how businesses may effectuate lasting change inside. Educational institutions and business training programs should prioritize the development of digital and sustainability literacy among present and future professionals.

CONCLUSION

The use of emerging technologies in sustainable entrepreneurship is a broad topic with several advantages. Big giants like Google, Microsoft, and Amazon are setting benchmarks for other firms to adapt these technologies and integrating AI with other technologies to minimize carbon footprints and wastage, improving supply chain, generating eco- friendly products in context to achieve entrepreneurship sustainability goals. Best practices and approaches for implementing AI in sustainable enterprises can be found with more research in this field.

Businesses have to embrace new technology quickly to adopt sustainable practices in the face of global issues like resource depletion and climate change. As this study makes clear, integrating these technologies is not just a business obligation; it's also a way to increase productivity, lessen environmental effect, and foster positive social change. By doing this, companies can not only prosper financially but also act as catalysts for significant change in the direction of a just and sustainable world.

REFERENCES

Adams, R., Bessant, J., & Phelps, R. (2006). Innovation management. *International Journal of Management Reviews*, •••, 1–45.

Aggarwal, K. K., & Agrawal, S. (2024). Artificial intelligence and its role in financial market. In *Global Financial Analytics and Business Forecasting*.

Aggarwal, K. K., Sharma, A., Kaur, R., & Lakhera, G. (2023). Algorithmic FinTech pioneering the financial landscape of tomorrow. In *Algorithmic Approaches to Financial Technology*. Forecasting, Trading, and Optimization., DOI: 10.4018/979-8-3693-1746-4.ch002

Bag, S., Gupta, S., & Luo, Z. (2020). Examining the role of logistics 4.0 enabled dynamic capabilities on firm performance. *International Journal of Logistics Management, 31*(3), 607–628. DOI: 10.1108/IJLM-11-2019-0311

Bressanelli, G., Adrodegari, F., Perona, M., & Saccani, N. (2018). Exploring how usage-focused business models enable circular economy through digital technologies. *Sustainability (Basel), 10*(3), 639. DOI: 10.3390/su10030639

Canton, H. (2021). Organisation for economic co-operation and development—OECD. In *The Europa Directory of International Organizations 2021* (pp. 677–687). Routledge.

Chang, V., Walters, R. J., & Wills, G. (2013). The development that leads to the Cloud Computing Business Framework. *International Journal of Information Management, 33*(3), 524–538. DOI: 10.1016/j.ijinfomgt.2013.01.005

Dao, V., Langella, I., & Carbo, J. (2011). From green to sustainability: Information Technology and an integrated sustainability framework. *The Journal of Strategic Information Systems, 20*(1), 63–79. DOI: 10.1016/j.jsis.2011.01.002

Dao, V., Langella, I., & Carbo, J. (2011). From green to sustainability: Information Technology and an integrated sustainability framework. *The Journal of Strategic Information Systems, 20*(1), 63–79. DOI: 10.1016/j.jsis.2011.01.002

Delmas, M. A., & Burbano, V. C. (2011). The drivers of greenwashing. *California Management Review, 54*(1), 64–87. DOI: 10.1525/cmr.2011.54.1.64

Elkington, J. (1997). The triple bottom line. *Environmental Management: Readings and Cases, 2*, 49–66.

Geissdoerfer, M., Savaget, P., & Evans, S. (2017). The Cambridge business model innovation process. *Procedia Manufacturing*, *8*, 262–269. DOI: 10.1016/j.promfg.2017.02.033

Jayawardhana, A., & Colombage, S. (2020). *Does Blockchain Technology Drive Sustainability? An Exploratory Review*. Emerald Publications. DOI: 10.1108/S2043-052320200000015002

Kumar, V., & Vidhyalakshmi, P. (2012). Cloud Computing for Business Sustainability. *Asia-Pacific Journal of Management Research and Innovation*, *8*(4), 461–474. DOI: 10.1177/2319510X13481905

Li-Hua, R., Wilson, J., Aouad, G., & Li, X. (2011). Strategic aspects of innovation and internationalization in higher education: The Salford PMI2 experience. Journal of Chinese Entrepreneurship .

Muntean, M. (2018). *Business Intelligence Issues for Sustainability Projects. Knowledge Management, Innovation and Big Data: Implications for Sustainability*. Policy Making and Competitiveness.

Nishant, R., Kennedy, M., & Corbett, J. (2020). Artificial intelligence for sustainability: Challenges, opportunities, and a research agenda. *International Journal of Information Management*, *53*, 102104. DOI: 10.1016/j.ijinfomgt.2020.102104

Nižetić, S., Šolić, P., González-de-Artaza, D. L.-I., & Patrono, L. (2020). Internet of Things (IoT): Opportunities, issues and challenges towards a smart and sustainable future. *Journal of Cleaner Production*. PMID: 32834567

Parmentola, A., Petrillo, A., Tutore, I., & Felice, F. D. (2021). Is blockchain able to enhance environmental sustainability? A systematic review and research agenda from the perspective of Sustainable Development Goals (SDGs). *Business Strategy and the Environment*.

Saberi, S., Kouhizadeh, M., Sarkis, J., & Shen, L. (2019). Blockchain technology and its relationships to sustainable supply chain management. *International Journal of Production Research*, *57*(7), 2117–2135. DOI: 10.1080/00207543.2018.1533261

Sharma, A., Aggarwal, K. K., & Malik, M. (2023). IoT integration with Big Data and Cloud Computing: A Multidisciplinary Approach. *2023 International Conference on Communication, Security and Artificial Intelligence, ICCSAI 2023*, 139–144. DOI: 10.1109/ICCSAI59793.2023.10420987

Székely, F., & Knirsch, M. (2005). Responsible Leadership and Corporate Social Responsibility: Metrics for Sustainable Performance. *European Management Journal*, *23*(6), 628–647. DOI: 10.1016/j.emj.2005.10.009

Tohanean, D., Buzatu, A.-I., Baba, C. A., & Georgescu, B. (2020). Business Model Innovation Through the Use of Digital Technologies: Managing Risks and Creating Sustainability. *Amfiteatru Economic*, •••, 758–774.

Chapter 6
Asia Pacific Perspectives on Digital Innovation and Integrated Frameworks in Aerospace Manufacturing

Yashaswi Agrawal
https://orcid.org/0009-0008-3853-8224
National University of Singapore, Singapore

ABSTRACT

As a result of digital transformation, aerospace manufacturing in the Asia Pacific region is changing dramatically. This book chapter examines how digital innovation is being incorporated into aerospace manufacturing through the use of an integrated framework approach. The book chapter emphasizes that successful adoption of digital technologies in manufacturing requires not only advanced technological infrastructure, but also alignment with organizational strategy, the development of workforce skills, and cross-industry collaboration. This book chapter adopts a socioeconomic and technical perspective. This chapter is based on a conceptual framework that highlights the key dimensions of technological development, organizational development, people, and stakeholder communication. A study of current literature, industry trends, and practical applications has been conducted by the author to illustrate how Asia Pacific aerospace manufacturers and policymakers may leverage this integrated framework to enhance performance efficiency, quality, and innovation.

DOI: 10.4018/979-8-3373-1082-4.ch006

INTRODUCTION

Digital innovation is driving profound changes in the aerospace manufacturing industry. Modern digital technologies are redefining how aircraft components are designed, produced, and maintained. Aerospace companies around the world use a wide range of technologies, including robotic sensors that facilitate real time monitoring of machines and products, artificial intelligence (AI) data capabilities that maximize complex processes, and machine learning and automation that enhance accuracy and productivity. Manufacturing sector has played an important role in the development of Asia-Pacific countries and has contributed to regional economic growth and sectoral transformation (Guisan & Exposito, 2015). The graph of manufacturing competitiveness in Asia has been quite influenced by both domestic industrial policies and global economic pressures (Jomo, 2003). Digital manufacturing in combination with flexible assembly technologies makes it possible for aerospace production systems to become highly reconfigurable and adaptable (Jackson, Efthymiou, & Borton, 2016). It is expected that these innovations will be of significant benefit to the industry, including the reduction of production costs, the enhancement of quality control, the production of new aircraft models at a faster rate, and the ability to customize products according to customer needs. In recent times the competition for skilled manufacturing talent in Asia has intensified (Zheng, Soosay, & Hyland, 2007). For aerospace manufacturing to maximize the benefits of digital technologies, more than ad hoc adoption of technology and gadgets is required. Asian manufacturing has gone through a significant structural transformation in the late twentieth century, also influenced by rising productivity and technological advancement across several economies (Timmer, 2000). People create new technologies and industries to meet human needs more effectively and at a lower cost. Innovation is a major agent of progress (Frosch & Gallopoulos, 1989). The implementation of a strategy for manufacturing requires a framework that aligns technological capabilities with business strategy, workforce readiness, and supply chain interconnections. Foreign direct investment (FDI) in manufacturing has been a strong driver of economic growth in Asian economies, supporting industrial expansion and technology transfer (Wang, 2009). The future of manufacturing is based on redefining its basic support blocks and raising awareness about how advanced production paradigms will integrate digital, intelligent, and green innovations (Wang, 2018). Organizations invest continually in process improvements because such enhancements support production but also lay the groundwork for future product success (Wheelwright & Hayes, 1985). The aerospace manufacturing industry involves a wide range of material processing technologies that are critical to the production of aircraft components (Saha, 2016). Manufacturing has to an extent increased resource consumption and accelerated technological development

(Young, Byrne, & Cotterell, 1997). Manufacturing is now more of a dynamic and interdisciplinary field, increasingly shaped by advancements in emerging technologies that enable innovative production approaches (Kalpakjian, 2001).

Manufacturing research and practice are increasingly focusing on integrated digital transformation frameworks. It is becoming apparent that advances in the design and manufacturing of large aerospace composite structures have significantly emphasized the use of polymeric materials in order to enhance the performance and reduce the weight of these structures (Harris, Starnes, & Shuart, 2002). Companies are increasingly approaching the adoption of digital technology holistically rather than as isolated projects, integrating them into every aspect of their business operations and ensuring they are interoperable. The importance of an integrated approach is especially evident in aerospace manufacturing, where design, production, and maintenance activities are highly interdependent, involving a wide network of suppliers and partners. As manufacturing processes become increasingly complex, they depend increasingly on well-developed material supply chains to enhance operational efficiency (Bales, Maull, & Radnor, 2004). Design specifications must be changed rapidly by a digital twin model and transmitted to production tooling and supplier processes via a connected data thread; quality data collected on the assembly line should also be fed back into field maintenance programs and design improvements. An integrated digital platform, cross-functional coordination, and a strategic vision spanning the entire product lifecycle are necessary to achieve such integration. Using digital twin technology, it is possible to create real-time virtual models of aerospace and manufacturing systems, which significantly improves simulation accuracy and operational efficiency (Phanden, Sharma, & Dubey, 2021).

As promising as the outlook may seem, integrating digital innovation into aerospace manufacturing is a complex journey filled with a number of challenges. There are many complicated aerospace products, such as passenger jets, military jets, and satellites, which require a rigorous certification process in order to comply with stringent safety and quality requirements. It is imperative for manufacturers to ensure that any new digital technology or workflow meets the industry's stringent requirements for reliability before rolling it out. As part of the aerospace manufacturing process in Asia Pacific, local firms and established global aerospace companies are often in collaboration, resulting in digital integration involving different organizational cultures, systems, and sometimes even different levels of technological sophistication, which can be challenging. As a result, different issues need to be addressed, such as data sharing between partners. These issues include the protection of sensitive intellectual property while connecting legacy digital systems and upgrading production lines while ensuring no disruption to ongoing programs. Through an integrated framework approach, these issues can be addressed in a structured manner by integrating digital innovations with existing processes and stakeholder relationships.

BRIEF OVERVIEW OF THE ASIA PACIFIC AEROSPACE INDUSTRY

There has been a considerable amount of research conducted on the topic of digital transformation in manufacturing within the last decade, which is part of a larger industry-wide focus on the topic. Throughout the past few years there has been an increasing interest in what is called Industry 4.0, referred to as the concept of what is sometimes referred to as the fourth industrial revolution, which consists of smart factories and cyber-physical systems. There are several digital technologies which are being integrated with traditional industrial operations as part of the Industry 4.0 strategy, such as artificial intelligence, 3D printing, and extended reality. These technologies are often referred to as being capable of enabling a shift from isolated, linear production lines to highly interconnected, intelligent manufacturing ecosystems in the literature. Aerospace manufacturers are faced with multiple unique challenges due to the strict regulatory requirements they must adhere to, along with the typical pressures such as competition, cost reduction, and the continuous development of products (Abollado, Shehab, & Bamforth, 2017).

It is becoming more and more common for machine operations to be converted into digital data streams, which can be monitored and controlled in real-time by using technology. A handful of AI algorithms have been developed for optimizing workflows or predicting equipment failures based on large datasets from productions. With advanced robotics, tasks can be performed with a precision and adaptability that exceed the capability of humans. In the context of digital convergence, scholars and practitioners agree that it can have many benefits, including greater efficiency, better and error-free products, high level customization, enhanced flexibility, and new business models, including predictive maintenance services and on-demand manufacturing, among other things (Tiwari, 2025).

Taking into account the Asia-Pacific region as a context for examining digital innovation in aerospace manufacturing is particularly interesting and dynamic. This region represents not only one of the fastest-growing markets for air travel, but also one of the fastest-growing centers for the development of new aircraft and aerospace technology. Over the past two decades, many countries such as Australia, China, India, Japan, the Philippines, Singapore, South Korea, and Thailand have invested heavily in the development of aerospace manufacturing capabilities, establishing factories, research centers, and joint ventures that are contributing significantly to the global aerospace supply chain. In line with this growth, there has been a strong interest in leveraging cutting-edge digital technologies to boost competitiveness. According to industry survey results, Asian-Pacific manufacturers are among the companies that have adopted digital operations and incorporated new technologies more quickly than their counterparts in North America and Europe with respect to certain aspects of digital maturity, often outpacing their counterparts in North America and Europe.

A regional enthusiasm for Industry 4.0 is accompanied by government policies and initiatives that support this enthusiasm. For example, Japan's vision of Society 5.0 promotes the integration of cyber-physical systems into industry and Singapore's Smart Industry initiatives support the adoption of Industry 4.0 technologies in high-value sectors such as aerospace and precision engineering, respectively. Several initiatives have been made by the aerospace firms in the region in an effort to create an enabling environment for them to experiment with and implement integrated digital solutions both in their factories and across their supply networks. There is one underlying theme that has repeatedly emerged during the author's research, which is that it is not enough for technology to be the sole driver of a successful digital transformation in manufacturing. According to early findings, the installation of new machines or software without adjusting organizational processes and culture tends to produce limited results compared to the installation of existing machines or software. An increasingly wide range of stakeholders are recognising the need for integrated frameworks to govern digital innovation. This includes factors such as the organizational structure, the strategy, the people, as well as stakeholders in the process. It seems that the concept of a socioeconomic and technical framework is relevant here, i.e., it is necessary to manage technological change in conjunction with social change, e.g., changes in processes and people, in order to make sure that information is shared and accessible. A common term used to describe this is aligning the three dimensions of change "people, processes, and technology". As manufacturing companies increasingly embrace digital transformation, they invest just as much time and money on training their workforce, redesigning their workflows, and evolving their corporate strategies as they do on investing in new technologies. Industry ecosystem and recent insights in commercial aerospace show the increasing complexities and integration challenges faced across the aviation value chain (Strube, Eloot, Griessmann, Dhawan, & Ramaswamy, 2016).

The importance of integrated change in the aerospace industry and in the context of Asia pacific is amplified by the complexity and the high stakes that are associated with the sector. A history of innovation has traditionally been located at the cutting edge of the aerospace industry when it comes to adopting technology in manufacturing. Aerospace companies were among the first companies to use computer-aided design and engineering tools for the design of complex aircraft components and to simulate the performance of these complex components years ago. In a continuous effort to improve quality, aerospace companies have also successfully incorporated technology such as mass manufacturing and kaizen in order to improve quality and reduce waste as part of their continuous improvement process. Earlier waves of innovation laid the foundation for a disciplined and precise process as a result of these earlier waves of innovation. The digital transformation builds upon and extends upon these foundations by enabling a far greater level of connectivity and intelligence than has ever been possible before. By automatically identifying inefficiencies in real-time and dynamically adjusting production schedules and resource allocation in response to those inefficiencies, digital tools can go beyond what lean manufacturing was aiming for by eliminating waste by systematically improving processes. There is a great opportunity for aerospace companies, in particular small and medium-sized manufacturers, to transition from traditional production approaches to lean manufacturing in order to become more efficient and flexible in their operations (Garre, Bharadwaj, Shashank, Harish, & Dheeraj, 2017).

DESCRIPTION OF INTEGRATED FRAMEWORK

A cohesive framework that incorporates technology, process, organization, and human resources will be the most effective guide to digital innovation in aerospace in the Asia Pacific region. Given the complexity and exacting standards of the aerospace industry, it is obvious that such integration is a must, given its complexity and exacting standards. With its proactive stance and unique conditions, Asia-Pacific is a unique area to study how these integrated frameworks are being implemented in practice because of its proactive stance and unique conditions. Taking this context into account, the next part of the chapter introduces a conceptual framework that captures the key dimensions of using digital technologies in aerospace manufacturing. It follows, via a diagram, a framework that will serve as the base for understanding the examples and analysis.

Figure 1. A variable matrix for analysis of organizational development and change management, using digital technologies in the manufacturing industry

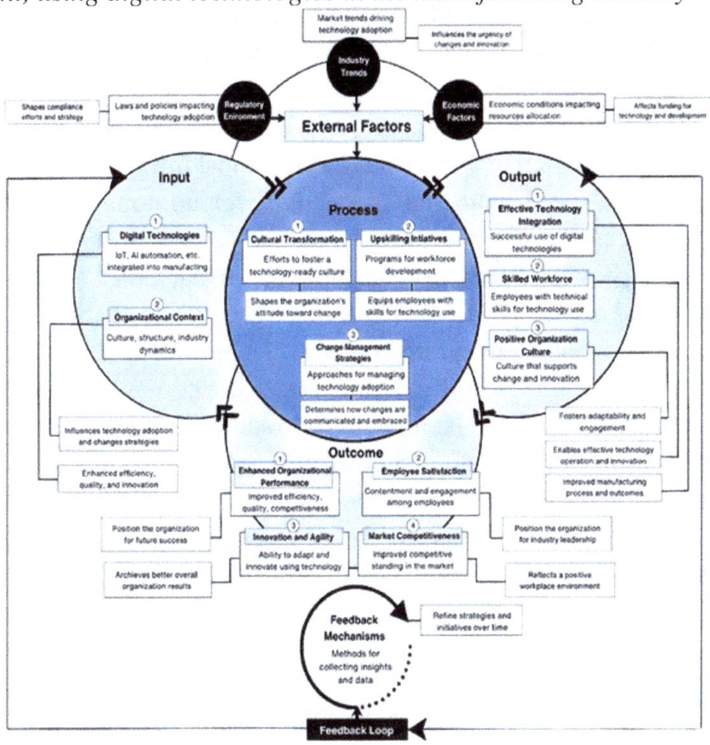

Figure 1 illustrates how innovative digital technology can be successfully integrated into aerospace manufacturing. In order to do this, one must take a holistic approach that addresses multiple interconnected dimensions in order to be successful. Figure 1 also presents a conceptual model titled using digital technologies in the manufacturing industry that encapsulates the core components necessary for effective digital transformation to be carried out in a manufacturing setting. Integrated frameworks such as this one emphasize not only the need to deploy technological tools in conjunction with organizational strategy and human resource development, but also the need to collaborate with stakeholders. The use of digital technologies can be seen as situated within a broader socioeconomic and technical system. In the framework, four key dimensions are outlined that must be coordinated in order to successfully implement digital transformation in the aerospace ecosystem with reference to Asia Pacific, which can be described as follows, which are divided into four categories: (1) Technology capabilities, deriving efficiencies, infrastructure, (2) Strategic alignment, future organizational direction, (3) Human capital, training and skills, and (4) Integration and alignment of stakeholders ecosystem, collaboration.

Digital technologies can be embedded into manufacturing processes in a holistic way by integrating and enhancing each of these dimensions, creating a holistic approach to the integration of digital technologies. There is a foundational layer which forms the basis of this framework which consists of technical capabilities and infrastructure. A significant component of this dimension is the hardware, software, and data architecture that support advanced manufacturing processes. An aerospace factory setting in Asia pacific might have an array of technical infrastructure components, which include high-speed industrial connectivity, cloud computing platforms or data centers that can manage and store the massive amounts of data generated by the factory, and specialized systems that can process that data.

The concept of digital threads has been regarded as one of the most important and relevant components of integrated frameworks in aerospace manufacturing literature for a long time now. Using a digital thread as a communication framework, a continuous strand of data can be fed into the traditional silos of product lifecycle design, production, quality, and maintenance in order to create an interconnected web of data that connects these traditionally siloed elements of product design, production, quality, and maintenance. There is a potential for a digital thread to allow data from every phase of construction of an aircraft to become accessible and usable throughout the lifecycle of the aircraft. The seamless flow of information helps ensure that changes or insights in one domain are propagated to other domains as soon as possible. Engineers can trace back through the thread if operational data indicates that a certain component is failing more frequently than others and adjust either the design or the production process of the component to solve the problem. Developing a digital thread can be challenging and will require integration of a variety of software systems, including product lifecycle management platforms that are specific to the aerospace industry, standardized data formats, and tightly regulated access to and security of data.

Technology combined with process integration connecting the various phases of the aerospace lifecycle and bringing together teams that collaborate across traditional borders is a major positive for the aerospace industry in Asia Pacific. A number of Asia Pacific aerospace companies also communicated with one another about their strategic objectives, operational processes, and workforce development efforts in conjunction with the digital transformation of the industry. Many companies that lag behind tend to have isolated pilot projects or lack a clear, overarching digital strategy when compared to those that lead. The common pitfall is not integrating the new digital solutions with existing systems and processes, which can lead to duplication, data silos, and user resistance due to the lack of integration. While interacting with aerospace industry experts, the author has also noted that a major pitfall of establishing an integrated digital roadmap is the absence of top-management vision and support. Without top-management vision and support, various departments might pursue digital initiatives which are at odds or redundant with one another.

IDENTIFYING AND ADDRESSING CHALLENGES AND OPPORTUNITIES

Despite a solid framework and a strong regional support, integrating digital innovations into aerospace manufacturing remains a complex task, even with a solid framework and strong regional support. There are a number of challenges that the Asia-Pacific aerospace sector is facing that need to be navigated carefully. As a result of implementing an effective digital transformation process, it is also left with significant opportunities for capitalizing on. This section of the book chapter will be about identifying and describing the key challenges and opportunities that aerospace businesses in Asia Pacific region face in order to succeed. Several of these challenges and opportunities are of the same nature, which means that overcoming one opens the door to the next, as they are similar in nature. Our integrated approach takes a holistic approach, addressing the technical, organizational, human, and ecosystem factors in a way that is aligned with our holistic perspective.

Key Challenges

1. The investment in advanced digital infrastructure can be a very costly and uncertain process at the beginning, as it involves new technologies, upgrading to new platforms, and updating precision equipment that require substantial upfront investments. This is a significant cost for many Asian companies, especially for the smaller suppliers in the region. The fact that government funding can sometimes be provided in the aviation sector is not a guarantee that the investment will return a positive return on investment (ROI). Due to the long project timelines and strict quality requirements involved in the aerospace manufacturing industry, the benefits of digital technology, such as fewer defects and a faster production process, may not be apparent in the short term. A decision-maker may become cautious as a result of this. For a multi-million-dollar smart factory upgrade for instance, the business case often requires projecting the benefits over a number of years with a view to proving the business case. In the absence of clear quantifications of these benefits or if past pilot projects have not demonstrated clear wins, companies may hesitate or begin to implement half-measures in the absence of clear quantifications. A successful solution to this challenge will require careful planning and phased implementation that will demonstrate incremental successes so as to build confidence and secure further investments.

2. There are legacy and siloed systems as well as complexity involved with integration. There are many established aerospace manufacturing facilities which have developed legacy systems that are a patchwork of older computers, machines with proprietary control software, and databases that are not designed to communicate with one another but work together as one system. The integration of these legacy systems into a modern, unified digital framework is a technical challenge that needs to be addressed. The old software system used on some older assembly lines to track production may have been developed in the 1980s, whereas the newer machines on the floor utilize a different protocol. To combine data from both of these sources, one may have to write custom middleware or devote a great deal of time to engineering to be able to do this. The majority of aerospace companies in Asia use specialized software for designing, analyzing, and performing other related tasks, so there is a possibility that data may not be able to be exchanged between systems easily. There is a real concern about the threat of production downtime when replacing legacy systems outright. It can be quite risky and disruptive. There is a great deal of difficulty in gradually connecting or upgrading legacy assets in order to become part of the digital thread rather than isolated islands that are part of the digital thread. The importance of standards and open architecture in this area is apparent, but achieving those goals may require working closely with multiple vendors or undertaking significant integration projects in order to do so. Asia-Pacific has a bigger problem with this issue because there are many corporations that are younger, and it is less of a problem for older companies or when new companies acquire old facilities through acquisitions, which is why it is a big issue for older companies.
3. Asia Pacific's aerospace industry faces a number of skills gaps and workforce readiness challenges, which pose a number of challenges. In order to transform an organization, it is often the human capital component that proves to be the most challenging. Psychological well-being can be fully managed in AI-integrated operational environments, as long as human factors and technology management strategies are considered at the same time (Fahrudin, Tiwari, & Rahmi, 2025). A skill shortage exists in the aerospace industry with regards to both domain knowledge and digital technologies. An aerospace manufacturing engineer, for example, might not initially be familiar with the process of precision engineering, and a seasoned manufacturing engineer might not be proficient in coding or data analytics in the first place. The task of bridging this gap is a challenging one. There are many cases in which the pace of technological change in Asia-Pacific countries has outpaced the pace of curriculum updates and on-the-job training that are available in those countries, despite the fact that there is a youth and educated labor force in those countries. Automation and artificial intelligence can be viewed as a threat to many workers in the aerospace indus-

try. This is mainly due to their apprehension that these technologies will cause them to lose their jobs or that it will cause them to change the nature of their work in an unpleasant way. A successful management of this change requires effective communication and the demonstration that digital tools can be used to complement rather than replace workers' roles, emphasizing that robots will be able to handle repetitive tasks, leaving workers to focus on tasks that require them to be more skilled. There is a need for comprehensive training programs and perhaps new hiring strategies in the manufacturing industry that will bring hiring software engineers and data experts. Even the best technologies could not be used fully if the skill gap was not addressed, or they could be implemented incorrectly if this gap was not addressed.

4. Among the challenges faced by the aerospace industry in Asia is cybersecurity and data protection. The move toward digitalization in aerospace manufacturing has also resulted in it being more vulnerable to cyber threats. Having machines and systems connected to the Internet is one of the characteristics of a fully integrated digital factory. There is a risk that a breach or malware incident could disrupt production or lead to the theft of valuable intellectual property such as design files for a new aircraft. Cybersecurity is also one of the most vital factors when it comes to the aerospace sector, since military aerospace projects are often directly linked to national security. Due to the fact that Asia-Pacific companies often collaborate internationally, data may move across borders between partners and therefore create further exposure for companies from Asia-Pacific. An effective cybersecurity strategy involves investing in a secure network architecture with encryption, access controls, and a continuous monitoring process in order to remove any intrusions that may occur. The challenge arises from the fact that it adds another layer of complexity and costs to the process. Aside from data sovereignty laws, there are other laws such as the cybersecurity law, which requires certain data to be stored locally, as well as other regulations. Global data integration can pose a number of challenges. Having to navigate different regulations while maintaining an integrated system is a challenge for companies today. There is the possibility that a cyber incident, in which manufacturing data is exposed due to a supplier's inadequate security, can severely erode trust among the stakeholders. Stakeholder integration is an issue that needs to be considered as an integral part of the framework's stakeholder integration dimension, to the extent that it is part of this issue for the industry.
5. Integration only brings its full benefits if the systems are able to communicate seamlessly with each other. Standardization and inter-polarity play an important role in the aerospace sector in Asia, where different companies and different parts of the value chain may use different standards or platforms, and that is

where standardization and inter-polarity come into play. The key to making sure that digital systems are working well together is to make sure they speak the same language. While this is true in theory, it is the case in practice that different companies use different tools, systems, or formats, so there is no easy way for everything to work together seamlessly. Delays, misunderstandings, and extra work are created as a result of this. There has been a great deal of effort being made in the Asia-Pacific region in the aerospace industry to bring everyone on the same page, but it will take a while. There might be a reluctance among companies to change what already works for them if it is working for them. When it comes to organizing common ways of working, governments and industry groups need to take a step forward and encourage everyone to come to an agreement. To make the systems connect until this happens, businesses often have to devise unique solutions just to make the systems work together, and these solutions often prove to be quite expensive and not always reliable. This is like trying to put puzzle pieces together that weren't designed to be put together and it makes no sense. This slows down the production process and increases the chances of errors occurring during the process. A small supplier is usually the one that struggles the most, as they lack the resources required to keep up with continuous changes in the system. The strength of a collaborative process depends not only on the availability of technologies, but also on the development of trust and cooperation among all stakeholders involved in the manufacturing process. It is crucial to understand that real progress is achieved when the benefits of sharing in the benefits outweigh the inconveniences caused by individual decisions.

6. Digital transformation can disrupt long-established ways of working that have been in place for a very long time and changing the way we work can only be achieved with the support of culture. The idea of allowing younger tech teams and external advisors to contribute to corporate culture can be very challenging, especially in corporate cultures where hierarchy is strong. It is not uncommon for some heads of departments or IT executives to believe that their traditional methods are superior or at least adequate to the digital tools they use and view them as unnecessary or even as a fad. There is a subtle challenge involved in overcoming this mindset. The first step is to build a culture that promotes innovation and openness within the organization. Asia-Pacific is a region that has a diverse range of cultures and cultures play an important role in addressing this challenge. An organization with a culture that highly values experience and seniority, for instance, might find that its younger, more digitally savvy employees find it difficult to be heard when they propose changes to the company. As opposed to this, if there was a more tech-driven culture, perhaps that would lead to an overshadowing of the shop-floor wisdom, ultimately leading

to friction between both parties. Embracing new digital technologies while respecting proven aerospace practices is a delicate balance. Having the top understand the importance of digital integration and that new skills are required is more likely to reduce resistance to the change. An organization can use several methods to shift mindsets, such as demonstrating how one team has improved its performance after adopting a new system.

Opportunities for the Future

1. It is possible to achieve significant efficiency and productivity gains through the integration of digital technology. This is the primary reason driving the digital integration of technology. Using real-time data and automated processes, aerospace manufacturers are able to reduce the waste of materials and time in the production process. A well-integrated production monitoring system can, for instance, greatly reduce idle time of machines in an organization simply by better scheduling and a quick response to any issues that arise. With advanced analytics, a company will be able to identify processes that can be re-ordered, modified, or tweaked to reduce the amount of time it takes to manufacture each unit. As a result of these efficiencies, it may be possible to produce more aircraft parts with the same or fewer resources, which represents a tremendous competitive advantage given that the aerospace industry has high capital costs. As the demand for aerospace products is high in Asia-Pacific, the opportunity for you to become a reliable source for high-quality aerospace products is tantalizing because the potential is there. Among the many benefits of taking advantage of digital tools is the fact that they often notice defects earlier, or even prevent them from occurring altogether, and this leads to a higher yield and a reduction in rework. A result of this is not only a reduction in costs for an organization but also a considerable boost to their reputation as a company with their customers as a result. By mastering these efficiencies in the global supply chain, companies can position themselves to be preferred partners on the global stage.
2. Innovation and the development of new products are both vital for the future of aerospace companies, which will be one of the biggest opportunities in the industry. The development of digital technologies has opened up doors for designing and manufacturing of products that were previously not feasible in the past. Taking advantage of this opportunity will be helpful for Asia-Pacific aerospace companies to innovate beyond typical approaches in the field. A significant opportunity for aerospace is also generative design and additive manufacturing in order to enable new forms and lighter components. A product like an aerospace product can also benefit from the use of artificial intelligence to optimize

complex systems like flight control software or maintenance scheduling, which can offer an additional selling point in addition to the ability to utilize artificial intelligence. There would be a strong differentiation advantage if a company developed a subsystem of a plane that includes a feature powered by artificial intelligence that monitors the health of its aircraft, this would give them a significant advantage over their competitors. There has been an increasing desire by Asia-Pacific countries that aim to manufacture their own aircraft products such as regional jets, drones, advanced satellites, to become more digitally adept so that they can accelerate the development cycle and keep up with or complement the giants in the industry. There have been rapid developments in the aerospace sector as well as advanced air mobility electric vertical takeoff and landing vehicles in this region as a result of the rapid development of aerospace startups. There are also a number of entrepreneurial opportunities emerging as a result of new companies adopting digital technologies to run their businesses in order to stay competitive. Using digital technologies to develop and prototype novel aerospace concepts quickly is one way to use technology for swiftly prototyping and iterating new ideas. The concept of integrated frameworks enables such innovation to happen faster and more reliably, since they allow design to be connected to testing and manufacturing in a tightly controlled manner.

3. A new market access and collaboration opportunity exists in Asia for Aerospace. The digital operation of a company can collaborate more easily with international partners when it is well integrated. Creating deep integrations with global aerospace programs is a big opportunity for Asian-Pacific manufacturers, as well as a chance to win contracts that might have been awarded elsewhere if they had been able to integrate deeper.
4. There is a strong correlation between the greater the ability of a supplier to connect digitally with a manufacturer, providing transparency, responsiveness, and solid data-driven assurances of quality, the more likely it is that the manufacturer will do business with the supplier. Integration of digital technology can be a powerful signaling tool. It proves a company's modernity, efficiency, and ability to meet the stringent requirements that aerospace prime contractors require. A regional increase in market access will also be possible as the aerospace industry in Asia-Pacific nations is often reliant on local participation for the ordering of defense and national airline equipment. If a local company is able to partner with a big aerospace company with advanced digital capabilities, they will stand out from the crowd. A company that is ahead of the curve can be an advantage as standards for digital integration become part of certification criteria or procurement criteria in the future. An organization may reach a point in the future where the level of digital maturity of the company will be beyond its control. Regulations will evaluate the ability of suppliers to integrate and

trace their processes as well as to guarantee that critical aerospace components are produced to the highest quality when they certify the production quality. The benefits of integrating frameworks early in the process will therefore be more readily accessed by those who adopt them at a later stage.

5. There are many opportunities for the Asian aerospace industry in terms of resilience and risk management. Whether due to a pandemic or other global disruptions in the supply chain such as natural disasters or trade wars, the aerospace industry has been exposed to a wide range of vulnerabilities. Developing more resilient operations is possible with the help of digital integration. It has been demonstrated that factories that have a strong digital backbone are able to perform remote monitoring and even remote operations, which have proven invaluable when physical access is restricted. It has been demonstrated that companies with digital workflows and cloud-based collaboration suffered fewer disruptions due to disruptions. During times of travel restrictions or when employees were required to work from home, aerospace engineers were able to access designs and simulations remotely through dashboards, and manufacturers could track factory status through software dashboards. When the information is integrated with the supply chain, companies are able to simulate what-if scenarios in the supply chain in the event that one supplier goes down and can quickly find alternatives, since they have the data and connections available to them. When it comes to the aerospace industry in Asia-Pacific, where natural disasters such as typhoons and earthquakes can occur, and where supply chains span many borders, there are many natural hazards to deal with. In a world where operations are perhaps more susceptible to political risk than ever before, the ability to dynamically reconfigure them with the least amount of downtime is a real advantage. There is a growing trend in which digital twin models of factories or supply chains are used by planners to test their systems and create contingency plans in case of an incident. There are many companies that invest in this in order to be prepared for future crises, helping them gain trust from customers by being able to deliver on time despite challenges, and possibly even benefiting by being able to capture business from their competitors when they are unable to deliver.
6. The Asian aerospace industry has a great opportunity to unleash its potential through global workforce empowerment and talent attraction. The lack of skilled employees has been a challenge for aerospace manufacturing over the years, but on the other hand, a successful digital transformation can make the industry more attractive to talent in the future. Many young professionals are looking for opportunities to work in a technologically advanced, forward-looking environment. The use of cutting-edge tools and technologies is likely to attract engineers, data scientists, and technicians who are eager to work with high-

tech technology at an aerospace company known for its digital innovation and smart factory. There is a high level of talent competition across a wide range of industries in Asia-Pacific, which is particularly relevant. As manufacturing transforms from a less glamorous industry to a high-tech one, historically seen as less glamorous than software or services. Manufacturing in the aerospace industry can draw some of the brightest minds into the industry who may otherwise decide to work for companies in the other fields. Embedding a digital environment into the organization allows for the capabilities of the existing workforce to be empowered since they can be freed from drudgery and have rewarding analytical or creative tasks at hand. There is a possibility that a machine operator who was previously just following set procedures could now be part of a team that experiments with parameters based on data insights, which could give him or her a more fulfilling role than before. With the use of digital systems, employees can see their suggestions implemented in a quick and efficient manner. The idea that a culture of continuous improvement can be established can be achieved by employees and systems observing bottlenecks, making recommendations for tweaks, and ensuring that the data system confirms that changes have been made. Morale and productivity are boosted as a result of this empowerment. Organizations that emphasize the importance of digital tools in augmenting human capabilities and turn a potential point of resistance into an enthusiastic adoption, as workers realize they are being elevated and not sidelined, can ultimately increase adoption as employees realize they are being augmented.

7. The author of this book chapter believes that contributing to sustainable goals will be both an excellent opportunity and a necessity for the aerospace industry in Asia in the next decade. Taking part in global sustainability efforts is a good way to align with local efforts. There is growing pressure on the aerospace industry to reduce its carbon footprint, not only during aircraft operations, but also during manufacturing processes. With the help of digital technologies, machines can be optimized in order to reduce their energy consumption through optimization. A smart energy management system for factories, a manufacturing process that minimizes scrap material through precise fabrication and better first-time quality, and a process that supports the use of greener and more recyclable materials, such as designing parts for recycling or reusing. Several Asian-Pacific countries are parties to international climate treaties and have committed themselves to sustainable development. As a result, aerospace firms located in these regions can lead by example when it comes to green manufacturing practices. A great way to improve sustainability is by using digital technologies in a measurable manner through the use of digital technologies. A factory can reduce its power consumption by X% if it implements energy monitoring and AI optimization

in order to lower its energy usage and then reduce the cost of its equipment. Sustainability not only benefits the planet and enhances the public image of a company, but it can also meet the requirements of new regulations for sustainable supply chains, as well as meet customers' demands for them. It is possible that this can also become a competitive edge as airlines and defense departments begin evaluating the environmental impact of their supply chain in order to differentiate themselves from their competitors. The opportunity here is to get ahead of those demands and ensure that the aerospace industry is leading the way with sustainable manufacturing in the future.

It has become evident that the path forward for Asia-Pacific aerospace manufacturing will require a combination of careful but bold action, in light of these challenges and opportunities. Technology solutions, policy support, and human centric management are all necessary in order to address the challenges that we face. A challenge associated with skill development can be tackled by industry-academy programs, which work as a team. A number of government programs are in place in the region that are aimed at developing a pipeline of engineers with a high level of digital fluency. Such specialized aerospace technology programs are in fact being funded by governments in the region as part of their public policies. It is also important to address issues of standardization and interoperability, which can be challenging in some cases. In an area where Asia-Pacific collaboration can yield fruit, regional forums and partnerships can be beneficial in terms of developing common frameworks that can serve as a complement to global standards for local implementation guidelines, thereby mitigating some of the risks. There are two things that companies need to be willing to do if they want to be able to capture efficiencies and innovation on the opportunity side. There is even a possibility that their business models may be affected. The traditional mindset may have to be shifted in order to make this happen. The aerospace industry is able to leverage the power of a digital ecosystem to benefit its customers through the help of a digital ecosystem of partners and suppliers.

CONCLUSION

As the aerospace industry in Asia-Pacific evolves, it will be necessary for each organization to chart its own course. Those journeys will be aided by recognizing these common challenges and opportunities, as well as by sharing knowledge and expertise with each other. In this book chapter, an integrated framework is presented which serves as a guide while guiding the reader through the journey, ensuring that while pursuing high tech solutions, the journey stays on track. The strategic alignment, the people, and the partnership components of companies do not get lost in the shuffle. The integration of digital innovation through such frameworks carries a great deal of significance in the region, and the implications for the future of aerospace manufacturing in the region are certainly positive. The aerospace manufacturing sector in Asia-Pacific will have to take advantage of its coherence if it is to succeed in the digital era as it moves forward. There are obstacles along the way, and the journey is not without challenges. There are several challenges that the aerospace industry must address, including ensuring cyber security, overcoming talent gaps, investing wisely, and integrating different systems in order to operate efficiently.

As a result, there may be economic benefits, such as greater efficiency, innovation, connectivity of the markets, resilience, empowerment of the workforce, and sustainability, which may even serve as further reasons for persevering and innovating solutions to these challenges. It is likely that the organizations in the aerospace industry which treat digital innovation as more than a gadget that needs to be acquired, but as an integrated change in its culture, structure and relationships that will define the future of the industry will emerge as the leaders who will shape it. Asia-Pacific is a region of increased ambition, strong market momentum, and advancements in technology that is well positioned to set benchmarks in this domain because of its combination of ambition, momentum, and technological prowess. The aerospace industry is poised to achieve unprecedented levels of efficiency, develop more innovative and safe aircraft, and respond adeptly to the evolving demand for innovative and safe aircraft that is affecting the global aerospace industry as a whole. Asia-Pacific aerospace manufacturers and their partners can achieve digital transformation by adopting integrated frameworks to enable them to improve their digital capabilities. There is more to this evolution, as explained throughout this book chapter, than just adopting new tools; it involves reimagining the aerospace manufacturing enterprise as a digitally empowered enterprise. There is a need for aerospace professionals to be flexible, intelligent, and collaborative. In the coming years, key decisions will be made that will have a major impact. Concepts are now being translated into practice at a much larger scale, and lessons are being learned here that will be relevant not only regionally, but for the entire aerospace industry worldwide, as everyone strives to build the aircraft and spacecraft of tomorrow with the technologies available today. Aerospace manufacturing processes are significantly improved by measurement-assisted assembly methods because their precision and accuracy are significantly improved (Drouot, Zhao, Irving, Sanderson, & Ratchev, 2018).

As the digital aerospace manufacturing industry gains market share in Asia-Pacific, many factors are fueling the growth of this industry. This can be attributed to a variety of factors, such as a growing market demand, favorable government policies, as well as an inherent ambition to ascend the value chain of global aerospace. Based on the findings of the author of this chapter, it can be concluded that Asia-Pacific companies are not just followers in this domain but are often leading in terms of adoption rates and innovation in regard to implementations of this technology. Using digital twins and artificial intelligence to design and maintain smart factories on greenfield sites is one of the latest trends in manufacturing. As a result of the diverse nature of a region like Asia, there cannot be one template that applies to all of its people. Designed to meet the needs of both high-tech production lines and suppliers seeking to meet the standards of international markets, this integrated framework is flexible enough to be able to be adapted as required so as to satisfy both sets of needs.

REFERENCES

Abollado, J. R., Shehab, E., & Bamforth, P. (2017). Challenges and benefits of digital workflow implementation in aerospace manufacturing engineering. *Procedia CIRP*, *60*, 80–85. DOI: 10.1016/j.procir.2017.02.044

Bales, R. R., Maull, R. S., & Radnor, Z. (2004). The development of supply chain management within the aerospace manufacturing sector. *Supply Chain Management*, *9*(3), 250–255. DOI: 10.1108/13598540410544944

Drouot, A., Zhao, R., Irving, L., Sanderson, D., & Ratchev, S. (2018). Measurement assisted assembly for high accuracy aerospace manufacturing. *IFAC-PapersOnLine*, *51*(11), 393–398. DOI: 10.1016/j.ifacol.2018.08.326

Fahrudin, A., Tiwari, S. P., & Rahmi, K. H. (2025). Psychological Well-Being of Human Content Moderators and Wellness Strategies in AI-Driven Content Moderation Organizations. In *Content Moderation in the Age of AI* (pp. 221–250). IGI Global Scientific Publishing. DOI: 10.4018/979-8-3373-0335-2.ch009

Frosch, R. A., & Gallopoulos, N. E. (1989). Strategies for manufacturing. *Scientific American*, *261*(3), 144–153. DOI: 10.1038/scientificamerican0989-144

Garre, P., Bharadwaj, V. N., Shashank, P. S., Harish, M., & Dheeraj, M. S. (2017). Applying lean in aerospace manufacturing. *Materials Today: Proceedings*, *4*(8), 8439–8446. DOI: 10.1016/j.matpr.2017.07.189

Guisan, M. C., & Exposito, P. (2015). Manufacturing and development in countries and areas of Asia-Pacific, 2000-2010. *Regional and Sectoral Economic Studies*, *15*(2), 203–208.

Harris, C. E., Starnes, J. H.Jr, & Shuart, M. J. (2002). Design and manufacturing of aerospace composite structures, state-of-the-art assessment. *Journal of Aircraft*, *39*(4), 545–560. DOI: 10.2514/2.2992

Jackson, K., Efthymiou, K., & Borton, J. (2016). Digital manufacturing and flexible assembly technologies for reconfigurable aerospace production systems. *Procedia CIRP*, *52*, 274–279. DOI: 10.1016/j.procir.2016.07.054

Jomo, K. S. (2003). *Manufacturing competitiveness in Asia*. Taylor & Francis.

Kalpakjian, S. (2001). *Manufacturing engineering and technology*. Pearson Education India.

Phanden, R. K., Sharma, P., & Dubey, A. (2021). A review on simulation in digital twin for aerospace, manufacturing and robotics. *Materials Today: Proceedings*, *38*, 174–178. DOI: 10.1016/j.matpr.2020.06.446

Saha, P. K. (2016). *Aerospace manufacturing processes*. CRC Press. DOI: 10.1201/9781315367965

Strube, G., Eloot, K., Griessmann, N., Dhawan, R., & Ramaswamy, S. (2016). Trends in the commercial aerospace industry. In *Supply Chain Integration Challenges in Commercial Aerospace: A Comprehensive Perspective on the Aviation Value Chain* (pp. 141–159). Springer International Publishing.

Timmer, M. (2000). The dynamics of Asian manufacturing: a comparative perspective in the late twentieth century. In *The Dynamics of Asian Manufacturing*. Edward Elgar Publishing. DOI: 10.4337/9781035304103

Tiwari, S. P. (2025). Building a Content Moderation Framework to Provide a Safe and Secure Experience for Omnichannel Commerce. In *Content Moderation in the Age of AI* (pp. 55–84). IGI Global Scientific Publishing. DOI: 10.4018/979-8-3373-0335-2.ch003

Wang, B. (2018). The future of manufacturing: A new perspective. *Engineering (Beijing)*, *4*(5), 722–728. DOI: 10.1016/j.eng.2018.07.020

Wang, M. (2009). Manufacturing FDI and economic growth: Evidence from Asian economies. *Applied Economics*, *41*(8), 991–1002. DOI: 10.1080/00036840601019059

Wheelwright, S. C., & Hayes, R. H. (1985). *Competing Through Manufacturing*. Harvard Business Review Case Services.

Young, P., Byrne, G., & Cotterell, M. (1997). Manufacturing and the environment. *International Journal of Advanced Manufacturing Technology*, *13*(7), 488–493. DOI: 10.1007/BF01624609

Zemskov, A. D., Fu, Y., Li, R., Wang, X., Karkaria, V., Tsai, Y. K., . . . Li, P. (2024). Security and privacy of digital twins for advanced manufacturing: A survey. *arXiv preprint arXiv:2412.13939*.

Zhang, C., Xu, Q., Yu, Y., Zhou, G., Zeng, K., Chang, F., & Ding, K. (2025). A survey on potentials, pathways and challenges of large language models in new-generation intelligent manufacturing. *Robotics and Computer-integrated Manufacturing*, *92*, 102883. DOI: 10.1016/j.rcim.2024.102883

Zheng, C., Soosay, C., & Hyland, P. (2007). Manufacturing to Asia: Who will win the emerging battle for talent between Dragons and Tigers? *Journal of Manufacturing Technology Management*, *19*(1), 52–72. DOI: 10.1108/17410380810843453

Zhu, M., Liang, C., Yeung, A. C., & Zhou, H. (2024). The impact of intelligent manufacturing on labor productivity: An empirical analysis of Chinese listed manufacturing companies. *International Journal of Production Economics*, *267*, 109070. DOI: 10.1016/j.ijpe.2023.109070

Chapter 7
Cognitive Ergonomics in Smart Manufacturing:
Aligning Technology With Human Capabilities

Mohd Dahlan Abdul Malek
https://orcid.org/0000-0003-2960-4134
Universiti Malaysia Sabah, Malaysia

Ida Shafinaz Mohamed Kamil
https://orcid.org/0000-0002-8229-0464
Universiti Malaysia Sabah, Malaysia

ABSTRACT

The rise of smart manufacturing under Industry 4.0 demands a harmonious integration between advanced technology and human cognitive abilities. This paper explores the role of cognitive ergonomics in ensuring that complex systems, automation, and artificial intelligence are designed to support, rather than overwhelm, the human operator. By focusing on cognitive load, decision-making processes, mental models, and user interfaces, this study emphasizes the importance of human-centred design in smart work environments. Through a review of empirical research and case studies, the paper highlights the psychological challenges and opportunities arising from human-technology interaction in manufacturing settings. Practical recommendations are provided for optimizing task design, reducing mental fatigue, and improving performance and well-being in technologically advanced workplaces. The paper argues that aligning technological innovation with psychological capabilities is critical to achieving sustainable productivity and employee satisfaction in smart manufacturing systems.

DOI: 10.4018/979-8-3373-1082-4.ch007

1. INTRODUCTION

Smart manufacturing is reshaping the global industrial landscape, driven by the convergence of automation, data analytics, and interconnected systems collectively termed Industry 4.0. In Malaysia, this technological revolution is guided by initiatives such as the National Policy on Industry 4.0 (Industry4WRD), which aims to transform the manufacturing sector into a digitally driven and people-centric industry. While much of the focus has been placed on technological innovation and automation, an equally critical factor, the cognitive capacity of human operators has often been underemphasized.

Cognitive ergonomics, a key subfield of human factors psychology, offers a lens through which the interaction between humans and complex systems can be examined and optimized (Wickens, 2008). It emphasizes the design of technologies and environments that align with human mental processes such as perception, memory, decision-making, and attention (Vicente, 2006). In high-stakes manufacturing environments, these mental processes are often taxed by the pace and complexity of operations, leading to cognitive overload, performance errors, and psychological fatigue.

Malaysian manufacturing industries, particularly in Penang, Selangor, and Johor, have seen rapid adoption of smart technologies, including robotics, automated quality inspection, and IoT-enabled machinery. While these innovations boost efficiency, they also demand a high degree of cognitive agility from operators who must interpret complex data, make time-sensitive decisions, and maintain situational awareness in digital control environments. For example, a case study from an electronics plant in Penang revealed that machine operators struggled with interpreting multi-screen dashboards, resulting in increased error rates and delayed response times (Mayer, Schneider & Braun, 2020).

This paper argues that the sustainable success of smart manufacturing depends on how well these systems are aligned with the cognitive capabilities of their users. Neglecting the human component risks undermining technological investments and jeopardizing both productivity and well-being. The integration of cognitive ergonomics into manufacturing practices ensures that technological complexity does not exceed human limitations and that workplaces are designed to enhance mental performance rather than diminish it.

Accordingly, the objectives of this paper are threefold:

1. To define cognitive ergonomics and its relevance in the context of Malaysian smart manufacturing.
2. To examine the psychological challenges and opportunities arising from human-technology interaction.

3. To propose evidence-based strategies, including system design, training, and policy recommendations that align manufacturing systems with human mental capacities.

Through a combination of empirical evidence, theoretical insights, and industry case examples, this study underscores the need for a paradigm shift in how smart manufacturing systems are conceptualized, moving from machine-centric to human-centred innovation. Ultimately, by integrating cognitive ergonomics into Malaysia's Industry 4.0 journey, we can create smarter, safer, and more resilient workplaces. This topic is increasingly critical as smart manufacturing evolves, requiring greater synergy between human cognition and intelligent systems.

Definition and Scope

Cognitive ergonomics in smart manufacturing refers to the interdisciplinary study and application of psychological principles and cognitive science to design manufacturing systems, technologies, and environments that align with the mental capabilities and limitations of human operators. It focuses on optimizing human-machine interactions in Industry 4.0 contexts by considering factors such as perception, memory, attention, decision-making, and mental workload. The goal is to ensure that advanced technologies such as artificial intelligence, robotics, and the Internet of Things (IoT) are integrated in ways that enhance human performance, reduce cognitive strain, and support psychological well-being in complex, automated production settings.

In other words, cognitive ergonomics in smart manufacturing is a branch of human factors and ergonomics that specifically focuses on optimizing the interaction between human mental processes and advanced technological systems in modern manufacturing environments. It addresses how cognitive functions such as perception, attention, memory, reasoning, problem-solving, and decision-making are influenced by the design of machines, software, work tasks, and environments. The aim is to ensure that technologies used in Industry 4.0 not only increase efficiency and automation but also align with the cognitive strengths and limitations of human workers.

This alignment is essential in smart manufacturing systems, which often involve interconnected machines (IoT), autonomous robotics, cyber-physical systems, and real-time data analytics. These systems can overwhelm workers if not designed with cognitive ergonomics principles, leading to mental fatigue, errors, stress, and even accidents. This topic is increasingly critical as smart manufacturing evolves, requiring greater synergy between human cognition and intelligent systems

Key Elements of Cognitive Ergonomics in Smart Manufacturing:

1. **Mental Workload Optimization:** Manufacturing tasks often require multitasking, monitoring complex dashboards, and quick decision-making. Cognitive ergonomics ensures these systems distribute workload appropriately to avoid overload.

   Example: In a Malaysian electronics factory in Penang, a redesign of machine interfaces was conducted to present only the most relevant information to operators during shift changes, significantly reducing mental workload and transition errors.
2. **Human-Machine Interface (HMI) Design:** Interface design that supports intuitive navigation, clear data visualization, and timely feedback helps operators make informed decisions without confusion.

   Example: At a glove manufacturing plant in Selangor, operators initially encountered challenges in interpreting data from the new digital dashboards. Through user-centered design workshops, the interfaces were refined using color-coded alerts and clear visual hierarchies, leading to more effective and timely responses.
3. **Decision Support and Error Prevention:** Systems should assist users in making decisions by guiding them through complex processes and preventing common errors through automation or alerts.

   Example: In Johor, an automotive component supplier implemented predictive analytics tools that alert technicians when parts deviate from production tolerance. This decision aid reduced inspection-related errors and improved product quality.
4. **Training and Cognitive Skill Development:** Training programs should reflect the cognitive demands of new systems, using simulation, augmented reality (AR), or virtual reality (VR) to build mental models.

   Example: A Sabah-based aerospace firm uses VR simulation to train technicians on assembling sensitive components, enabling workers to visualize step-by-step instructions, thus reducing onboarding time and improving knowledge retention.
5. **Cognitive Task Analysis (CTA):** CTA helps understand the cognitive steps involved in a task, revealing potential mismatch between system complexity and user capability.

   Example: UiTM researchers used CTA to evaluate decision points in a petrochemical plant's control room. Their findings led to redesigning alarm systems to prioritize critical alerts and suppress non-urgent notifications.

Cognitive ergonomics in smart manufacturing is about designing intelligent systems that *think with*, not *for*, humans. When machines are engineered to accommodate how people perceive, understand, and act, the result is a more resilient, productive, and psychologically healthy workforce. For Malaysia, embracing this discipline will be key to ensuring that its transformation into a high-tech manufacturing hub does not come at the expense of human capacity and well-being.

Historical Foundations of Cognitive Ergonomics in Smart Manufacturing

To appreciate the contemporary significance of cognitive ergonomics in smart manufacturing, it is essential to understand its historical evolution. The development of cognitive ergonomics has paralleled major shifts in industrial technology and psychological science, each wave of advancement demanding a deeper understanding of the human mind in relation to complex systems.

Origins in Human Factors and Industrial Psychology. Cognitive ergonomics emerged as a distinct discipline in the mid-20th century, particularly after World War II, when aviation and military systems required operators to interpret high-density data in real-time. This led to foundational research on mental workload, attention, and decision-making under pressure. Human factors research expanded during the 1950s and 1960s, emphasizing the importance of aligning system demands with human capabilities, especially in high-risk environments such as air traffic control, nuclear plants, and military command centres.

During this era, psychologists such as Paul Fitts and Donald Broadbent laid the groundwork for cognitive ergonomics by investigating information processing, human error, and decision latency. Their work shifted the focus from merely physical ergonomics concerned with posture and muscle strain to cognitive aspects like mental fatigue and attention span.

Industrial Evolution and the Emergence of Automation. With the rise of computer-integrated manufacturing (CIM) in the 1970s and 1980s, cognitive ergonomics became more relevant. Factory floors started incorporating computerized control panels, requiring operators to process digital feedback and troubleshoot automated systems. Researchers began applying models such as Wickens' Multiple Resource Theory (2008) and Rasmussen's Skill-Rule-Knowledge framework (1983) to explain operator behavior in complex, fast-paced environments.

Cognitive Ergonomics and Industry 3.0. The third industrial revolution (Industry 3.0) further solidified the role of cognitive ergonomics. As manufacturing systems became more computerized, human-machine interaction became a key concern. In Malaysia, this was the period when electronics manufacturing began to dominate the export economy, creating demand for ergonomic interventions to address repetitive stress, attention fatigue, and error-prone tasks in assembly lines.

Transition to Industry 4.0 and the Rise of Smart Manufacturing. The 21st century ushered in Industry 4.0 marked by interconnected devices, autonomous robots, and cyber-physical systems. With this shift, the cognitive demands on human operators dramatically increased. Workers now interact not only with machines but with complex digital ecosystems that require continuous monitoring, adaptation, and real-time decision-making.

In response, cognitive ergonomics has evolved to encompass neuro-ergonomics (integrating neuroscience and ergonomics), wearable cognitive sensors, and predictive analytics. It now addresses mental load in smart control centres, cognitive fatigue in robotic supervision, and decision-support design in AI-driven production. This topic is increasingly critical as smart manufacturing evolves, requiring greater synergy between human cognition and intelligent systems.

Malaysian Context: Local Adaptation and Research In Malaysia, cognitive ergonomics research began gaining momentum in the early 2000s, primarily in the context of human-computer interaction (HCI) and manufacturing safety. Institutions like Universiti Teknologi MARA (UiTM), Universiti Teknologi Malaysia (UTM), and Universiti Malaysia Sabah (UMS) have since conducted research into mental workload in high-tech industries.

The Malaysian Productivity Corporation (MPC) and the Department of Occupational Safety and Health (DOSH) have also started integrating cognitive considerations into workplace assessment frameworks, particularly under the Industry 4WRD initiative.

From its roots in military aviation to its current application in intelligent manufacturing systems, cognitive ergonomics has continuously adapted to technological progress. Its history reflects a broader shift toward valuing human cognition as a critical asset in system design. As Malaysia advances its smart manufacturing agenda, embracing the historical lessons of cognitive ergonomics ensures that future innovations remain not only efficient but human-centred and sustainable.

2. FOUNDATIONS OF COGNITIVE ERGONOMICS

Cognitive ergonomics refers to the scientific discipline concerned with understanding mental processes that affect interactions among humans and other elements of a system. These mental functions include perception, memory, reasoning, motor response, decision-making, and situational awareness. In the context of smart manufacturing, where operators interact with intelligent systems, the ability to process information accurately and efficiently becomes critical.

Theoretical models such as Wickens' Multiple Resource Theory (2008) help explain how individuals allocate attention across tasks and why multitasking in complex environments can lead to mental fatigue. Rasmussen's Skill-Rule-Knowledge Framework (1983) also identifies different levels of cognitive control that operators use when handling familiar, routine, or unfamiliar tasks (Rasmussen, 1983). These models are particularly relevant in smart factories, where rapid switching between control interfaces, dashboards, and real-time alerts places enormous demand on the worker's cognitive system.

In Malaysia, the application of cognitive ergonomics is gradually gaining traction in sectors such as electronics, automotive, and oil and gas manufacturing. However, there is still limited integration of cognitive models into workplace design, training, or human-machine interaction protocols. Bridging this gap requires an understanding of how operators think, learn, and make decisions within the constraints of high-technology environments.

3. HUMAN-TECHNOLOGY INTERACTION IN SMART MANUFACTURING

Human-technology interaction is a defining characteristic of Industry 4.0. Technologies such as industrial IoT, cyber-physical systems (CPS), big data analytics, and autonomous robots promise increased productivity, but they also shift the nature of work from manual labor to cognitive coordination. This transformation is not without challenges.

In Malaysia, the implementation of smart systems has increased exponentially, particularly in high-tech zones like Bayan Lepas (Penang), Senai (Johor), and the Klang Valley. In one example, a Malaysian semiconductor facility introduced a smart dashboard integrated with predictive maintenance alerts. While effective in reducing machine downtime, the system initially led to cognitive overload among technicians unfamiliar with interpreting layered data streams. An internal assessment found that errors in system interpretation contributed to avoidable equipment delays.

Such examples highlight the importance of designing systems that account for the limitations of human cognition. Poorly designed interfaces, non-intuitive data visualizations, and alert fatigue are common issues that arise when human factors are not adequately considered.

4. PSYCHOLOGICAL IMPACTS OF COGNITIVE MISMATCH

Cognitive mismatch occurs when the cognitive demands imposed by a task exceed the mental resources of the human operator. This phenomenon is particularly relevant in smart manufacturing environments, where operators are frequently required to manage complex, real-time information, operate advanced machinery, and make rapid decisions under pressure. Such environments, while technologically sophisticated, often fail to accommodate the limitations of human cognitive architecture.

In practical terms, cognitive mismatch can manifest in several detrimental ways:

- **Delayed decision-making in high-pressure situations:** When presented with overwhelming volumes of data or unfamiliar system interfaces, operators may experience decision paralysis, hesitating or second-guessing actions that should be routine.
- **Difficulty interpreting real-time data from multiple sources:** Smart manufacturing systems often include multiple data dashboards and visual alerts. Without intuitive design, these can overload visual and working memory capacities, leading to misinterpretation of critical metrics.
- **Increased absenteeism due to mental exhaustion:** Chronic cognitive overload contributes to psychological fatigue and stress-related absenteeism. In Malaysia, HR records from several industrial zones in Selangor have indicated a rise in cognitive burnout symptoms among technical staff following automation upgrades.
- **Human-machine system breakdowns due to over-reliance on automation:** Operators may become overly dependent on automation, leading to skill degradation and a reduced ability to intervene effectively during system anomalies or malfunctions.

A particularly illustrative case comes from electronics manufacturing plant in Shah Alam. The facility implemented a digital twin system designed to mirror and simulate real-time production conditions. However, the system's interface lacked proper user segmentation and layered control logic. Workers, insufficiently trained and overwhelmed by the interface complexity, made numerous operational errors. Within the first three weeks, task accuracy declined by 18%, and production delays spiked by over 12%. A follow-up ergonomic audit recommended simplification of user interfaces, phased training programs, and cognitive readiness assessments for all machine operators.

To effectively mitigate cognitive mismatch, proactive strategies must be implemented before and during system deployment (Parasuraman, Sheridan, & Wickens, 2008):

- **Pre-deployment assessments:** Use validated tools such as NASA Task Load Index (NASA-TLX) to quantify perceived workload across mental demand, physical demand, temporal demand, effort, and frustration.
- **Simulation-based testing:** Introduce new systems in simulated environments where operator feedback and behavioural data can be collected to adjust design elements.
- **Cognitive readiness programs:** Evaluate individual cognitive traits such as working memory span, problem-solving style, and attentional flexibility to better match workers with appropriate technology interfaces.
- **Interface modularity:** Design systems with modular or role-specific dashboards that allow operators to only access information relevant to their task, reducing unnecessary cognitive strain.

In the Malaysian context, many factories still rely on a one-size-fits-all approach to technology adoption, overlooking individual differences in cognitive load tolerance. A tailored, cognitive ergonomics-informed approach can significantly reduce the psychological burden on employees and ensure that technological advancement is matched by cognitive accessibility. Government agencies such as the Department of Occupational Safety and Health (DOSH) and the Malaysian Productivity Corporation (MPC) should also advocate for national guidelines that include cognitive workload evaluations in industrial ergonomics audits.

5. DESIGNING FOR COGNITIVE ALIGNMENT

Cognitive alignment in smart manufacturing hinges on designing systems that respect the mental boundaries of human operators while enhancing their capabilities. This section outlines three pivotal strategies: user-centred interface design, adaptive systems with cognitive support tools, and comprehensive training through simulation and decision aids.

User-Centred Interface Design

User-cantered design (UCD) focuses on tailoring interfaces to the needs, limitations, and cognitive processes of users. In manufacturing, this means creating dashboards, control panels, and software that are intuitive, uncluttered, and logically organized. Design principles such as minimizing visual clutter, prioritizing high-frequency tasks, and applying consistent visual cues can reduce cognitive load and improve response times. Engaging end-users early in the design process fosters systems that align closely with real-world workflows and expectations.

Adaptive Systems and Cognitive Support Tools

Adaptive systems dynamically adjust their behaviour based on user performance or context, enhancing usability under varying cognitive conditions. For instance, if an operator exhibits slower-than-normal responses, the system may simplify tasks or provide additional prompts. Cognitive support tools such as checklists, heads-up displays, and context-aware assistance reduce memory load and decision fatigue. Augmented reality (AR) has proven effective in guiding complex assembly tasks by overlaying digital instructions directly onto physical components.

Training, Simulation, and Decision Aids

Comprehensive training plays a crucial role in strengthening mental models and equipping operators for high-stress, rapidly changing industrial environments. Simulation-based learning provides a safe and immersive platform for workers to rehearse real-world scenarios, improving situational awareness, decision-making under pressure, and response confidence. These virtual environments mirror actual operational conditions, enabling personnel to internalize protocols and develop adaptive strategies in a controlled setting.

In addition, decision-support tools such as predictive analytics systems and diagnostic algorithms facilitate swift, evidence-based judgments, significantly reducing the risk of human error during critical moments. When integrated effectively, these strategies form a robust framework for designing cognitively optimized systems that enhance, rather than hinder, human performance.

6. BEST PRACTICES AND STRATEGIES

Effectively applying cognitive ergonomics in smart manufacturing environments requires a structured approach encompassing organizational policy, technology design, and workforce development. The following practices have shown to yield significant benefits in both Malaysian and global settings:

Practical Ergonomic Interventions: Manufacturing firms in Malaysia such as those in the Kulim Hi-Tech Park have adopted ergonomic principles to redesign workstations based on cognitive flow. For instance, a solar panel manufacturer optimized visual controls and task layout to match operator routines, which led to a 17% reduction in task completion time. Ergonomic lighting, colour coding, and minimized switch-tasking also contribute to reduced mental effort.

Cognitive Task Analysis (CTA) for Smart Workstations: CTA breaks down the knowledge and thought processes required for a task. A study conducted by Universiti Teknologi MARA (UiTM) applied CTA to assess control room operations in a petrochemical facility. The analysis revealed bottlenecks in decision points and prompted the redesign of the HMI to better match operator expectations. Such interventions improve decision quality, reduce stress, and promote operational resilience.

Creating Psychologically Supportive Environments: The psychological climate in a manufacturing facility plays a critical role in shaping cognitive performance. Organizations like Western Digital Malaysia have implemented mental health programs and mindfulness-based stress reduction (MBSR) workshops. These interventions not only address cognitive fatigue but also enhance emotional regulation and job satisfaction. Managers must encourage a culture where cognitive concerns are recognized and addressed without stigma.

7. CASE EXAMPLES FROM INDUSTRY

The application of cognitive ergonomics in smart manufacturing is best illustrated through real-world examples that demonstrate the effectiveness of aligning technology with human mental capabilities. This section presents selected case studies from Malaysia and international settings, highlighting successful implementations of ergonomic strategies and the key lessons derived from these experiences. This topic is increasingly critical as smart manufacturing evolves, requiring greater synergy between human cognition and intelligent systems.

Selected Malaysian and International Case Studies This topic is increasingly critical as smart manufacturing evolves, requiring greater synergy between human cognition and intelligent systems. Malaysian Case Study: Ergonomic Work Systems in Electronics Manufacturing (Penang) This topic is increasingly critical as smart manufacturing evolves, requiring greater synergy between human cognition and intelligent systems.

Selected Malaysian and International Case Studies

Malaysian Case Study: Ergonomic Work Systems in Electronics Manufacturing (Penang)

Major electronics manufacturing facility in Penang undertook a cognitive ergonomics intervention to address operator errors in a high-speed production line. By analysing task complexity and interface usability through cognitive task analysis (CTA), the company introduced simplified control panels, real-time visual alerts, and structured task flow designs. The result was a 30% reduction in error rates and a 25% improvement in operator satisfaction. This initiative, supported by the Malaysian Productivity Corporation (MPC), demonstrated that small design changes could yield significant psychological and productivity benefits. (Lim & Tan, 2021).

A global automotive manufacturer in Germany integrated Radio Frequency Identification (RFID) systems into its smart production chain to improve inventory tracking and task sequencing. Operators were previously burdened by frequent data entry tasks, which interrupted workflow and increased cognitive load. With RFID automation, part identification became seamless, allowing workers to focus on value-added activities. Subsequent assessments revealed a 40% decrease in task time variability and improved employee focus. The company also reported a decrease in stress-related absences.

Success Stories of Ergonomic Design in Smart Manufacturing

Adaptive Human-Machine Interfaces (HMIs)

At a Japanese robotics firm, engineers deployed adaptive HMIs capable of detecting user expertise through interaction patterns. Novice users received more detailed prompts, while experienced operators viewed streamlined interfaces. This adaptive approach not only shortened the learning curve for new hires but also reduced cognitive overload in critical task sequences. Productivity gains of up to 20% were recorded, along with fewer training hours.

Collaborative Robots (Cobots) in European Assembly Lines

In Denmark, a consumer electronics company integrated collaborative robots (cobots) alongside human workers on an assembly line. These cobots handled repetitive, precision-based tasks, while humans were assigned to roles requiring judgment and adaptability. The ergonomic redesign reduced musculoskeletal complaints by 45% and allowed for extended focus time during complex operations. The project received acclaim for its success in balancing automation and human-centric values (Bogue, 2019).

Lessons Learned: Advancing Cognitive Ergonomics in Smart Manufacturing

1. **Holistic Integration is Essential for Sustainable Design**

One of the most significant lessons emerging from smart manufacturing initiatives is the need to integrate physical, cognitive, and organizational ergonomics into a unified system. Addressing these domains in isolation often leads to disjointed or inefficient solutions. For example, a well-designed workstation (physical ergonomics) may still cause stress or overload if the cognitive demands are not managed effectively.

Similarly, even the most intuitive interface may fail if organizational workflows do not support decision autonomy or task clarity. A holistic approach ensures that system design supports users not only in terms of body mechanics but also in perception, cognition, and social organization. This alignment enhances productivity, reduces error, and contributes to psychological well-being.

2. **Technology Must Minimize, Not Increase Cognitive Load**

Although digital tools are intended to improve operational efficiency, poor implementation can inadvertently raise mental workload. Technologies such as RFID tracking, adaptive HMIs (Human-Machine Interfaces), real-time dashboards, and collaborative robots (cobots) are most effective when they are intuitive and aligned with users' mental models. For instance, adaptive systems that adjust interface complexity based on user experience or context can significantly reduce cognitive fatigue. Conversely, systems that flood users with non-prioritized alerts or overly complex analytics may undermine decision-making, especially in high-pressure environments. Therefore, usability testing and mental workload assessments must be embedded into the design phase to ensure that technology supports, rather than burdens, human cognition.

3. **Operator Participation Enhances Ownership and Effectiveness**

Direct involvement of frontline workers in the design, development, and evaluation of technological systems has proven to be a game-changer. Participatory ergonomics where users contribute their experiential knowledge ensures that systems are grounded in actual workflows and practical challenges. When operators help shape their tools, they are more likely to adopt new technologies, follow protocols accurately, and provide meaningful feedback for further improvement. Furthermore, involving workers promotes psychological ownership, builds trust in automation, and improves job satisfaction. This approach not only enhances usability but also encourages a culture of collaboration and continuous improvement (Hollnagel, 2014).

4. **Continuous Assessment and Feedback Loops Drive Long-Term Value**

Ergonomic systems are not static; they must evolve alongside technological, organizational, and human changes. Organizations that implement continuous assessment mechanisms such as real-time performance dashboards, user satisfaction surveys, and ergonomic audits are better equipped to detect and address emerging issues. Feedback loops enable rapid adaptation, ensuring systems remain aligned with user needs over time. This is particularly important in dynamic manufacturing environments, where production demands, workforce demographics, and technological capabilities are constantly shifting. Continuous learning from frontline data allows for the fine-tuning of systems and supports a culture of resilience and innovation.

5. **Cognitive Ergonomics is a Strategic Asset, Not Just a Compliance Tool**

Lastly, a broader cultural shift is necessary: organizations must begin to see cognitive ergonomics not merely as a safety requirement, but as a strategic enabler of innovation and competitiveness. Just as Lean and Six Sigma methodologies became standard operational frameworks, cognitive ergonomics should be institutionalized as a core pillar of smart industry. When organizations invest in the mental capacity, comfort, and confidence of their workers, they unlock higher levels of engagement, creativity, and adaptability qualities essential for thriving in Industry 5.0 and beyond.

These case studies reinforce the premise that cognitive ergonomics, when applied effectively, not only improves operational efficiency but also promotes psychological well-being and workplace resilience in smart manufacturing.

8. APPLICATION OF COGNITIVE ERGONOMICS IN EDUCATIONAL SETTINGS: IMPLICATIONS FOR UNIVERSITIES

While cognitive ergonomics has gained traction in industrial contexts, its application in educational settings particularly in universities offers immense potential for enhancing teaching, learning, and student cognitive performance. As smart education technologies become more embedded in higher education systems, there is a growing need to design these systems to align with students' cognitive processes. The same principles used to optimize human-technology interaction in smart manufacturing can be adapted to foster more effective, inclusive, and human-centred learning environments.

Cognitive Load Management in Smart Learning Environments

Modern classrooms increasingly rely on digital platforms, learning management systems (LMS), and real-time collaboration tools. Cognitive ergonomics can be applied to structure course content and interfaces in a way that reduces extraneous mental effort and promotes efficient learning.

Example: In Malaysian engineering programs, smart classrooms can be designed to present lecture materials using information hierarchy and visual cues. For instance, Universiti Teknologi Malaysia (UTM) uses modular e-learning platforms where core concepts are prioritized visually and secondary details are layered via hyperlinks. This scaffolding approach mirrors industrial dashboard design, helping students process information incrementally and effectively.

Human-Machine Interface (HMI) in Educational Tools

Just as poorly designed interfaces hinder workers in manufacturing, confusing educational interfaces can obstruct student learning. Cognitive ergonomics promotes intuitive navigation, consistency, and feedback in educational technologies such as online labs, simulations, and assessment tools.

Example: Universiti Malaysia Sabah (UMS) introduced a redesigned interface for its online technical drawing module, reducing cognitive friction by using icons, step-by-step animations, and simplified toolbars. Students reported faster task completion and less confusion during assignments.

Adaptive Learning Technologies

Educational AI platforms can incorporate cognitive ergonomics by adapting to each student's pace and capacity. These platforms can monitor student progress, identify cognitive bottlenecks, and dynamically adjust task complexity or provide supportive hints.

Example: An adaptive digital platform in use at Universiti Sains Malaysia (USM) tracks how long students spend on problem-solving tasks in a mechanical engineering module. If students demonstrate signs of struggle, the system introduces guided hints, video explanations, or reduces task complexity to sustain engagement and prevent overload.

Simulation and Scenario-Based Learning

Cognitive ergonomics also supports the design of simulations that replicate high-pressure or cognitively demanding environments. These tools are especially effective in preparing students for real-world manufacturing scenarios.

Example: In a final-year capstone course at UiTM, students operate a virtual smart factory simulation using augmented reality (AR). They must make production decisions based on real-time data while managing alerts and system constraints. This setting mimics cognitive demands encountered in actual industrial roles and builds resilience and decision-making capacity.

Cognitive Readiness Assessments in Curriculum

Universities can incorporate formal assessments of cognitive skills such as working memory, attention control, and reasoning. These metrics can be used to personalize instruction and prepare students for complex problem-solving in smart environments.

Example: In a human factor engineering course, UTM students use tools like NASA-TLX and dual-task experiments to assess their own workload during technical tasks. This reflective practice enhances metacognitive awareness and links theory to industrial applications.

Inclusive Cognitive Design for Neurodiversity Learners

Smart educational technologies should be designed to accommodate students with diverse cognitive profiles, including those with dyslexia, ADHD, or autism. Universal Design for Learning (UDL), grounded in cognitive ergonomics, promotes flexible content delivery and assessment.

Example: Multimedia lecture content at Universiti Malaya (UM) includes transcript overlays, chunked explanations, and color-coded visuals to support students with different learning preferences. These approaches mirror ergonomic practices used to support diverse operators in automated factories.

Smart Campus Infrastructure with Cognitive Ergonomics Principles

Beyond courseware, cognitive ergonomics can be applied to campus design. Lighting, ambient noise, digital signage, and classroom layout all influence mental performance and stress levels.

Example: Universiti Malaysia Sabah (UMS) piloted smart lighting systems in study areas that automatically adjust based on circadian rhythms and task intensity. This aligns with cognitive fatigue research and promotes sustained concentration.

Interdisciplinary Cognitive Ergonomics Courses and Research

Universities can foster cognitive ergonomics research by introducing interdisciplinary courses that integrate psychology, industrial design, and engineering. Collaborative projects allow students to apply theoretical models in developing ergonomic interfaces or systems.

Example: A cross-faculty elective titled "Cognitive Ergonomics in Industry 4.0" could combine cognitive science and engineering design. Students would analyse case studies, conduct usability tests, and propose redesigns of smart educational or manufacturing tools.

The integration of cognitive ergonomics into university settings ensures that educational innovation keeps pace with student cognitive needs (Vicente, 2006). By aligning learning environments with the principles of mental efficiency and accessibility, institutions can:

- Improve student engagement and knowledge retention
- Reduce cognitive fatigue and academic stress
- Enhance inclusivity and support for neurodiversity learners
- Prepare graduates to work effectively in cognitively demanding smart manufacturing systems

This approach positions universities as incubators of not only technological talent but also cognitive-aware professionals equipped to lead in the age of Industry 4.0

Cognitive Efficiency and Inclusive Excellence

In the era of digital transformation and cognitive-intensive work environments, universities play a pivotal role in shaping learners' readiness to navigate complexity. Integrating cognitive ergonomics into higher education is a strategic imperative that ensures learning environments are not only accessible and technologically current, but also attuned to the cognitive capacities and limitations of diverse student populations. This integration supports the broader mission of education not merely to transfer knowledge but to develop adaptive, mentally agile graduates who can thrive in cognitively demanding fields such as smart manufacturing, artificial intelligence, and digital engineering.

Cognitive ergonomics focuses on how individuals perceive, interpret, store, and use information. In the context of university education, this includes how students interact with lectures, digital platforms, reading materials, collaborative tools, and assessment formats. When learning environments are poorly designed whether due to information overload, misaligned pacing, unclear visual organization, or inflexible teaching methods students are at risk of cognitive fatigue, disengagement, and even burnout. Conversely, cognitive-friendly educational design enhances mental clarity, promotes deeper engagement, and fosters sustained academic achievement.

Key Benefits of Applying Cognitive Ergonomics in Universities

1. Improved Student Engagement and Knowledge Retention

Cognitive ergonomics encourages the use of instructional design techniques that align with how the brain naturally processes and retains information. Methods such as chunking (breaking content into smaller, manageable units), dual coding (presenting verbal and visual information simultaneously), and retrieval-based learning (frequent low-stakes quizzes) are grounded in cognitive science. These strategies help students form stronger memory connections and maintain focus during complex or abstract material. Moreover, adaptive learning platforms that adjust content delivery to students' pace and comprehension level have shown significant improvements in retention and satisfaction.

2. **Reduced Cognitive Fatigue and Academic Stress**

University students often juggle dense coursework, multiple deadlines, and technology-mediated communication. Poorly designed learning interfaces or excessive multitasking can overload working memory and drain attentional resources. By reducing extraneous cognitive load for example, simplifying navigation in learning management systems, limiting redundant or conflicting information, and allowing task sequencing educators can help preserve students' mental energy. Integrating cognitive recovery strategies such as breaks, guided reflection, and structured peer collaboration also mitigates mental fatigue and promotes long-term academic sustainability.

3. **Enhanced Inclusivity and Support for Neurodiverse Learners**

Modern classrooms are neurodiversity spaces. Students with ADHD, dyslexia, autism spectrum conditions, and other cognitive differences often struggle in rigid, traditional educational setups. Cognitive ergonomics offers a blueprint for universal design for learning (UDL) an approach that proactively considers variability in cognition. Features such as customizable interface settings, multi-sensory instructional modes, flexible submission formats, and scaffolded task structures allow all learners to access and engage with content effectively. Importantly, such inclusive design not only benefits neurodiversity students but also enhances the experience for the broader student body.

4. **Enhanced Readiness for Smart Manufacturing and Cognitive-Intensive Workplaces**

As graduates transition into smart manufacturing or Industry 4.0 environments, they are expected to work with real-time data systems, advanced automation, and cyber-physical infrastructure. These workplaces require high cognitive flexibility, rapid decision-making, and technological fluency. Embedding cognitive ergonomics in university learning environments fosters the development of these competencies. Exposure to simulation-based learning, problem-based modules, and interactive digital platforms prepares students to function effectively under cognitive load and to manage complex tasks independently or as part of interdisciplinary teams.

The integration of cognitive ergonomics into university settings ensures that educational innovation keeps pace with student cognitive needs. Cognitive ergonomics recognizes that just like workers in smart factories, students are also cognitive actors who must process information, make decisions, and solve problems often within high-pressure, technology-rich learning environments. When educational systems and technologies are designed with mental efficiency in mind, learning becomes not only more accessible but more sustainable in the long term.

By aligning learning environments with cognitive ergonomics principles, institutions can achieve the following:

- **Improve student engagement and knowledge retention:** Learning tools designed for cognitive flow minimize distractions and help students focus on core concepts. Structured content delivery via layering, cueing, and modular formats enhances comprehension and long-term memory.
- **Reduce cognitive fatigue and academic stress:** Ergonomic lighting, smart content pacing, and streamlined interfaces lower mental workload. By re-

ducing unnecessary cognitive effort, students are less likely to experience burnout and more likely to maintain consistent academic performance.

- **Enhance inclusivity and support for neurodiversity learners:** By applying Universal Design for Learning (UDL) principles, universities can ensure that students with varying cognitive styles including those with ADHD, autism, and dyslexia can fully participate in and benefit from digital learning ecosystems.
- **Prepare graduates to work effectively in cognitively demanding smart manufacturing systems:** Students trained in environments that mirror the decision-making, information processing, and interface interaction challenges of Industry 4.0 are better equipped for real-world industrial roles. Exposure to simulations, adaptive learning, and ergonomic design fosters critical thinking and cognitive agility.

Moreover, universities that implement cognitive ergonomics are not just optimizing the student experience; they are future-proofing their graduates. As Malaysian industries adopt more AI, robotics, and cyber-physical systems, there will be a premium on human talent that can adapt quickly, manage complexity, and remain cognitively resilient. Universities must evolve from being centres of information transfer to becoming ergonomic ecosystems for cognitive development.

In this context, cognitive ergonomics is not just a supportive feature it is a strategic imperative for higher education in the age of smart manufacturing.

9. FUTURE DIRECTIONS

As smart manufacturing continues to evolve, new frontiers in cognitive ergonomics are emerging that promise deeper integration between human cognitive functions and advanced technology. This section discusses future avenues for enhancing human-technology synergy, focusing on neuro-ergonomics, ethical considerations, and pressing research needs.

Integrating Neuro-Ergonomics and Wearable Technology

Neuro-ergonomics, an emerging subfield that merges neuroscience with ergonomic design, aims to optimize human-system performance by studying brain activity in real-world environments. This integration is particularly critical in smart manufacturing contexts, where cognitive demands are high and system complexity can induce mental fatigue, inattentiveness, and human error. By leveraging insights into neural functions, neuro-ergonomics enhances the ability of systems to adapt to the cognitive states of workers in real time.

Technologies such as electroencephalography (EEG), functional near-infrared spectroscopy (FNIRS), and eye-tracking tools enable the continuous monitoring of neural activity (Sarter & Woods, 1991; Endsley, 1995). These tools can measure brainwave patterns, blood oxygenation in the prefrontal cortex, and gaze fixation all indicators of mental workload, focus, and cognitive stress. In smart manufacturing settings, these data streams can be used to trigger adaptive responses. For example, if a worker's EEG data indicate cognitive overload, the system can initiate alerts, reduce task complexity, or recommend short breaks. This dynamic feedback mechanism aligns operational demands with the brain's current processing capacity, ensuring sustained performance and safety.

Wearable technologies further extend these capabilities by providing unobtrusive, real-time monitoring of physiological and psychological states. Smartwatches, sensor-embedded clothing, and headbands equipped with biometric sensors can track:

- **Heart rate variability (HRV):** An indicator of stress resilience and autonomic nervous system balance.
- **Skin conductance (EDA):** A marker of arousal and emotional responses to workload.
- **Body posture and motion:** Used to detect signs of physical fatigue and repetitive strain.

In Malaysia, pilot initiatives have been launched where factory workers wear biometric wristbands that collect HRV and movement data during shifts. The aggregated data is analysed to identify peak stress periods, prompting managerial interventions such as rotating workstations, offering cognitive microbreaks, or redesigning workflows. This has proven effective in reducing injury rates and improving worker morale.

The integration of wearable data with cognitive models also supports predictive analytics. Over time, patterns of cognitive fatigue and strain can be used to develop personalized ergonomic profiles, guiding job design, training needs, and workload distribution. These insights also inform broader health and safety strategies by identifying workers at risk of burnout or injury (Hollnagel, 2014)

Importantly, ethical considerations must be addressed. Continuous biometric monitoring raises concerns about data privacy, consent, and surveillance. To ensure ethical compliance, organizations must (Cheng et al., 2011):

- Establish transparent policies on data use and storage
- Secure informed consent from all employees

- Limit data access to health and safety professionals
- Ensure that monitoring serves supportive not punitive functions

Looking ahead, the convergence of neuro-ergonomics and wearable technology holds transformative potential. It enables the creation of smart manufacturing systems that not only monitor cognitive states but also evolve based on them. This human-centred approach ensures that technological innovation enhances, rather than diminishes, worker well-being, efficiency, and adaptability in Industry 4.0.

Ethical Considerations in Human-Machine Collaboration

As cognitive data collection and human-machine interactions become increasingly sophisticated, ethical challenges in smart manufacturing intensify. With the integration of neuro-ergonomics and wearable technologies, organizations now have the capacity to monitor workers' mental and physiological states in real time. While these innovations can enhance safety and productivity, they also raise significant ethical questions related to privacy, autonomy, consent, and algorithmic bias.

1. **Privacy and Consent** Biometric monitoring tools such as EEG headsets, smartwatches, and eye-tracking devices collect deeply personal data that may reveal not only mental workload but also emotional states, stress levels, and cognitive health. Without proper safeguards, this information could be misused to penalize workers, determine job suitability, or inform performance appraisals unfairly. It is imperative that organizations obtain explicit, informed consent from workers before collecting any cognitive or biometric data. Consent must be ongoing, not a one-time checkbox, and employees should retain the right to withdraw without consequences.
2. **Data Ownership and Governance** A major concern in cognitive data monitoring is the question of data ownership. Who controls the data the employer, the employee, or the technology vendor? (Parasuraman et al., 2008). To ensure transparency and fairness, organizations must develop data governance frameworks that define clear boundaries for data use, storage, access, and sharing. Workers should have visibility into what data is being collected, how it is interpreted, and for what purpose it is used. Anonymization techniques and strict access control protocols should be standard.
3. **Algorithmic Bias and Discrimination** The AI and machine learning systems that interpret cognitive and biometric data are not immune to bias. If these algorithms are trained on non-diverse datasets, they may misinterpret signals from workers of different genders, ethnicities, or neurological profiles. This could result in unfair treatment or exclusion from certain roles. Therefore, ethical

human-machine collaboration must include algorithm audits, diverse training datasets, and cross-disciplinary oversight involving ethicists, psychologists, and engineers.

4. **Surveillance and Worker Autonomy** There is a fine line between supportive monitoring and invasive surveillance. Workers may feel constantly watched, which could lead to anxiety, reduced trust, and disengagement. To avoid creating a "digital panopticon," ethical design frameworks must prioritize autonomy and dignity. Monitoring systems should be opt-in rather than mandatory, and their purpose must be supportive focused on health, safety, and ergonomic improvement not disciplinary enforcement.
5. **Participatory Design and Employee Involvement** Involving workers in the development and implementation of monitoring protocols can enhance trust and ethical compliance. Participatory design approaches empower employees to voice concerns, co-develop policies, and tailor monitoring systems to real needs. This inclusiveness fosters a culture of shared responsibility and ensures that technological tools align with human values.
6. **Cultural and Legal Considerations** In Malaysia and many other countries, labour laws and workplace privacy regulations are evolving to keep pace with digital transformation. Organizations must align their practices with local legal frameworks such as the Personal Data Protection Act (PDPA) 2010 in Malaysia, and ensure international compliance if operating across borders. Ethical considerations must be integrated into organizational policies, training modules, and internal review mechanisms.

Ethical integration of cognitive monitoring and human-machine collaboration is not merely a legal compliance issue it is a matter of organizational integrity and long-term sustainability. Companies that embrace ethical principles such as transparency, consent, fairness, and inclusivity will not only avoid reputational and legal risks but also cultivate a high-trust work environment. As Industry 4.0 evolves, ensuring the ethical alignment of cognitive technologies with human values will be as critical as their technical sophistication.

Research Gaps and Opportunities

Despite commendable advances in the application of cognitive ergonomics within smart manufacturing, there remain critical gaps that need to be addressed to fully realize its potential. Understanding these gaps allows researchers, policymakers, and industry leaders to prioritize strategic interventions that align technological innovation with human capability.

1. **Longitudinal Studies on the Impact of Cognitive Interventions:** Most existing research focuses on short-term outcomes of ergonomic implementations, such as immediate productivity gains or error reduction. However, there is a significant lack of longitudinal studies that track the sustained cognitive, emotional, and operational impacts of such interventions over months or years. These studies are essential for evaluating the durability of cognitive enhancements, retention of training, and the evolution of mental workload over time. In Malaysia, long-term ergonomic research remains underfunded, particularly in small and medium enterprises (SMEs), which are the backbone of the national manufacturing sector.
2. **Cross-Cultural Validation of Cognitive Ergonomics Tools:** Many cognitive workload assessment tools, such as NASA-TLX or SWAT (Subjective Workload Assessment Technique), were developed in Western contexts and may not fully capture cultural nuances in decision-making, communication, or stress perception among Southeast Asian workers. Validation studies must be undertaken to ensure these tools are culturally appropriate, especially in diverse multicultural settings like Malaysia, where language, educational background, and socio-cultural norms can influence ergonomic outcomes.
3. **Cost-Benefit Analyses of Neuro-Ergonomic Investments:** Although neuro-ergonomic technologies such as EEG wearables and eye-tracking systems show promise in workload measurement and real-time adaptive feedback, their adoption remains limited due to high costs and perceived complexity. There is a need for detailed cost-benefit analyses that quantify ROI in terms of reduced errors, improved training outcomes, and enhanced safety. Such analyses can support evidence-based budgeting for cognitive enhancements, especially in resource-constrained environments.
4. **Frameworks for Evaluating Cognitive Readiness in Industry 4.0 Roles:** As manufacturing roles evolve from manual to cognitive-centric tasks, traditional competency models are becoming obsolete. There is a pressing need to develop cognitive readiness frameworks that include metrics such as attention control, problem-solving flexibility, and digital literacy. These frameworks can guide recruitment, training, and career progression. For instance, the Malaysian Skills Certification System (Sijil Kemahiran Malaysia) could integrate cognitive performance benchmarks into technical training syllabus.
5. **Interdisciplinary Collaboration and Integration:** Solving ergonomic challenges in smart manufacturing requires collaborative research across disciplines. Psychologists provide insights into mental processes, engineers translate those insights into system design, and data scientists optimize decision-support algorithms. Ethicists ensure that innovations respect worker rights and dignity. Malaysia's public universities and government-linked companies (GLCs) should

foster interdisciplinary consortia that promote applied research and pilot studies in industrial settings.

6. **Inclusivity in Ergonomic Design:** Inclusive cognitive ergonomics means designing systems that accommodate workers across a wide spectrum of cognitive abilities, including aging workers and those with neurodiversity profiles (e.g., autism spectrum, ADHD, or dyslexia). These populations are often underrepresented in ergonomic research and risk being marginalized in increasingly automated work environments. Universal design principles must be embedded into digital systems to ensure accessibility and equity in the workplace.

By addressing these research gaps, the field of cognitive ergonomics can mature into a more inclusive, culturally sensitive, and technologically integrated discipline. For Malaysia, this presents an opportunity to position itself not just as a manufacturing hub but also as a leader in human-centred industrial innovation. Creating work environments where cognitive excellence, inclusivity, and psychological well-being are prioritized will be critical to achieving sustainable growth in the age of smart industry.

10. CONCLUSION

This paper has presented a comprehensive exploration of cognitive ergonomics as a foundational pillar in the evolution of smart manufacturing. The shift from manual labour to knowledge-driven, technology-assisted work has increased the cognitive demands placed on human operators. Smart factories rely heavily on real-time data, advanced automation, and interconnected systems, which, while increasing efficiency, introduce significant mental workload and complexity for workers. Cognitive ergonomics addresses this challenge by designing systems that harmonize technological functions with human mental capacities.

Key findings from the paper indicate that the integration of cognitive principles such as human-centred interface design, adaptive systems, and cognitive task analysis leads to measurable improvements in operator performance, workplace safety, and psychological well-being. Operators perform more effectively when systems are intuitive and when mental models are supported through simulation, training, and feedback loops. Real-world case studies, both from Malaysia and abroad, validate the tangible benefits of cognitive alignment, such as reduced human error, lower mental fatigue, improved response times, and more efficient onboarding.

The strategic importance of cognitive ergonomics cannot be overstated. As smart technologies become more deeply embedded into manufacturing ecosystems, the human component must be viewed not as a liability but as an integral asset. Neglecting the human factor in automation leads to operational inefficiencies, increased safety risks, and a deterioration in workforce morale and mental health. In contrast, systems designed with cognitive ergonomics in mind promote both resilience and adaptability among workers.

Implications for Policy, Practice, and Research:

- For policymakers, cognitive ergonomics should be incorporated into national industrial development frameworks such as Malaysia's Industry4WRD strategy. Ergonomics guidelines should not focus solely on physical comfort but also include cognitive criteria for evaluating technologies and workflows.
- For industry leaders, investments should prioritize workforce cognitive readiness through continuous professional development, ergonomic audits, and the adoption of user-centred design practices in new technology rollouts.
- For engineers and system designers, cognitive ergonomics offers actionable frameworks for reducing mental complexity. Interface simplicity, alert prioritization, and user feedback mechanisms must be standard design requirements.
- For psychologists and educators, there is a critical need to collaborate with industry stakeholders to assess, design, and evaluate cognitive demands across diverse operator roles. This also extends to university curricula, which should include courses and certifications in applied cognitive ergonomics.

The Role of Education and Young Generations in Cognitive Alignment

Young people today are digital natives, often fluent in navigating smartphones, digital interfaces, and online platforms. However, this fluency in technology does not inherently translate into deep cognitive engagement, critical thinking, or readiness for complex decision-making in dynamic, high-stakes environments like smart manufacturing. Familiarity with digital tools often leads to multitasking, fragmented attention, and surface-level learning. Cognitive ergonomics in education plays a crucial role in bridging this gap by aligning educational design with the ways the brain naturally learns and processes information.

Cognitive ergonomics in education aims to create learning environments that support attention, memory, reasoning, and executive functions. This involves reducing unnecessary cognitive load, presenting information in digestible formats, and offering scaffolded challenges that build higher-order thinking. When implemented effectively, this approach not only improves academic performance but also prepares students to function effectively in cognitively demanding roles.

To align educational ecosystems with cognitive ergonomics principles, institutions must take several strategic steps:

- **Integrate cognitive assessments and feedback mechanisms into smart learning platforms:** Learning analytics can be used to track cognitive engagement indicators such as time-on-task, error rates, and attention shifts. By using tools like real-time quizzes, cognitive load ratings, and adaptive question sequencing, educators can provide timely feedback and personalize instruction. For instance, an engineering design course can include embedded prompts that assess decision-making strategies and recommend targeted interventions to support cognitive development.
- **Design simulations that mimic real industrial cognitive challenges:** Simulations are highly effective in preparing students for real-world environments by replicating stressors, multitasking requirements, and rapid decision-making scenarios. These tools can be particularly beneficial in technical education fields, such as robotics, manufacturing, and logistics. A smart factory simulation using augmented reality (AR), for example, can expose students to fluctuating production schedules, machinery alerts, and interdependent tasks, fostering mental flexibility and resilience.
- **Promote interdisciplinary learning that combines engineering, psychology, and user experience:** Courses that cross traditional disciplinary boundaries encourage students to understand the human side of technological systems. A program combining cognitive science, ergonomics, and software design can train future professionals to create user-centred systems. Capstone projects can involve real-world design problems where students evaluate cognitive demands and recommend ergonomic improvements to existing platforms.
- **Foster inclusive environments that support neurodiversity learners:** Cognitive ergonomics supports the design of educational environments that respect and respond to diverse cognitive profiles, including those of students with ADHD, autism spectrum disorder (ASD), dyslexia, and other neurodevelopmental conditions. This can include offering multimodal content (text, audio, visuals), allowing flexible pacing, and incorporating assistive technologies. Universal Design for Learning (UDL) principles, embedded in digital course platforms, can ensure that all learners access content in ways that suit their cognitive strengths.

Furthermore, student mental health must be viewed through a cognitive ergonomics lens. Excessive multitasking, digital overload, and poorly designed instructional tools can lead to stress, disengagement, and burnout. Cognitive ergonomics calls for the design of humane learning experiences ones that balance challenge and support, foster intrinsic motivation, and promote self-regulation.

In preparing the younger generation for Industry 4.0, cognitive readiness is as essential as technical proficiency. Universities must ensure that their graduates possess not only knowledge and technical skills, but also the cognitive agility, emotional regulation, and ethical awareness required to navigate smart systems. By embedding cognitive ergonomics into every layer of educational design from curricula to campus infrastructure Malaysia can cultivate a workforce that is not only tech-savvy but cognitively resilient, adaptable, and future-ready.

Graduates who are trained in cognitively ergonomic learning environments will not only understand the operational systems they manage but will also be equipped to contribute to their design and improvement. This positions them as future leaders in smart manufacturing sectors.

Ultimately, the success of Industry 4.0 will not be determined solely by the sophistication of its machines, but by how well these systems are integrated with the cognitive realities of the human operators who supervise, maintain, and make decisions within them. Cognitive ergonomics is no longer optional it is a strategic imperative for creating smarter, safer, more sustainable, and more human-centric industrial ecosystems.

Integrating cognitive ergonomics into university education is not just a pedagogical improvement, it is an investment in the cognitive capital of the next generation. As industries increasingly demand workers who can synthesize information, solve complex problems, and adapt in real-time, universities must evolve into cognitive-friendly ecosystems. By doing so, they not only foster academic success but also support lifelong learning, mental resilience, and equitable access to future-oriented career paths.

REFERENCES

Bogue, R. (2019). Collaborative robots: A review of capabilities, applications and market prospects. *The Industrial Robot*, *46*(4), 404–408. DOI: 10.1108/IR-05-2019-0092

Cheng, B., Wang, M., Yang, S., Kinshuk, , & Peng, J. (2011). Acceptance of competency-based workplace e-learning systems: Effects of individual and peer learning support. *Computers & Education*, *57*(1), 1317–1333. DOI: 10.1016/j.compedu.2011.01.018

Endsley, M. R. (1995). Toward a theory of situation awareness in dynamic systems. *Human Factors*, *37*(1), 32–64. DOI: 10.1518/001872095779049543

Hollnagel, E. (2014). *Safety-I and Safety-II: The past and future of safety management*. Ashgate Publishing.

Lim, C. H., & Tan, S. Y. (2021). *Ergonomic work system implementation in electronics manufacturing: A Malaysian case study*. Ministry of Human Resources Malaysia.

Mayer, A., Schneider, K., & Braun, D. (2020). Reducing complexity in operator dashboards: A case study. *International Journal of Industrial Ergonomics*, *76*, 102935. DOI: 10.1016/j.ergon.2020.102935

Parasuraman, R., Sheridan, T. B., & Wickens, C. D. (2008). Situation awareness, mental workload, and trust in automation: Viable, empirically supported cognitive engineering principles. *Theoretical Issues in Ergonomics Science*, *8*(4), 285–301. DOI: 10.1080/14639220701754741

Rasmussen, J. (1983). Skills, rules, and knowledge; signals, signs, and symbols, and other distinctions in human performance models. *IEEE Transactions on Systems, Man, and Cybernetics*, *13*(3), 257–266. DOI: 10.1109/TSMC.1983.6313160

Sarter, N. B., & Woods, D. D. (1991). Situation awareness: A critical but ill-defined phenomenon. *The International Journal of Aviation Psychology*, *1*(1), 45–57. DOI: 10.1207/s15327108ijap0101_4

Vicente, K. J. (2006). *The human factor: Revolutionizing the way people live with technology*. Routledge.

Wickens, C. D. (2008). Multiple resources and mental workload. *Human Factors*, *50*(3), 449–455. DOI: 10.1518/001872008X288394 PMID: 18689052

Chapter 8

The Human Engine of Industry 4.0: A Psychological Framework for Digital Transformation in Manufacturing

Adi Fahrudin
https://orcid.org/0000-0001-7140-6006
Universitas Bhayangkara Jakarta Raya, Indonesia

Kus Hanna Rahmi
https://orcid.org/0009-0005-1990-7440
Universitas Bhayangkara Jakarta Raya, Indonesia

Fahmi Ilman Fahrudin
https://orcid.org/0000-0003-3028-6399
Chiang Mai University, Indonesia

Siddhartha Paul Tiwari
https://orcid.org/0000-0003-1596-1501
Google Asia Pacific, Singapore

ABSTRACT

This study explores the psychological factors influencing the adoption of Industry 4.0 technologies in manufacturing. Moving beyond a purely technological perspective, it introduces a psychological framework based on the Precaution Adoption Process Model and Resource Dependence Theory. The study identifies seven key psychological barriers: acceptability, resistance, readiness, motivation, cognizance, funding, and commitment to change. Findings show that while organizational resources affect

DOI: 10.4018/979-8-3373-1082-4.ch008

perceived competence, this does not necessarily reduce risk perception. The paper emphasizes the importance of human-centric design, emotional intelligence, and employee engagement in facilitating digital transformation. It highlights the role of cognitive ergonomics, motivation strategies, and adaptive skill development in shaping workforce readiness. Addressing challenges such as technological anxiety and change resistance requires inclusive leadership, continuous training, and feedback mechanisms.

1. INTRODUCTION

Through its ability to integrate core production, core service and development processes, the implementation of Industry 4.0 technologies provides competitive advantages. This study aims to develop an understanding of the interplay among the psychological barriers affecting the adoption of Industry 4.0 in the manufacturing sector. Grounded on the precaution adoption process model and resource dependence theory, a research model is proposed to assess the interplay between seven psychological barriers. Data obtained from 305 manufacturing organisations was analysed using the partial least squares structural equation modelling approach. The findings show that both the availability and the assessment of organisational resources affect the existence of perceived competences. Surprisingly, perceived competences do not hinder risk perception. This study contributes to the literature by revealing new insights into the adoption of the Industry 4.0 phenomenon from a psychological barrier's perspective. The findings also provide implications for both researchers and practitioners, who can benefit from an articulate understanding of the interplay between the psychological barriers of the adoption of Industry 4.0 (Mahmood et al., 2021). In the case of successful adoptions of the Industry 4.0 phenomenon, the production engineering design will combine intelligent products, smart machines, knowledge-based processes, and fully connected "digital" supply chains. On this new stage, new digital products and services, greater integration of automation equipment, new design methodologies and technologies, new organisation of logistics, production and VAL processes, as well as new forms of collaboration and business models will arise. As a result, the perceived benefits with respect to operational excellence, strategic reorientation, and sustainable development will be achieved. These, in turn, will have strong implications on competitiveness. (Kapp, 2018) concludes that the technological and institutional transition towards Industry 4.0 differs by regional conditions and presents global implications for public policy. In reply to the question of "What would Industry 4.0 be?" indicates that digital technology advances, cyber-physical systems, internet of things, and big data analytics are four notable aspects of this phenomenon.

2. UNDERSTANDING INDUSTRY 4.0

The first industrial revolution started at the end of the 18th century with mechanization powered by water and steam. This was followed by the second industrial revolution that started in the late 19th / early 20th century, marked by mass production by the introduction of electricity. The third industrial revolution began in the late 1970s, depicted by automation and electronics. Industry 4.0 can be contemplated as new avenues of production such as the prevalent expansion of information and communication technologies, learning machines, autonomous robots, and the complete digitization of the supply chain (Mahmood et al., 2021).

This fourth industrial revolution can also be described as smart factories, cloud computing, cyber-physical systems, and the internet of things. The concept of smart factories first arose in 2011 along with the commencement of the German Government's digital manufacturing project, applying the phrase Industry 4.0 draw attention worldwide. Industry 4.0 has emerged as a prominent solution achieving sustainability and efficiency along the whole life cycle of processes, products, and resources from product design to product disposal. However, the shift from an embedded to cyber-physical system presents technological, organizational, and managerial challenges. Transitioning towards Industry 4.0 with higher intangible resources like the interconnectivity of information technology and production systems is a new and difficult challenge for many manufacturing organizations. Future production systems will require new competencies in the production of good quality that fit customer specification requirements: some processes will simply be further automated, while others become even more complex with higher skill-requirement. Organizations are also increasingly searching for ways to be agile to cope with rapidly changing product life cycles and consumer expectations. However, this concept still appears relatively understudied in the research literature, and most organizations lack a formal strategy to become more agile (Huber et al., 2022). This technological thrust calls for an exploration of the literature to facilitate transition and take advantage. The definition vocabulary for Industry 4.0 is very broad and has many opportunities and barriers, e.g., poor value-chain integration, uncertainty of economic environment, lack of required infrastructures, job disruptions, lack of skills to work in the new way, organized resistance to change reluctance to change, data security, internet security, standard privileges, mistrust of data bas, lack of competitive advantages, high investment cost.

2.1. Definition and Overview

Industry 4.0 is often described as the cyber-physical transformation of industries by and between several advanced technologies. This definition appropriately combines the technology and business goals within the description scope. To gain a better understanding of Industry 4.0, it is essential to break down this definition into its components. Industry 4.0 contains both cyber-physical and industrial transformation as processes. To keep the object of the analysis broad enough to include different industrial sectors, a wider notion of industry that comprises manufacturing, assembly, and processing is adopted. The cyber-physical transformation is understood as the refinement from an embedded sensing/actuating system to an autonomous linguistic/affective/cognitive entity containing an embedded sensing/actuating system.

As expected in theory development, a thorough literature review identifies core components/constitutive elements of the two component processes of Industry 4.0. Cyber-physical as processes of industry 4.0 embrace system concepts and technologies. Industrial transformation as processes of industry 4.0 celebrates digital transformation in sectors, leading to new driver-intermediary-vehicle-value structures. These components of Industry 4.0 are sufficiently clarified through formal and informal discussions and illustrative examples.

Shrunken to a neat framework, a comprehensive framework of Industry 4.0 is proposed, giving a synthetic overview of the transformation of industries. This framework naturally organizes diverse Industry 4.0 development efforts in descriptive dimensions. Acting as a problem-oriented analytical framework, the scheme enables a systematic composition of detailed modeling of particular business cases by selecting the components of interest.

2.2. Technological Advancements

Industry 4.0 encompasses an influential grouping of digital transformations in smart manufacturing and industrial contexts. Concurrently, it encapsulates and invigorates a persuasive coalition among research, business, and politics, nurturing testbed concepts, tangible solutions, and sprawling workfields (Ramzi et al., 2019). Industry 4.0 is a conceptual umbrella term embracing market contexts, research communities, and political policies. Presenting a strong persuasive coalition impacting profoundly the nature of manufacturing and industrial contexts, it cohesively bundles transformative concepts, radically disruptive influences, sociotechnical phenomena, and ecological consequences on the macro level.

Technologically, Industry 4.0 encompasses a trendy combination of key digital technologies inducive of competitive advanced industrial emergence. It encapsulates an influential family of Cyber-Physical Systems (CPS) empowering novel interconnectedness and assimilation of physical manufacturing systems, including both legacy machines, and advanced amassing apparatus augmented with on-board computing, memory, and communication capabilities (Mahmood et al., 2021). Specifically, although abundantly coined in conjunction with smart manufacturing contexts, CPS presides well over any and all industries and industrial contexts. As an integral class of CPS, Industrial Internet of Things (IIoT) enlarges the emphasis on smart computing and interconnectivity among industrial physical systems incorporating diverse digital interfaces, exposing the functionalities of a broader range of manufacturing systems and transcending the boundaries of traditional industrial concerns. Enabling two-way data streams and mobilizing the outcome of new types and forms of Artificial Intelligence (AI) and big data technologies over ample time intervals, IIoT supports higher levels of autonomous intelligence and enhances extensive intelligence over intricate and distributed geographic areas.

2.3. Impact on Manufacturing Processes

The manufacturing processes for products in digitalized manufacturing environments change significantly. Machinery and manufacturing equipment in digital environments, known as cyber–physical systems, connect through the internet of things to form a collaborative network that handles written data. Production systems were previously implemented in a hierarchical organizational structure, where one single machine was controlled in the front-end of production planning and operations management. In contrast, in digitalized environments, machines operated by control programs form a collaborative network to process written messages and do not need assistance from a central supervisor. These adjustments also extend to the relationships between products and production systems. Products evolve into a shared information space with cyber twin digital models, allowing them to interactively present 'user situations' to production systems that are able to comprehend their own capabilities, thus creating a digital value chain loop.

Digitalized manufacturing processes require wider participation in the planning and execution of operations beyond machine and product roles. On the one hand, this allows the introduction of completely new conditions for product and production functions. On the other hand, it requires the human functions of a digitalized plant to adapt to changes. Some tasks will become simpler or even automated, while others may become embedded in machinery and therefore require a higher level of competence in troubleshooting problem situations ((Mahmood et al., 2021)). In data-driven production systems, it remains uncertain whether the human operator can understand the overall operation status of the machines. The difficulty of knowing which machine and process is working as intended may increase. AI models are unfamiliar and cannot be comprehensible, nor controllable, by the operator.

Consequently, new working styles and psychological factors will arise in the human adaptations and emotional engagement with changes in the operation of manufacturing processes. These drawbacks in production organizations need to be assessed considering the related questions.

3. THE ROLE OF HUMAN FACTORS

The digital transformation of manufacturing industries — often termed Industry 4.0 — has predominantly emphasized technological capabilities such as automation, artificial intelligence, and the Internet of Things. However, as organizations progress in integrating complex digital ecosystems, it becomes increasingly evident that human factors play a pivotal role in ensuring the success, sustainability, and ethical alignment of these innovations. Workers are no longer passive operators but active collaborators in intelligent systems, necessitating a human-centered approach to design, engagement, and skill development.

3.1 Human-Centric Design

Human-centric design in manufacturing emphasizes tailoring digital systems to match the cognitive, physical, and emotional capabilities of workers. Instead of expecting employees to adapt to rigid technological frameworks, the systems themselves must evolve to accommodate human needs. Cognitive ergonomics becomes a critical pillar here, ensuring that information displays, decision-support systems, and machine interfaces align with natural human perceptual and decision-making processes (Norman, 2013). Moreover, user experience (UX) principles are increasingly vital even in industrial settings. Poorly designed digital platforms can lead to errors, stress, and resistance, while intuitive designs can enhance productivity, safety, and worker satisfaction. Human-machine collaboration models, such as cobots (collaborative robots), illustrate successful human-centric innovation by combining machine precision with human judgment and adaptability. Incorporating employee feedback during technology development stages (co-design) is crucial. As the complexity of manufacturing processes increases, designing technologies that account for human variability and strengths can be the difference between project success and failure.

3.2 Employee Engagement and Motivation

Digital innovation is not merely a technical upgrade but a psychological journey for the workforce. Research consistently shows that employee engagement is a strong predictor of innovation success (Deci & Ryan, 2000). Highly engaged employees demonstrate proactive learning, resilience to technological disruptions, and greater collaboration in digitally enhanced environments. Self-Determination Theory (SDT) offers valuable insights into maintaining engagement. Workers need to feel autonomy (control over their tasks), competence (capability to meet new challenges), and relatedness (connection with teams and organizational goals) to embrace change. Motivation strategies must thus move beyond monetary incentives, focusing on intrinsic motivators such as mastery, purpose, and belonging.

Importantly, leaders must recognize and manage resistance to change, a natural psychological response when faced with unfamiliar technologies. Open communication, participation in decision-making, and clear articulation of personal and collective benefits help mitigate fears and foster commitment. Organizational culture plays a key role: cultures that celebrate learning, experimentation, and adaptive thinking will fare better in digital transformations.

3.3 Skill Development and Training

Successful digital transformation demands a continuous evolution of skills among employees. Workers must not only acquire technical competencies but also develop adaptive expertise, emotional intelligence, and problem-solving skills tailored for complex, dynamic environments. Modern approaches to skill development integrate blended learning methods, including e-learning platforms, virtual reality (VR) simulations, augmented reality (AR) guides, and on-the-job mentorships. Cognitive apprenticeship models, where experienced workers guide novices through real-world tasks with scaffolded support, are particularly effective in industrial contexts.

Upskilling (enhancing current roles with new capabilities) and reskilling (preparing employees for entirely new roles) are both essential strategies. Psychological theories such as Bandura's Self-Efficacy Theory suggest that belief in one's ability to learn new skills is a strong predictor of successful adaptation (Bandura, 1997). Training programs must therefore build not just knowledge, but confidence and growth mindsets. Moreover, lifelong learning cultures need to be embedded organizationally. Workers must perceive continuous learning not as a threat but as a valued part of their professional identity. Organizations that invest in human-centered development frameworks will not only retain talent but also unleash innovation from within.

Since the human factor is regarded as one of the main prerequisites for any industrial development, the focus is placed on how mindfulness and mental paradigms change and transform on the way to the smart factory of the future. Modern industrial resources are subject to an increasing evolution, which is driven by various frameworks. While initial drives of the transformation originated from the development of genetics, biomedics, and new materials, the recent human industry forum is strongly associated with the fourth industrial revolution, with the term "Industry 4.0" (i4.0) purposely created for that purpose.

In the emerging i4.0 paradigm, it is realized that producers will be no longer passive participants of the production – "the wheels of industry" but active individuals producing either products or services tailor-made to the individual needs of consumers as "active riverbed." The impact on the consumer's interface is likewise profound and encompasses mental paradigms changes in the way humans enter the world of production. Thus, i4.0 focuses on people in this technological landscape as an innovation factor and uncertainty mitigator. As the perspective of both producers and consumers intertwine and mutually affect each other, the technological or product paradigm of one domain may result in a mental paradigm change in the other domain, as illustrated in Fig. 1.

Consequently, I4.0 or Industry 4.0 is considered the general name encompassing many frameworks and paradigms, including digisticated manufacturing, AI & machine learning, 5G networks, vir•tual reality, & mixed reality, digital twin, cyber threat, blockchain, and various human factors. The many frameworks may have address influence on the way in which the i4.0 approaches impact the human domain. As to the producers' workplace, the highlighted changes are focused on behavioural and physiological needs and intrinsic motivation adjustment. Understanding of the individual's motivations in the i4.0 workplace would allow possible scenario evaluations for valid instruments to promote mental health and individual development in this transformed environment (Mahmood et al., 2021).

4. COGNITIVE LOAD THEORY

Cognitive Load Theory is one of the most important constructs to be considered in the field of instructional design and Human-Computer Interaction. Therefore, it is essential to identify the effects of cognitive load on a user's performance during relative Work Tasks such as AR assembly or disassembly scenarios, in order help future Research in Identifying ways to assist users in reducing excess cognitive load. Over the years, there have been multiple different definitions of the Human Cognitive Load (working memory load), but they all mainly revolve around the same core areas. It can be defined as the total amount of mental activity imposed on working memory of the cognitive system, which holds information for limited amount of time and capacity (Lagomarsino et al., 2021). In the literature, the concept of Cognitive Load has been handled in a two-folded manner, illuminated by the distinction between intrinsic, extraneous and germane cognitive load. Working memory can only handle a limited amount of information a second, which makes a model of human memory storage, a solid building block for designing training tools in any instructional design task. It can be structured as a three-part model which consists of the sensory register, memory, and the long-term memory. Studies have shown, that a user's cognitive load can not only be implied by using self-assessment scales, but can also be estimated by using Observational Measures. More sustainable measures are available, which can make use of the User's Behaviour during Task Execution. The subtle differences in the user's head, eye, and body movements can be classified with Machine Learning methods to show associations with the user's cognitive load. Recent works have adapted this research area with AR Support Scenarios in Industrial Settings. As Cognitive Load takes the central place in the proposed Methodological Framework, it is necessary to bridge the gap between Cognitive Load Theory and psychological underpinnings of Cognitive Load. This is essential, because it defines how Cognitive Load is impacting a User's Performance, which is critical for research in evaluating situations when Cognitive Load is exceeding the Attended Threshold. If Production is to be designed in such a way, that human agents can effectively assist in Digital Transformation, there is no use in uncritically immersing workers in AR environments without understanding the relationship between cognitive load, motivation, attentional focus, and AR user performance.

4.1. Behavioral Change Models

Behavioral change models can be generally classified as voluntary compliance models and acceptance models (Mahmood et al., 2021). The Theory of Planned Behavior (TPB) has been used extensively to study the predictor variables influencing behavioral change. The TPB posits that intention is the best predictor of actual behavior, and that intention can be predicted from three antecedents or exogenous variables: attitude, subjective norms, and perceived behavioral control (PBC). Attitude refers to the individual's evaluation of the target behavior as positive or negative, and the evaluation is in turn predicted by a set of beliefs. Subjective norms refer to what people with social contact to the individual, such as family or friends, perceive about the target behavior. This belief is also predicted by a set of beliefs. Perceived behavioral control is the most recent addition in the model. It is defined as the perceived constraints and opportunities in order to perform the target behavior. It can refer to possession or lack of skills, money and time, as well as external barriers in the environment. PBC is assumed to have both direct and indirect influences on behavior, i.e. it is supposed to influence behavior directly, and at the same time it is supposed to influence intention as well, which also affects behavior, with intention and PBC being the only two predictors of behavior.

Two-box, five-degree scales were used to score behavioral change variables (Ramzi et al., 2019). The definite positive intention scales fully correspond to the Becker's scales of two boxes and five degrees, the second box with completely positive intentions and the best states of bridging gap and fully engagement. Full voluntary compliance of behavioral change is expressed through scoring. It is conditioned on an ACTS-compliance gap model composed of four viewpoints, incidental viewpoints, and three behavioral gap views, and compared to the Becker's process model. The Becker's scales of completing behavioral change cover a wide extent of scopes and ten 0–4 scoring degrees. The scoring system of degrees 1-4 appears to correspond or bridge the Becker's scoring systems, but careful efforts under extended degrees out of five boxes were made to differentiate also three other views, and incidental behavioral change was modeled more reflective to the Becker's model.

4.2. Emotional Intelligence in the Workplace

Its nature is so much more dispositional than cognitive. Emotions and the processes that ignites them are an integral part of the human engine. Although they may differ in intensity, emotions are aroused in all of us. Everyone is emotional to a greater or lesser extent. Nevertheless, there can be significant differences between people within the same profession as well as within the same organization. The complexity and variability in human behaviour are more generally captured by the concepts of emotional and social intelligence. Consequently, when assessing the merits of a candidate or a member of the manufacturing industry, it is equally important to probe their emotional and social intelligence. The term emotional intelligence refers to the ability to perceive and express emotion, to use emotion to facilitate thought, to understand and reason with emotion, and to regulate emotion to promote emotional and intellectual growth. Previous studies suggest connection between emotional intelligence and communication levels especially in contexts within the call centre industry. Recommendations to improve communication levels revolve around training interventions focusing on motivation and emotional competency development (Kapp, 2018). Coaching and mentoring are also offered.

Emotional intelligence is a set of competencies that enable us to analyze our social networks, make smart social decisions and maintain healthy interpersonal relationships, particularly in the face of stress. Emotional intelligence competencies impact interpersonal behaviours in two main ways: interpersonal perception effects, which determine the accuracy of social perceptions, and interpersonal influence effects, which determine whether a social influence strategy is successful. Deficits in interpersonal perception might prevent leaders from recognizing when and how to intervene in misunderstandings and conflicts, while deficits in interpersonal influence may result in ineffective strategies for resolving conflict or leading change. In the face of stress, leaders with low interpersonal perception may exhibit behaviours that heighten tensions and inhibit effective change, although they may often be unaware of these effects on social processes. Importantly, higher emotional intelligence is linked to a greater awareness of the effects of emotional expressions in oneself and others. It is, therefore, plausible that leaders with lower emotional intelligence may experience a cycle of social failure that further impacts their social behaviour.

5. CHALLENGES IN DIGITAL TRANSFORMATION

Several challenges obstruct the process of digital transformation for businesses looking to leverage manufacturing advancements (Mahmood et al., 2021). At a general level, the most severe issues relate to fostering an innovative corporate culture and understanding how to use the technology to put a vision into practice. The lack of skilled people, alongside the gap between concept and implementation, falls in this category. Although technologies can deliver outstanding capabilities, they cannot do much if no clear goals are in place to leverage them. Such a vision and mission are the prerogatives of the company's top management. The tendency to focus solely on technology is particularly damaging to smaller companies, where the information gaps threaten their profitability. Specific issues arise increasingly in accordance with the level of maturity of the processes that need attention. Often, critical investments have not been made in organizational processes, leading to high volumes of non-value-added work in conjunction with dissipative tools that people would like to work with. Therefore, specific initiatives are needed to improve the way of working before advanced productivity-boosting technologies can be leveraged. The manufacturing sector is increasing the model variability of manufactured products at an uncontrolled pace, threatening to make the system utterly dysfunctional. The whole digital supply chain, product lifetime, and business sustainable model indicate uncontrolled complexity. By bringing relatively cheap and highly flexible communication, service, computation, and logistics elements in a new overall architecture, the structure complexity of manufacturing can be kept in control at a semantic level. It is assumed that the implementation of such an architecture will lead to a curbing of the information, process, and structural complexity, which may threaten the operation of the manufacturing system ahead. A significant risk lies in the absence of a reliable technology stack. Given examples hint at potential exponential growths in productivity along with this. However, building a responsible and reliable technology stack that ensures safety, security, and intended use before uncontrolled proliferation occurs is crucial. A knowledge gap lies in how writing conversational agents can be designed and deployed to support the economic and sustainable growth of the most crucial manufacturing companies. The implementation of such systems can disturb the way work gets done and the skills and abilities of the employees that are involved. Considerable knowledge about potential design elements exists, and initial investigations show promising business opportunities. However, more knowledge is needed regarding how the designs can be optimal both in terms of the chatbots' well-being and in supporting the intended transformation of the companies.

5.1. Resistance to Change

Manufacturing organizations are on the road to implementing Industry 4.0 technologies and operational strategies making their workplaces safer, faster, and more effective, which means fewer accidents, shorter response and manufacturing times, lower errors and costs, and higher quality products, leading to more sales and finally higher returns. Such workplace transformations require large investments, which are not easily justified without considering the human issues that all industries must deal with. This is even harder because the technologies involved are already poorly understood or even prematurely developed. As a result, many of the cost-justification or feasibility analyses conducted for the implementation of Industry 4.0 tend to disregard the human side. Unfortunately, there is a strong belief that automation will remove human operators completely from manufacturing systems, with often overstated productivity and flexibility/robustness claims. Automation and management may lead to increased efficiency, but the lessons from past paradigm shifts and industrial crises also show that such disruptions can and will blindside organizations that discard their human factors.

The experience and learning acquired by human operators in the mass production era and the related mass distribution have in many ways been neglected by early 21st century cyber-physical intelligent production systems. As a result, similar problems as in the late 20th century are re-emerging. Rather than humans being removed from objectionable tasks, there is a need to keep humans in many areas of responsibility and develop their competencies for successful human-centered smart machines and collaborative robotics in technological processes for value-added production services. Although such technologies can act independently at different levels in complex dynamic manufacturing environments, workers' skills and willingness to use their intelligence and creativity would still dictate the quality, efficiency, and sufficiency of their organization's offerings. The primary aim of this research is to develop a human-centered psychological framework for the successful adoption and transition to Industry 4.0/Smart Factory technologies and new operation strategies. The Psychology framework proposes as its primary framework elements seven psychological factors, namely: Acceptability of Change; Resistance to Change; Readiness to Change; Motivation for Change; Cognizance of Change; Funding for Change; and Commitment to Change. Each framework element has newsworthy characteristics and disciplines, both in general and Industry 4.0-wise; one often appearing in relevant disciplines and academic fields but combined in this work for Industry 4.0. Industry and academic/military sectors would be able to understand and adopt such concepts/tools as are considered a modular framework to pressing challenges/studies and the disciplines/variables/tools regarding the successful design and operation of world-best/smart manufacturing systems, services, and organizations.

5.2. Technological Anxiety

Although numerous benefits from digital transformation have been enumerated, the approach has been different across industries and countries. This leads to variation in the automation levels of different industry sectors. Higher levels of science and technology are also reflected in manufacturing outputs. At the same time, manufacturer-user perspectives have varied concerning understanding and adoption, attitude towards technologies, and the extent of changes in roles and work areas. Similarly, several human-centric, interdisciplinary, multidisciplinary, or transdisciplinary frameworks exist but none of them are industrial-sector agnostic. There are several psychological barriers that could affect the adoption of Industry 4.0 technologies. These include 1) lack of familiarity; 2) technological anxiety; 3) perceived risk; 4) perceptual incompleteness; and 5) perceived behavioural control.

Technological anxiety could be explained as fear or apprehension of the new innovation being excluded from a group, an uncontrolled event to malfunctioning of an ongoing activity, or ability to distinguish relevant from irrelevant information. Various workshops and surveys yielded insights on why and how psychological factors impede or facilitate technology adoption in manufacturing business units. Participants presented widespread concern on the interoperability of enabled Industry 4.0 technologies, the higher levels of abstraction in controlling processes, data processing, rigorous privacy-related issues, and ethical implications of affects-based modelling. Anticipation of job redundancy and less fulfilment began to unearth potential risk issues. Suggestive approaches to bolster a proactive stance on adoption included a persuasive vision that structured the needed level of innovation, a grown interest in augmented capabilities instead of changes in current own work, transparency at all levels to prevent negative outcomes, and gradual adaptation with an explanatory transitional phase. On the other hand, it was shared that no public participatory approach or knowledge-sharing platform has spread to exchange views preferably conducted by educational institutes or industrial associations. It was recommended to convert perceivable risks into comprehensible scenarios over technology relevance, noise filtering, cybersecurity protocols, and well-defined, economically feasible applications.

5.3. Workforce Adaptation Issues

Workforce adaptation to technology-induced change in manufacturing organizations is not straightforward as people and technology affect each other in complex ways. The workforce needs to adapt in terms of its skills, mindsets, and behavior, which is determined by other people and systems as well. Hence, the focus is on the frameworks and approaches that can be used by industrial organizations to proactively enact the required adaptations of the workforce. The required adaptations are based on an analysis of the changes in manufacturing organizations caused by the ubiquitous introduction of new digital, autonomous, and connected technologies, and their implications for the workforce (Whysall et al., 2019). This results in the classification of three adaptation issues: (a) abilities: the workforce should have improved skills to use the new technologies, (b) work beliefs: the workforce should be open and proactive towards the new technologies and their impact on work, and (c) behavior: the workforce should behave independently and collaboratively when using the technologies together with other people. An analysis of the current approaches adopted by industrial organizations to address these adaptation issues reveals that they resort to ad-hoc initiatives (Ramzi et al., 2019). This casts doubt on their effectiveness and it remains unclear whether desired changes in the workforce occur. Hence, an evidential-driven approach is presented to sequentially decide on the workforce adaptation issues that need addressing, identify the appropriate frameworks to address them, and justify the subsequent selection of more specific interventions. The contribution consists of (1) a systematic classification of workforce adaptation issues for organizations in manufacturing domains, (2) an analysis of the state-of-the-art approaches to tackle these issues, and (3) an evidential-driven approach to select and justify the appropriate frameworks to tackle these issues.

6. STRATEGIES FOR SUCCESSFUL IMPLEMENTATION

A desire to latch onto a holistic vision for the sustainable, secure, reliable, and effective transformation of manufacturing industries has spurred the need to bring together a broad-range of experts and stakeholders from multiple disciplines. Interdisciplinary collaboration is required in order to articulate the potential of the digitally transformed manufacturing industries to deliver the vision of Industry 4.0 across multiple domains. This dialogue must address the short-term political and economic imperatives where evidence of potential gains and risk need to be shown. It must also promote societal awareness of the alienation from the systems of production, security, and privacy implications, loss of control over patterns of consumption, and ultimately, over social interaction impacts of the digital economy on our societies. For this scientific effort to have maximum impact, a framework that articulates the various domains, timings, impact, and extent of transformations is required. In this regard, central to Industry 4.0 is the human technology interaction at the foundation of Industry 4.0 and the relevance of existing human-centered cyber-physical systems of systems concepts. Hints of a decision framework governing the transformation of the digital economy are also identified (Ronell Leo, 2018).

Organizations are always in search for ways to stay ahead of competition, gain agility to responding to evolving changing patterns with product life cycles, product and service variety, consumer wealth, value addition, expectations of consumers and customers, increase productivity, decrease cost, and to achieve smooth and fast turnaround. This desire probably spans from the primitive point in time when an individual picked a stone and remained busy in finding ways to utilize that stone for self-benefitting activities. As the times passed and primitive individuals turned into primitive societies, they somehow discovered economies of scale through division of a large stone into small pieces. The quest for more benefits and productivity couturialized their attentions toward contriving replacement for stones and pebbles, which democratized the use of tools, setups, machines, and ultimately industries of better and more processes for higher value addition (Mahmood et al., 2021).

6.1. Leadership and Vision

In addition to technology and strategy, successful digital transformation critically depends on management and leadership. Properly oriented and highly skilled leaders who inspire and motivate others are arguably the most important business capitals needed for successful Digital Transformation in general. Such leaders are considered drivers of change and growth in organizations. They have the ability to set a vision, anticipate challenges, provide guidance, champion crossdepartmental collaboration, inspire and motivate others, respond to crises, and act flexibly and emotionally. Great leaders score high in psychological competencies, such as self-awareness, emotional self-regulation, social awareness and interpersonal management. Current literature stresses the need for an adequate managerial culture paving the way for customization of Industry 4.0 initiatives and helping employees to embrace ongoing change, but research on specific leadership styles and behaviors is lacking.

Therefore, more attention should be paid to investigate the compatibility of Industrial 4.0 and management philosophies in first place.

Modern management combines various complementary, and sometimes conflicting, ideas and practices. As the modern workplace becomes more complex with increased professional mobility and stronger positions of employees in the knowledge era, less hierarchical, less directive styles based on morality, participation, virtuousness and mindfulness have come to the forefront. A broad system consisting of knowledge, competency and decision-making on all levels, flatness in organizational structure and cushioning of the leader's position is advocated. Furthermore, due to the extensive adoption of Smart 4.0 technologies, the role of management is changed in many respects. With availability of far more information than can possibly be processed, the role of control becomes subordinate to guidance and facilitation. Team collaborations, agile working and self-regulation require a supportive culture and emotional leadership (Mahmood et al., 2021).

6.2. Collaborative Work Environments

As Industry 4.0 (I4.0) systems become increasingly complex, new forms of collaborative work environments (CWEs) for connected human exploitation in decentralised manufacturing systems need to be developed. CWEs in I4.0 are characterised as collaborative, dynamic, non-hierarchical systems with fluid boundary conditions. Humans are considered flexible, decision-making resources that require a high degree of autonomy in their working methods. New roles of human operators must be analysed against the background of collaborative work for decision support, integrated design, and maintenance 3D modelling. The integration of humans into CWEs provides the potential for a human-centric I4.0, which has the potential to shift some value creation activities back to human operators. However, this also requires companies to provide more diverse tasks and a high degree of autonomy for employees. Significantly more organisational and technological consideration is needed to create the conditions for CWEs that are human-centric and that enable human operators to benefit from the opportunities presented by I4.0 in a way worthy of them.

The pandemic and the associated crisis have shown that companies increasingly need to concentrate on their core competencies and flexible, lean operational models in order to be successful and, at the same time, resiliently cope with disturbances in national economies and the global supply network. This also takes the form of CWEs for the linked orchestration of completely connected resource exploitation in decentralised manufacturing systems. Such systems can be unplanned, complex, and sophisticated. Fluctuating product demand of only a few pieces may require high variation in production with economically insignificant lot size. Very diverse production resources and heterogeneous activities enable the production of many different products. CWEs are strongly distributed with regard to different levels of decentralisation and autonomy. Resource nodes can behave independently from market firms and other resource nodes. CWEs may cross organisational and national boundaries. CWEs will have high fluidity of boundary conditions, continuously varying realisation of CWEs with the networked addition or decommissioning of resource nodes at different time scales (Libert et al., 2020). New roles of human operators must be analysed against the background of collaborative work for decision support, integrated design, and maintenance 3D modelling. The integration of humans into CWEs is strongly motivated by three factors.

6.3. Continuous Feedback Mechanisms

Industry 4.0 (I4.0) encompasses the evolution of manufacturing and means the ongoing automation of traditional manufacturing processes using modern smart technology, including the Internet of Things, cloud computing, and cognitive computing. It will transform what manufacturing is, how it operates, and the skills it demands. However, widespread adoption of new technologies, such as human-robot collaboration in assembly line work or data analysis for predictive maintenance, does not occur without integrating the process elements of digital transformation, encompassing not only advanced robotics but also accommodating the human element (Chen et al., 2022).

A generic digital transformation metaphor is developed focusing on the feedback between capabilities and architecture. The digital transformation dynamic model is composed of three first-order differential equations, representing the state equations of capabilities, architecture, and environment. The feedback circuit of digital transformation, comprising two feedback circles of continuous knowledge and continuous capabilities is characterized. Dynamic analysis reveals the market and maintenance potential feedback from architecture to capabilities can address strategic problems by fostering initiative innovation improving the store dynamic capability. The operational and innovative feedback from capabilities to architecture can meet uncertainty strategic needs and influence continuity of digital transformation via influencing architecture flexibility. This actionable knowledge of continuous feedback mechanisms in a dynamic model is beneficial to managers seeking to balance the increasingly complex composition of enterprise processes in the digital transformation journey of I4.0.

The digital industrial transformation is not limited to focused investment but rather encompasses continuous engagement, emphasizing the deleting transformation prioritization when switching from traditional capabilities. Complex networked information systems are needed to reconfigure and optimize across multi-level data and information and model the operation process to support process control subject to distributed and asynchronous operational switching.

7. CASE STUDIES IN MANUFACTURING

Industry 4.0 (I4.0) technology as an enabler for the fourth industrial revolution, manufacturers around the world have invested heavily to initiate digital transformation in their factories. However, many of these initiatives are either unimplemented or unsuccessful. Focusing on three case companies in the manufacturing sector, this paper argues that stronger attention needs to be paid to the human element of digitalization (M. Eriksson et al., 2022). Based on the framework outlined in a literature review, this paper highlights the importance and responsibilities of managers on three levels: strategists, architects, and implementers. It also discusses how the three companies differ in terms of their attention to and investments in human factors, gap between espoused values and actual values, and how companies can align their espoused values with those in practice. Further research is suggested both theoretically and practically.

The rapid spread of Industry 4.0 technology creates new possibilities for manufacturers to improve the efficiency and quality of their production. However, these technologies will only generate benefits if successfully adapted to local systems. There are thousands of different digital tools in areas such as production planning and control, which can potentially produce smarter production management processes. While this diversity of tools is one of the strengths of digitalization, it adds complexity to the selection process. Just entering the possibilities provided by data-driven tools does not automatically increase added value. On the contrary, it is essential to clearly understand the production system, identify needs and compatibility with these needs, and look at the feasibility of digital tools in the context of manufacturing (Ramzi et al., 2019). Otherwise, the decisions taken in environmental uncertainty may involve substantial investments in useless tools, ultimately leading to successful digitalization. Therefore, focusing on the human element of Industry 4.0 (I4.0) is critical for successfully digitalizing companies.

7.1. Successful Digital Transformation Examples

Globalisation, the emergence of new technologies and consumers shifting from mere usage to full-on co-creation are profoundly changing business models. Many organisations are struggling to deal with these drivers of change, especially in manufacturing and industry services, sectors that have heavily relied on stable, established business models. As a result, digital transformation is gaining traction across industries. Steps on the way to a more customer-facing business model in an industrial setting have been illustrated on the example of a global supplier to the automotive industry. Initialisation is particularly critical because of the issues mentioned above, but there are also opportunities: technologies are widespread, but still not ubiquitous. courage, leadership, persistence and willingness to learn are at least as important as technology, data and methods themselves in launching a corporate-wide digital transformation (Von Leipzig et al., 2017). Developing an interplay among the psychological barriers for the adoption of industry 4.0 phenomenon (Mahmood et al., 2021). Executives and policy makers around the globe are showing interest in how to accelerate the adoption of recent emerging technologies they refer as Industry 4.0 revolution. Organizations are more interested in searching for ways to be more agile to forthcoming changing patterns concerning product life cycles, variety, and consumer expectations. However, being a novel area, this concept is still understudied. The literature is scant, zero, or fragmented resulting in a difficulty for firms to know how to design a strategy to embrace this upcoming breakthrough successfully. The most organizations do not possess an organization-wide strategy to embrace it, delta in the literature still exist. This phenomenal technological thrust calls for investigation to understand and facilitate such a transition. The revolution has various opportunities and barriers, but little literature is available concerning it. Aspects, such as poor value-chain integration, economic benefits' uncertainty, lack of infrastructure, job disruptions, resistance to change, cybersecurity issues, standardization problems, lack of skilled workforce, expiration of existing business models, and organizational resistance at both employee and middle management level may be particular issues.

7.2. Lessons Learned from Failures

As with any technological evolution, firms can learn from the failures of others, both within (i.e., in the same industry) and outside their industry. In the three case studies discussed above, all companies felt the urgency of the Industry 4.0 movement but were all confounded as to how to respond. Efforts to react were often complex, with competing departments in firms (e.g., R&D versus Production) often developing fragmented, siloed responses. Organizations must take an integrated, company-wide approach to the problem (Mahmood et al., 2021). Moreover, while government organizations have provided assistance with grants and projects to encourage firms to adopt Industry 4.0 solutions, actors within these large organizations (e.g., bureaus of economics, R&D leaders, etc.) are often overwhelmed and unsure of how to proceed. Little has been done to provide a framework or path for individual actors or companies to follow, nor have they had the foresight to link other companies together within an industry or ecosystem.

The note of caution required in addressing the potential for aspirational phrases to, in the end, impinge on the very nature of Industry 4.0 and its potential benefits is, however, a double-edged sword. Such caution is fundamental as it speaks to the very heart of the matter that is the Industry 4.0 imperative. Such lexicons run the risk of discrediting applicants for future funding streams or belittling the spin outs of funding projects if, at the end of the day, firms are unable to financially justify the outcomes of such initiatives. More fundamentally, though, the very notion of 'solution' runs the risk of being interpreted at a more philosophical level, limiting the potential visioning of where Industry 4.0 might go in terms of its role and potential impact in reshaping society and the world in which we live. Would one take the position that we have already reached it, couldn't go further or might envision working in an Industry 6.0 world and what would that be like? What might be the implications at an individual firm level, industry level or even broader societal level?

8. FUTURE DIRECTIONS IN INDUSTRY 4.0

The advance of Industry 4.0 is spurring large-scale transformation within and across manufacturing sectors, on the shop floor and beyond (Briken et al., 2023). In parallel, considerable mega forces propel strategic and operational changes associated with decarbonization and thus sustainability. Industry 5.0, or the fifth industrial revolution, refers both to the potential of future non-linear uptake trajectories of these two streams of change, as well as the application of using digitalization for sustainable practices in manufacturing supply chains (SCs). The trajectory of this holistic vision of smart and sustainable production systems is uncertain as both streams of change are new and unprecedented. However, some early explicit or implicit agendas for Industry 5.0 are emerging. These lay out a series of high-level future directions organized both thematically and temporally. These future directions related to Industry 5.0 center on developing a stronger focus on the human factor in the transformation fueled by digital technologies and on finding solutions for more sustainable human-centric manufacturing.

The advent of Industry 4.0--the industrial transformation of interconnected technology in a digital and automated manufacturing system--implies great potential to improve performance in manufacturing. The emergence of "smart" factories has been linked to the broader concept of industrial "digital transformation." The new technologies introduced by the Industry 4.0 movement are far beyond what was previously offered by traditional automation such as numerical control. It is emphasized that given the increasing reliance on flexible adaptation of the production lines to fast-changing product demands and the customization of the products themselves, a decoupling of the geometric model of the product from its manufactured counterpart, known as a digital twin, will be crucial. Digital twins enable virtual real-time monitoring of the production process and proactive reaction to changes, ultimately allowing for process adjustment via intelligent decision-making to optimize production efficiency, product quality, and energy-consumption cost, amongst others (Mahmood et al., 2021).

8.1. Emerging Technologies

Upstream to the Factory: Analyzing Emerging Technologies in the Broad Context of Industry 4.0

This section provides a brief overview of emerging technologies in the broader context of industry 5.0. The discussion of indivisible pieces of technology focuses first on the entire idea or concept in a general manner and then moves to experts' analysis regarding the elements and/or components of the technologies. Industry 4.0 in its broadest definition is defined as the global phenomenon by which several concepts and technologies in information technology were brought to the factories where physical machines are being operated ((Kapp, 2018)). For the future of the industry, ideas such as industry 5.0 are being discussed. Nonetheless, industry 4.0 itself is still considered an emerging area for which nearly all broad consensus remains to be established. Currently, the non-academic interest in industry 4.0 is extremely high, but little general understanding exists. Several important concepts remain either completely or largely open. However, such open aspects still require consideration from business and at a higher level, broader governmental policy-making perspectives.

The concept of industry 4.0 (with lower case wording 'i') refers to the 4th industrial wave; that is, it is about what industrial players aim to accomplish in the future. This sense of the term is widely used in academic literature as well. In 2006, during the Hanover Trade Fair, the term 'Industry 4.0' was introduced in the public by representatives of the German government and industry opposed by engineers' group. The idea is to bring concepts and technologies such as the internet, the web, cloud services, big data, machine learning, deep learning, robotic process automation, etc. (collectively account as 'IT') which have been presented for public use in the outside industrial environment, into the factories where physical machines are being operated. At its simplest and most conceptual, the 'vision' of this idea is to digitize the entire factory processes and the objects moving within it, from raw materials to finished products, by connecting them all via the Internet ((Briken et al., 2023)). Tracing the history of the term, a crucial element is missing. The said broad definition on traces BE/CE technological progression, the conceptualized end-of-the-line robotization is phenomenal without any crossing backward event; there is no missed transition to speak of.

8.2. Evolving Workforce Dynamics

As technology continues to reshape industry, companies in the manufacturing sector are keen to understand how employees will be affected, and how production systems contribute to this impact. Regarding the evolving nature of employee tasks, a survey highlighted the rise of more complex tasks as the main factor for change in employee roles, although this appears to be industry dependent. Instead, the importance of mutual understanding on the production floor was a factor notable for its importance in these industries. With control applications becoming more complex in machine integrated systems, a significant percentage of employees expect varied control tasks to replace fixed control roles. Easy access to production systems is expected to have a large impact on the nature of employee tasks and the organization of the workforce. To meet expectations that employees will spend more time analyzing production, there is an anticipated reduction in routine analysis but with more time inspecting handling applications. Employees already possess machines and management dashboards, and methods to configure them are being implemented.

Recent discussions about employee-centric development within smart manufacturing systems have raised some questions. The aim of this study was to investigate how the manufacturing workforce will change with the evolution towards a digitized industrial environment. Qualitative interviews with employees of a European automotive production plant were conducted and complemented by group discussions. The analysis focused on employee expectations concerning the impact of evolving production systems on workforce stereotypes and dynamics. The research assesses the perception of impacts on the workforce concerning workforce stereotypes, structure and organization, work tasks, and knowledge and skills. Factors expected to impact technician, operator, and general employee stereotypes were identified.

8.3. Sustainability Considerations

The framework raised in Section 7 aims at addressing operative competitiveness and social sustainability considerations. The "human-computer integration" driver includes functionalities that increase the inter-operability and cognitive capabilities of workers in digitalized settings. Recommended functionalities are intuitive and perceptive Human-Machine Interfaces (HMIs) in the physical domain, such as augmented and virtual reality, and in the cognitive domain, such as Digital Twin assistants, generative design tools, and social bots. The "socio-organisational adjustment" driver encompasses functionalities that configure work practices, teams, and relations among actors in the CPS to enhance workforce performance and well-being. Functionalities are related to work system design, individual and team processes, sociability, and appreciation of the human factor in planning and operation. The "knowledge management" driver includes functionalities that allow managing knowledge accumulation and reuse in human-robot teams. It is recommended that companies adopt real-time knowledge capture and curation functionalities, such as Digital Twin-based knowledge graph, and interfaces for knowledge retrieval and transfer among human-robot collaborations, such as explanatory reasoning.

Dealing with anthropocentric sustainability considerations is one of the main challenges that manufacturing industries are facing in the coming years. The key lever to be addressed is that Human-Centric Cyber-Physical Production Systems (HC-CPPS) must re-think the technologies - and their adoption strategies - needed to ensure the sustainability of manufacturing processes. The direct result of such an introduction will be assisted and cooperative shop-floor activities supported by Artificial Intelligence (AI)/Machine Learning, Mixed Reality environments, and Human-Cyber-Physical production systems. While existing literature focused on the proposed technologies, however, manufacturers' processes, adoption conditions, and use must be analysed as well from the human point of view, thus representing knowledge gaps in the scientific literature, industry framework standards, and manufacturers' needs (Baptista Torres da Rocha et al., 2022). The proposed research follows the framework present in the ISO 26800: 2020 in terms of human-centric design principles to define how anthropocentric sustainability might be addressed (Pinzone et al., 2020).

9. CONCLUSION

The emergence of industry 4.0 is revolutionizing almost every aspect of life, particularly in industrial settings, with equivalent social, cultural, and economic implications. Industry 4.0 leverages digital technologies to capitalize on the IoT, the emerging ecosystem of interconnected things. Technological advances are reminiscent of the 3rd industrial revolution. Still, the key difference is that in the 4th revolution, there are radically new technologies such as the IOT, which enable the collection of data from connected devices in real time and the connectivity and interoperability of devices and systems (Mahmood et al., 2021).

Insights obtained through the application of these technologies across the entire system over time yield information and knowledge, which increase the self-organizing functionality of the system concerning the optimization of internal and external processes. Investment in advanced digital technologies can enhance the operability of the whole manufacturing system as technological innovations can reduce production costs, increase product customization, and ensure quality. The amount of technologically mediated operation will increase in terms of labor intensity in many large-scale and specialized production processes, becoming autonomous and highly individualized (Briken et al., 2023).

However, if revenues continue to stagnate, investment in advanced digital technologies will remain low, which will reduce competitiveness. Conversely, investment in advanced digital technologies can exacerbate the skill challenges faced by manufacturing companies. A low salaries/overwork test strategy, which increases the salary gap between the labor-intensive production process and the switching job, was sufficiently effective to successfully sustain the competitiveness of the labor-intensive production process. However, in terms of productivity enhancement, gig work-induced underemployment at first would impede the investments needed to steer the manufacturing company toward its future position.

REFERENCES

Bandura, A. (1997). *Self-Efficacy: The Exercise of Control.* W.H. Freeman and Company.

Baptista Torres da Rocha, A., Borges de Oliveira, K., Espuny, M., Salvador da Motta Reis, J., & José Oliveira, O. (2022). Business transformation through sustainability based on Industry 4.0

Briken, K., Moore, J., Scholarios, D., Rose, E., & Sherlock, A. (2023). Industry 5 and the Human in Human-Centric Manufacturing. *Sensors (Basel)*, *23*(14), 6416. DOI: 10.3390/s23146416 PMID: 37514710

Chen, S., Ren, Z., Yu, X., & Huang, A. (2022). A Dynamic Model of Evolutionary Knowledge and Capabilities Based on Human-Machine Interaction in Smart Manufactures. *Computational Intelligence and Neuroscience*, *8584888*, 1–10. Advance online publication. DOI: 10.1155/2022/8584888 PMID: 35515504

Deci, E. L., & Ryan, R. M. (2000). The "what" and "why" of goal pursuits: Human needs and the self-determination of behavior. *Psychological Inquiry*, *11*(4), 227–268. DOI: 10.1207/S15327965PLI1104_01

Eich, A., Klichowicz, A., & Bocklisch, F. (2023). How automation level influences moral decisions of humans collaborating with industrial robots in different scenarios. *Frontiers in Psychology*, *14*, 1107306. DOI: 10.3389/fpsyg.2023.1107306 PMID: 36968758

Eriksson, K. M., Carlsson, L., & Olsson, A. K. (2022). To digitalize or not? Navigating and merging human- and technology perspectives in production planning and control. *International Journal of Advanced Manufacturing Technology*, *122*(11-12), 4365–4373. DOI: 10.1007/s00170-022-09874-x PMID: 35968033

Gadekar, R., Sarkar, B., & Gadekar, A. (2022). Model development for assessing inhibitors impacting Industry 4.0 implementation in Indian manufacturing industries: An integrated ISM-Fuzzy MICMAC approach. *International Journal of System Assurance Engineering and Management*, •••, 1–26. Advance online publication. DOI: 10.1007/s13198-022-01691-5

Huber, R., Oberländer, A. M., Faisst, U., & Röglinger, M. (2022). Disentangling Capabilities for Industry 4.0 - an Information Systems Capability Perspective. *Information systems frontiers: a journal of research and innovation*, 1–29. Advance online publication. https://doi.org/DOI: 10.1007/s10796-022-10260-x

Kapp, J. (2018). The fourth industrial revolution: assessing the intelligences of engineers in the South African automotive industry. Master Thesis: Faculty of Economic and Bussiness, Nelson Mandela University

Krupitzer, C., Müller, S., Lesch, V., Züfle, M., Edinger, J., Lemken, A., Schäfer, D., Kounev, S., & Becker, C. (2020). A Survey on Human Machine Interaction in Industry 4.0. *ArXiv*. https://arxiv.org/abs/2002.01025

Lagomarsino, M., Lorenzini, M., De Momi, E., & Ajoudani, A. (2021). *An Online Framework for Cognitive Load Assessment in Assembly Tasks*. ArXiv., DOI: 10.1016/j.rcim.2022.102380

Libert, K., Mosconi, E., & Cadieux, N. (2020). Human-Machine Interaction and Human Resource Management Perspective for Collaborative Robotics Implementation and Adoption. *Hawaii International Conference on System Sciences*. Proceedings of the 53rd Hawaii International Conference on System Sciences. https://hdl.handle.net/10125/63805

Mahmood, A., Arshad Ali, A., Nazam, M., & Nazim, M. (2021). Developing an interplay among the psychological barriers for the adoption of industry 4.0 phenomenon. *PLoS One*, *16*(8), e0255115. DOI: 10.1371/journal.pone.0255115 PMID: 34339448

Meindl, B., & Mendonça, J. (2021). Mapping Industry 4.0 Technologies: From Cyber-Physical Systems to Artificial Intelligence. *ArXiv*. https://arxiv.org/abs/2111.14168

Nelson, P. J. B., Biddle, J., & Shapira, P. (2023). Applications and Societal Implications of Artificial Intelligence in Manufacturing: A Systematic Review, *ArXiv, abs/2308.02025*. https://doi.org//arXiv.2308.02025DOI: 10.48550

Norman, D. A. (2013). *The Design of Everyday Things (Revised and Expanded Edition)*. Basic Books.

Pinzone, M., Albueque, F., Orlandelli, D., Giovanna Barletta, I., Berlin, C., Johansson, B., & Taisch, M. (2020). A framework for operative and social sustainability functionalities in Human-Centric Cyber-Physical Production Systems. *Computers & Industrial Engineering*, *139*, 105132. DOI: 10.1016/j.cie.2018.03.028

Ramzi, B., Ahmad, H., & Zakaria, N. (2019). A Conceptual Model on People Approach and Smart Manufacturing. 8(4), 1102-1106. https://doi.org/DOI: 10.59160/ijscm.v8i4.3583

Rocha-Jácome, C., Carvajal, R. G., Chavero, F. M., Guevara-Cabezas, E., & Hidalgo Fort, E. (2022). Industry 4.0: A Proposal of Paradigm Organization Schemes from a Systematic Literature Review. *Sensors (Basel)*, *22*(1), 66. DOI: 10.3390/s22010066 PMID: 35009609

Ronell Leo, J. A. (2018). A model for smart factories in the automotive sector. Thesis. Faculty of Business and Economics Sciences, Nelson Mandela University.

Stadnicka, D., Sęp, J., Amadio, R., Mazzei, D., Tyrovolas, M., Stylios, C., Carreras-Coch, A., Merino, J. A., Żabiński, T., & Navarro, J. (2022). Industrial Needs in the Fields of Artificial Intelligence, Internet of Things and Edge Computing. *Sensors (Basel), 22*(12), 4501. DOI: 10.3390/s22124501 PMID: 35746287

Trstenjak, M., Opetuk, T., Cajner, H., & Hegedić, M. (2022). Industry 4.0 Readiness Calculation-Transitional Strategy Definition by Decision Support Systems. *Sensors (Basel), 22*(3), 1185. DOI: 10.3390/s22031185 PMID: 35161929

Von Leipzig, T., Gamp, M., Manz, D., Schottle, K., Ohlhausen, P., Oosthuizen, G., Palm, D., & Von Leipzig, K. (2017). Initialising customer-orientated digital transformation in enterprises. *Procedia Manufacturing, 8*, 517–524. DOI: 10.1016/j.promfg.2017.02.066

Whysall, Z., Owtram, M., & Brittain, S. (2019). The new talent management challenges of Industry 4.0. *Journal of Management Development, 38*(2), 118–129. DOI: 10.1108/JMD-06-2018-0181

Chapter 9
The Future of Manufacturing:
Integrating Digital Innovation and Integrated Frameworks

Siddhartha Paul Tiwari
https://orcid.org/0000-0003-1596-1501
Google Asia Pacific, Singapore

Adi Fahrudin
https://orcid.org/0000-0001-7140-6006
Universitas Bhayangkara Jakarta Raya, Indonesia

Eliseo A. Aurellado
Southeast Asia Interdisciplinary Development Institute, Philippines

Fahmi Ilman Fahrudin
https://orcid.org/0000-0003-3028-6399
Universitas Muhammadiyah Bandung, Indonesia

ABSTRACT

This book chapter examines the revolutionary impact of digital transformation on the manufacturing industry, as well as future applications. The book chapter provides corporates, academicians, consultants, and researchers with an overview of integrated frameworks and cutting-edge digital innovations that helps in the transformation of manufacturing. This book chapter explains how technological advances are reshaping traditional manufacturing methods. The book chapter contains real-world case studies, in-depth analysis, and expert insights that are intended to help readers in navigating and leveraging the shift to a more efficient, sustainable, and streamlined manufacturing environment. Throughout the manufacturing value chain,

DOI: 10.4018/979-8-3373-1082-4.ch009

automation is at the forefront of digital technological trends and innovation, which is why a growing number of companies are striving to enhance their operational and functional processes across their entire value chain in order to increase their operational and functional efficiency.

INTRODUCTION

Manufacturing companies are being forced to adapt to the new scenarios and processes for using digital technologies; this has caused a digital transformation as a result of the new processes for using digital technologies. Digital transformation, even though it is a highly relevant and recent theme in manufacturing, is being implemented in a design and study manner by manufacturing companies (Okano, M. T., Antunes, S. N., & Fernandes, M. E. 2021). There has been a great deal of digital transformation in the manufacturing industry, driving significant improvements in terms of quality control, efficiency, and product development, while at the same time reducing costs as well as improving environmental control. There is a global manufacturing revolution going on right now. Using the skills and strategies outlined in the Manufuture vision and the strategic agenda is a vital component of digital manufacturing (Westkämper, E. 2007). A significant part of this transformation is the integration of advanced technologies into manufacturing processes, such as the Internet of Things (IoT), artificial intelligence (AI), machine learning, and big data analytics based on big data.

Digital transformation has several significant benefits, and enhancing quality control is one of the most notable of these benefits. There are many manufacturing companies that have taken advantage of digital technologies to make rapid advances in their processes (Savastano, M., Amendola, C., & D'Ascenzo, F. 2018). Manufacturing companies can detect defects and inconsistencies early on in the production process by using real-time data and predictive analytics, which ensures that only the most high-quality products are shipped to the market at the end of the production process. Manufacturing companies that have made significant advances in their business by utilizing digital technologies to their advantage (Holmström, J., & Partanen, J. 2014). With this proactive approach to waste minimization, we are able to minimize costly rework and minimize waste.

Gaining efficiency is another important advantage to consider. Streamlining operations and optimizing resource utilization is possible with automated technologies and advanced analytics. Using the internet of things (IoT) devices in smart factories enables monitoring the performance of equipment in real time, predicting the need for maintenance, and preventing unplanned downtime. There are many benefits to this approach, including more efficient production cycles and maximum output.

There is also a positive effect of digital transformation on the creation of better products. By utilizing advanced modeling and simulation tools, manufacturers can ensure optimal performance and reduce the time-to-market of their products before they are actually manufactured by designing and testing them virtually before physical production takes place. Furthermore, as manufacturers are becoming more able to adapt to the changing demands of customers, customization becomes more feasible as well.

Among the many benefits of digital transformation, one of the most significant is cost reduction. The digitization of manufacturing can reduce the intensity of carbon emissions by improving the efficiency of the manufacturing process (Fang, H., Jiang, C., Hussain, T., Zhang, X., & Huo, Q. 2022). Manufacturers can reduce operational costs by improving the efficiency of their processes and reducing waste. Although the manufacturing industry showed a positive trend in terms of low-carbon development level and the use of digitalization, there was still a significant gap between regions and industrial sectors (Lyu, Y., Zhang, L., & Wang, D. 2023). Moreover, the industry's carbon footprint is reduced by using energy-efficient technologies and sustainable practices, further contributing to the industry's superior environmental controls. Across the manufacturing landscape, digital transformation has been reshaping the landscape, creating new opportunities for companies in a world that is changing rapidly.

Historically and globally, manufacturing has been undergoing rapid changes

Manufacturing has experienced numerous disruptive changes over the years as an industry with a rich and storied history. Among these disruptions is the COVID-19 pandemic, which has both short and long-term effects on manufacturing. Similar to its responses to past adversity, technology continues to help the industry in responding, recovering, and reimagining its future.

Manufacturing has undergone lasting changes as a result of global challenges throughout history. As the industry evolved, every significant event left an indelible mark.

During World War II, there was a significant period of transformation. A significant shift in the workforce landscape took place during this period, which revived manufacturing ingenuity. In the period 1940 to 1945, the number of women employed in the U.S. increased by 50 percent, reaching a peak of 20 million. The primary concern of the majority of these women was securing better wages and working conditions (Chafe, W. H. 1992).

Innovations in production techniques as well as a shift towards increased participation of women in the workforce resulted from the World War II effort, which required rapid innovation and adaptation. A milestone in gender diversity in manufacturing roles was set during this period that not only highlighted the industry's ability to adapt rapidly, but also illustrated the industry's capacity for rapid transformation.

An entirely different type of disruption was brought about by the 9/11 attacks in 2001. The 9/11 environment had many factors contributing to the need for increased security, and many of these factors persisted even after 9/11 (Makinen, G. 2002). The attacks elevated the level of security for strategic national assets and demonstrated the need for improved physical and cyber security in the industrial manufacturing sector. Security measures, both physical and digital, have been increasingly invested in as a result of increased awareness of vulnerabilities. With an increasingly interconnected world, it has become more important to protect critical infrastructure as well as maintain the integrity of manufacturing processes.

Figure 1. Manufacturing industry historic context vis-a-vis optimization and way forward with digital technologies

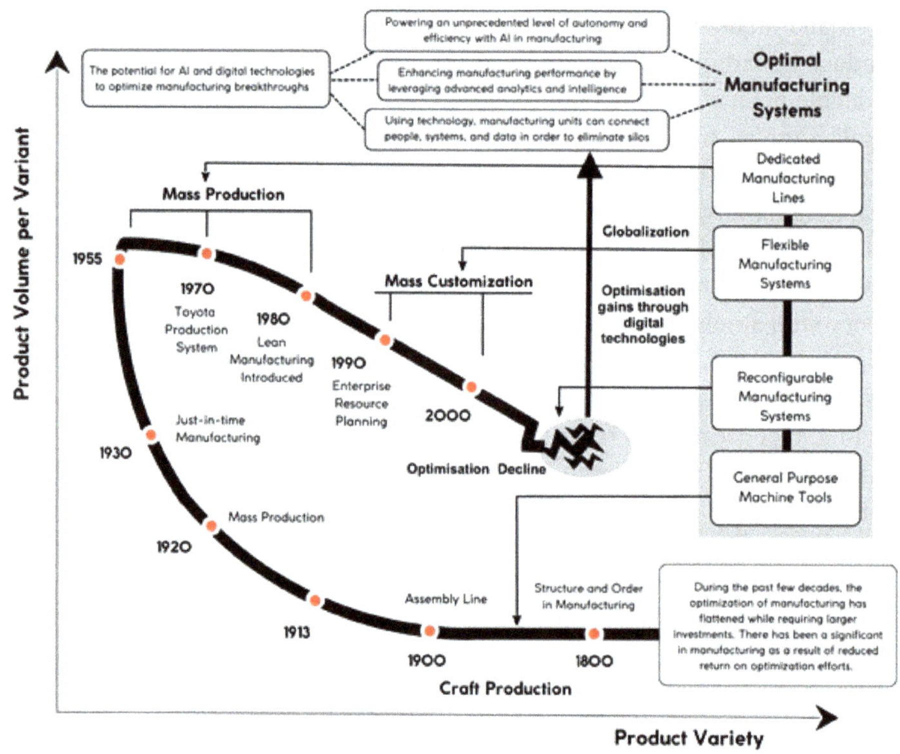

(Authors, 2024)

Manufacturing was once again challenged by the 2008 financial crisis. During the global financial crisis of 2008, a number of temporal and geographical trends were observed with regard to backshoring (Fratocchi, Luciano, Guido Nassimbeni, Marco Sartor, Alessandro Ancarani, Carmela Di Mauro, Andrea Zanoni, Paolo Barbieri, and Matteo Vignoli 2013).

Manufacturing companies were forced to reevaluate their strategies and operations as a result of the economic downturn. Companies have increased their reliance on global supply chains as a means of preserving profits and have engaged in production in low-cost countries in order to protect their profits. A difficult economic environment resulted in a greater utilization of outsourcing and offshoring during this time period. Flexibility and cost management were underscored by the crisis, which led to a shift towards a more globalized production network.

A significant disruption to the manufacturing industry has occurred as a result of the COVID-19 pandemic, which has forced manufacturers to rethink their speed and flexibility. The scope of disruptions caused by COVID-19 in manufacturing has been generally twofold: endogenous disruptions of manufacturing processes and systems, as well as exogenous disruptions in supply chains (Okorie, O., Subramoniam, R., Charnley, F., Patsavellas, J., Widdifield, D., & Salonitis, K. 2020). During the past few years, the pandemic has exposed vulnerabilities in global supply chains and highlighted the need for manufacturers to be more agile in their operations. Manufacturers have responded by establishing new lines, redirecting resources, and reconfiguring their global supply chains within a matter of months or even days. Automation, digitalization, and data analytics have all contributed to this rapid adaptation. Technological innovation has been crucial to overcoming disruptions as these tools have enabled manufacturers to quickly pivot and meet changing demands.

Though the pandemic presents immediate challenges, it is imperative to consider the long-term effects of the current disruption. The COVID-19 disease pandemic has disrupted manufacturing significantly, and has severe implications for production networks, as well as demand and supply chains that are impacted by this pandemic (Kapoor, K., Bigdeli, A. Z., Dwivedi, Y. K., & Raman, R. 2024). Supply chain resilience is one potential lasting change. There has been a great deal of emphasis placed on the risks associated with over-reliance on single sources of supply and the importance of diversifying supply chains in light of the pandemic. The manufacturing industry may continue to prioritize resilience by adopting strategies such as nearshoring, reshoring, and multisourcing in order to mitigate future risks.

Increasing adoption of digital technologies could also result in lasting changes. Digitalization has proved valuable in enabling remote work, enhancing communication, and optimizing operations during the pandemic. As manufacturers strive to improve efficiency and innovate, they may continue to invest in technologies such as the Internet of Things (IoT), artificial intelligence (AI), and cloud computing. Technology such as these enhances visibility, predictability, and real-time decision-making, enabling the industry to be more responsive and agile in the face of future disruptions.

This pandemic has also brought to light the importance of workers being able to adapt to changing conditions. There seems to be a resurgence in the focus on workforce development and training, just as there was during World War II with the influx of women into the manufacturing workforce. Depending on the industry, manufacturers may choose to invest in upskilling and reskilling programs to ensure their workforce is well equipped to cope with new technologies and job roles that are constantly evolving. There has been a considerable increase in stress and fear among employees of various manufacturing sectors as a result of the Coronavirus (COVID-19) spreading around the globe (Abdallah, A. 2020). By emphasizing the importance of human capital, the industry can be made more resilient and able to innovate more effectively. Manufacturing has a history of overcoming adversity and emerging stronger from disruptions than it was before the disruptions took place. There have been a number of challenges over the years that have shaped the industry's trajectory, including the COVID-19 pandemic. Technology, enhancing supply chain resilience, and investing in workforce development are all ways by which manufacturers can navigate the current disruption and position themselves to be successful in the future. The history of the manufacturing sector shows that every challenge brings with it the opportunity for growth and transformation, resulting in a more vibrant and innovative industry that is more resilient.

Using Digital Technoligies in the Manufacturing Industry

As technologies are becoming more and more sophisticated in manufacturing, they are dramatically altering how companies operate and how they give value to their customers. As a result of digital transformation, manufacturing companies are employing technology in an effort to digitize and automate operations in order to enable data-driven decisions to be made during production. Digital technology in manufacturing aims to increase efficiency, productivity, and profitability at the end of the project.

Digital transformation plays an important role in helping manufacturers stay competitive in the increasingly digital economy and fulfill changing consumer expectations at the same time. Digital technologies such as the Internet of Things (IoT), software, and cloud computing can help manufacturers improve operations, streamline the supply chain, decrease costs, boost customer satisfaction, and improve customer experiences by optimizing operations, the supply chain, and decreasing costs.

Digital transformation can be a powerful tool for enhancing operational efficiency and facilitating innovation as well. By implementing real-time data insights and analytics, manufacturers are able to adjust quickly to changes in the process conditions.

With the trend toward modernizing operations and improving competitiveness as a consequence of integrating digital technologies into the manufacturing process, digital technologies have become a critical component of the manufacturing process. It is important to emphasize that the implementation of new tools and systems is not the only aspect of this endeavor; rather, transform the entire organizational landscape in a holistic manner, not just make use of them and adopt new tools and systems in this instance. As the technology and organisational variables that affect the transformation are interconnected and multifaceted, a number of factors are playing a role in this transformation.

In their work, Frosch, R. A., & Gallopoulos, N. E. 1989) conclude with the proposal to make manufacturing strategies more sustainable and adaptable in order to make manufacturers more competitive and effective in order to make them more efficient and profitable. The authors argue for a move away from linear production models and towards more circular, sustainable processes, which will not only enhance efficiency, but also have a minimal impact on the environment as well. Several factors have been identified as having the potential of playing an important role in the future of manufacturing strategies, such as recycling, remanufacturing, and waste reduction, which Frosch and Gallopoulos highlighted in their 1989 study (Frosch & Gallopoulos, 1989). A holistic approach to manufacturing efficiency is proposed in this approach, which would include the consideration of environmental sustainability as an integral part of the equation when evaluating the efficiency of manufacturing processes when evaluating the efficiency of manufacturing processes.

It has been shown (Heshmati, A. 2003) that there is a direct relationship between productivity growth, efficiency, and outsourcing from the point of view of productivity growth in both manufacturing and service industries. Based on Heshmati's analysis of the outsourcing industry, it reveals that, when properly managed, outsourcing can result in significant improvements in the efficiency as well as productivity of businesses. Based on the findings of Heshmati (2003), global supply chains and specialization can enhance operational efficiency by leveraging global supply chains and specialization to optimize operations. A nuanced understanding of globalization and the strategic use of outsourcing could be added to our understanding of how globalization can be exploited to make manufacturing more efficient through the use of globalization. This perspective can be argued to add nuance to the understanding of globalization.

Manufacturing efficiency can be attributed to a number of factors, including technological innovation, strategic management, and global competitive pressures. There is a wide variety of factors which contribute to the pursuit of efficiency in manufacturing. It is clear from the insights provided by (Baily et al. 1995), (Beeson and Husted 1989), (Frosch and Gallopoulos 1989), and (Heshmati 2003) that comprehensive strategies are important. These strategies should not just be focused on operational efficiency, but should also take into account environmental sustainability, skill development, as well as the strategic use of global supply chains, in addition to operational efficiency. Despite the fact that the manufacturing sector continues to evolve, there is no doubt that these insights will continue to play an important role in guiding firms towards higher levels of efficiency and sustaining their competitiveness in a global marketplace as the manufacturing sector continues to grow.

Figure 2. Using digital technologies in the manufacturing industry conceptual framework

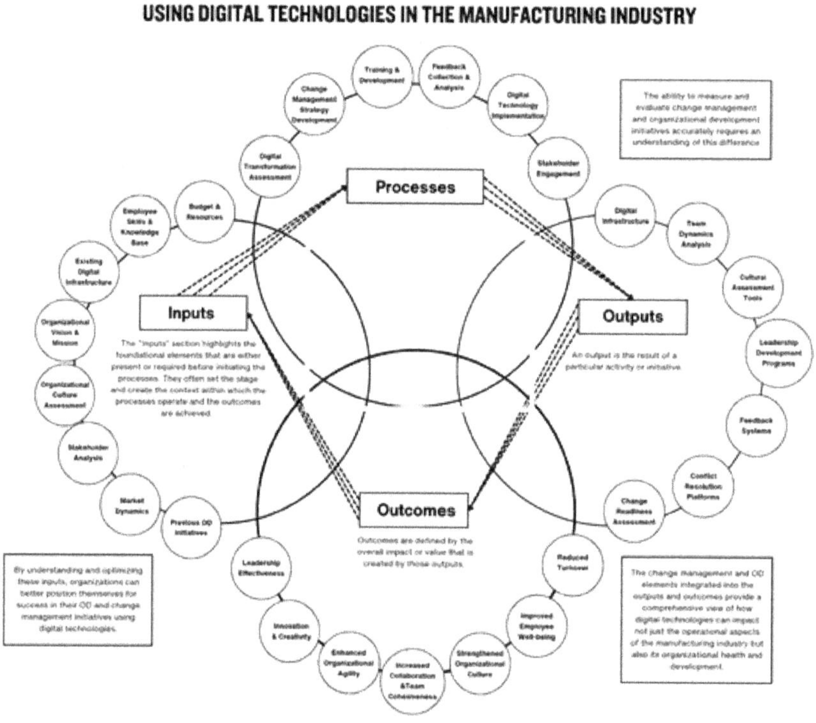

(Authors, 2025)

Technology at the core of this digital transformation is the integration of cutting-edge technology such as the Internet of Things (IoT), Artificial Intelligence (AI), automation, and data analytics that are all at the forefront of this digital transformation, which goes along with the Internet of Things (IoT). These new technologies have resulted in a fundamental reshaping of industrial processes in a way that cannot be ignored as a result of their impact on manufacturing processes. The use of IoT devices in manufacturing allows you to monitor and control the equipment in real time, AI algorithms can optimize production processes and help predict maintenance needs, automation improves production efficiency, decreases human error, and provides insights that can be used for making informed decisions, all of which can be accomplished through the use of data analytics. With the adoption of these technologies, manufacturers will be able to increase efficiency, improve quality, and increase flexibility in their production processes as a result of implementing these technologies into their processes.

Here is a selection of some of the more important points that were gathered by the authors on the topic of introducing digital technologies in manufacturing:

Manufacturing and production processes that rely primarily on manual processes, as opposed to automated processes or those that are supported by technological aids, are susceptible to a number of flaws, especially in terms of the consistency and quality of the output that is being produced. There are quite a few variations in human performance when it comes to manual processes, which are inherently reliant on human labor, and are therefore prone to variation in performance. It is important to note that there are a number of factors that can affect the production line, such as fatigue, distractions, as well as varying skill levels among the workers, which, in turn, can cause errors and inconsistencies in the process, which, in turn, may have a significant impact on the final product's quality. Aside from the lack of a robust quality control mechanism, a real-time monitoring system is one of the things that exacerbates this problem in manual processes. When a production environment is automated or technology-driven, a sophisticated system is often installed so that the production process can be continuously monitored so that deviations or defects can be identified in real time in order to take immediate corrective action if they occur. In automated or technology-driven production environments, a sophisticated system is frequently installed. This level of quality control ensures that the final product meets stringent quality standards on each and every occasion with the result that quality standards are met consistently.

It is critical for manufacturers to keep up with the latest technological developments in order to stay competitive in the marketplace. Technology advances such as advanced robotics, artificial intelligence, and additive manufacturing, in addition to streamlining production processes, present a world of opportunities for innovation as a result of the introduction of new technologies. Among the many benefits that manufacturers are able to take advantage of as a result of these technologies is their ability to explore and implement innovative materials, cutting-edge manufacturing processes, and innovative product designs. This will enable them to introduce new and improved products into the marketplace and increase profits by increasing their ability to introduce new and improved products. In environments where there are low adoption rates for technology, manufacturers are often at a significant disadvantage when it comes to adopting the technology. It is undeniable that the lack of technological resources within these organizations can seriously limit their ability to experiment with and apply new materials that have the potential of offering superior value or performance to the material they currently use. Depending on the type of composite material, innovative methods of handling and processing may, for example, be required which may not be feasible in the traditional manufacturing methods when working with advanced composite materials. They might not be used or realised as a result of this, which in turn could hinder their use.

A manufacturer without the aid of technology will find it difficult to adapt to changes in the market demand or customer preferences as a result of changes in the market conditions as a result of changing market conditions. As markets change and become more volatile, it becomes more challenging to achieve flexibility and a quick response to market shifts.

The introduction of digital technologies can have a significant impact on the way some manufacturing processes are conducted, as there are a number of safety concerns associated with them. Due to a lack of technological advancements in automation and safety measures, there may be an increase in the number of workplace injuries and accidents.

Without the aid of technology, manufacturers do not have access to many data insights that can help them make better decisions. The problem resides in the fact that manufacturers are not able to collect, analyze, and leverage data in order to optimize their processes, improve their decision-making, as well as determine where improvements are needed.

When it comes to competing globally, industries that do not adopt modern manufacturing technologies may find it hard to compete with their competitors who are able to take advantage of technological advances to improve efficiency, quality, and cost-effectiveness of their products.

Increasing demand for manual processes may make scaling them difficult as the number of people who need them increases. The advancement of technology has made it possible for manufacturers to meet increasing demands without having to increase labor costs proportionately, which is due to the advancement of technology that allows greater scalability.

The manual processes that are used in the production process tend to result in a prolonged production cycle due to the fact that they are manual. There is a possibility that the delivery of products to customers could be delayed as a result of this, which in turn could negatively impact both customer satisfaction and market share if this occurs.

There is a possibility that older manufacturing methods could have a negative impact on the environment since they consume a greater amount of energy and waste more resources, thus creating a negative public perception and causing regulatory challenges for the industry.

If manufacturers do not have access to technology, it may be difficult for them to connect with suppliers, distributors, and customers in an increasingly global market, and they may find it challenging to communicate with them on a regular basis.

Manufacturing workers currently face a skills gap that needs to be addressed in order for them to remain competitive in a rapidly changing technological environment. A lack of technological integration within an organization can result in a shortage of skilled workers that have the skills to operate and maintain advanced machinery as a result.

Various manufacturing technologies, such as automation and robotics, can be used to improve efficiency and innovation as a result of their potential to open up a number of opportunities. Without a proper understanding of these technologies, there may be a high risk that future breakthroughs will be missed.

Whenever skilled workers are in short supply or there is a sudden disruption to the labor force, relying solely on manual labor may be problematic for a business, especially in situations of shortages or disruptions in skilled workers.

FACTORS AND CHALLENGES IMPEDING THE ADOPTION OF DIGITAL TRANSFORMATION IN MANUFACTURING

Manufacturing companies are being forced to transform their businesses into comprehensive technological services and platforms in order to stay competitive in the marketplace as a result of the digital disruption affecting the manufacturing sector. There is no longer an option for manufacturers to simply provide connectivity; they now have to provide a range of technological services. It is the increasing demands of digital innovation, and the role that it plays in the industry, that is driving this change in the sector, which has led to a significant shift in the industry with this transition. In the past few years, there has been a strong emphasis on the importance of information technology (IT) strategy in the business sector. However, in the present day, C-suite executives are looking to make a much more substantial impact on their business with their investments in digital technology. There has been a rapid evolution of business transformation initiatives by manufacturers, with digital innovation becoming an integral part of their strategic goals, becoming the center of their business transformation initiatives over the years as they have evolved business transformation initiatives.

As much as there are many benefits that can be derived from digital transformation, there are also a number of obstacles that prevent manufacturers from fully embracing technology, adopting the "new normal," and, therefore, achieving breakthrough performance by embracing technology in the first place. The crucial factor for the success of the digital transformation agenda is to have a clear understanding of these barriers and how they can be overcome in order to drive forward the agenda. Having carefully researched and compiled a list of points that they believe to be most important, the authors have created a list of the most important ones that they found and have provided them below.

There is no doubt that manufacturing has a long, rich history, with the top 50 manufacturing companies by revenue having an average age of 85, which is among the oldest in the country. Many of these well-established businesses have deeply ingrained processes and cultures that make it difficult for them to change anything in the future. These companies find it difficult to drive breakthrough performance because they operate according to the traditional mindset and established ways of doing business. Managing change and fostering cultural transformation are tough challenges that require dedicated efforts and a visionary approach in order to overcome them.

The majority of manufacturers operate on a global scale with small and loosely integrated systems that are often supported by hand and operate on a local level. There are several languages that can be used by these systems as well as many revisions, which results in inefficiency and a lack of standardization. There is a fragmentation of systems across various regions of the world, which creates barriers to seamless data flow and integration, which are essential for the implementation of digital transformation across various regions. Creating a unified digital infrastructure and integrating these disparate systems is a daunting task that must be accomplished, but not without success.

One of the most significant barriers to efficient decision-making is the fact that data is siloed, strategies are multifaceted, and local politics are complex. As a result of inefficient and uneven decision-making, many manufacturing companies are unable to meet the needs of their customers. This leads to a lack of unified strategies and inconsistent decision-making. There are many data silos within organizations that prevent the free flow of information across the organization, which hinders the ability to conduct comprehensive analyses and make informed decisions. Taking advantage of the full potential of digital technologies requires breaking down these silos and nurturing a culture of data sharing and collaboration in order to unlock their full potential.

During the manufacturing process, manufacturers operate in highly competitive, capital-intensive environments that are fiercely competitive when it comes to both operational expenditures and capital expenditures. The decision to invest in a technology over a long-term technological innovation is often based on the priority of immediate operational capabilities, such as capacity, cycle time, and throughput. The focus on short-term gains at the expense of long-term strategic investments in technology creates a significant hindrance when it comes to digital transformation, for it is determined by short-term gains. To remain competitive in the long run, manufacturers need to maintain a balance between the requirements for operational excellence and the demands of digital innovation in order to remain viable.

The manufacturing industry is renowned for its high level of protection for its intellectual property (IP), with billions of dollars being poured into the research and development of new products every year. Manufacturing companies spend a significant amount of their R&D budgets on developing new technologies and products, creating an environment that makes them extremely risk-averse to any threats they may face to their intellectual property. It is common for the emphasis on security to trump optimization in favor of a protective stance, which often inhibits innovation. Digital transformation can only be achieved by striking a balance between protecting intellectual property while fostering a conducive environment for innovation to flourish.

The manufacturers usually grow through mergers and acquisitions (M&A), but without proper integration of the data acquired through the mergers and acquisitions (M&A), this can result in further data silos and inefficiency. During the acquisition process, best practices and innovations from acquired companies may be ignored or buried, adding another layer of complexity to decision-making and operational processes that were already complex enough to begin with. The integration of acquired companies and their systems requires an effective approach in order to streamline operations and leverage the full benefits of M&A activities to their full potential.

The manufacturers have over the years implemented an intricate patchwork of custom-built systems that were designed for particular purposes, resulting in a significant amount of technical debt. This heterogeneous environment of outdated and fragmented systems impedes the implementation of modern digital solutions in an environment that tends to be complex and heterogeneous. Modernizing the IT infrastructure and dealing with technical debt is a significant challenge that requires significant investments and strategic planning in order to be able to overcome.

Several areas of the manufacturing value chain have reached the point of "peak optimization," where further improvement yields decreasing returns with greater efforts. As an unfortunate consequence of the existing data silos and fragmented systems, optimizing one part of the value chain often leads to sub-optimizing another part. This can result in optimization initiative fatigue in which companies are reluctant to invest in further improvements that do not show any sign of delivering a significant return on investment. An end-to-end view of the value chain must be taken in order to overcome this problem, which means focusing on end-to-end improvements rather than isolated program improvements.

EMBRACING AI AND DIGITAL TECHNOLOGIES IN MANUFACTURING

A culture of continuous improvement is both essential and essential for organizations in order to fully take advantage of the advantages of digital integration, in order to do so, they must commit to a culture of continuous improvement. The best way to accomplish this is to regularly evaluate the effectiveness of the digital tools and processes that are in place, to keep abreast of the latest technological advancements, and to be willing to make modifications based on real-world results and feedback from customers. Such an approach entails one of the advantages of providing real-time information, products, and services that enable the business organization, a workforce, and a broader market environment to respond effectively to the changing needs of the organization, workplace, and market environment. The commitment to continuous improvement ensures that we will optimize operational efficiencies and the quality of our products, and that our customers will be more satisfied with our products, for years to come. A company's digital strategy can be used to better position itself to be in a better position to respond to any competitive pressure, market demands, or the rapid pace of technological innovation by proactively adapting their digital strategy to the changing circumstances. Digital integration can be viewed as an ongoing process if viewed from the perspective of perpetual improvement, as it is an ongoing process that, when approached with a mindset geared towards continuous improvement, can lead to sustained improvements in the factory performance and can lay the foundation for a sustained business success for many years to come.

A multifaceted approach is required to overcome the challenges and maximize the benefits of the use of artificial intelligence and digital technologies within the manufacturing industry. There is a need to implement a phased modernization strategy, because technical debt and legacy systems are both serious problems for manufacturers. There are several ways to achieve this, including incrementally upgrading existing infrastructure over time and integrating new technologies in a way that ensures compatibility and minimizes disruption as much as possible. Providing the necessary scalability and flexibility to support these efforts can be achieved through cloud-based solutions.

Data integration and management plays an essential role in breaking down silos and using AI and advanced analytics more effectively by enabling effective data integration and management. Investing in data integration platforms should be considered by manufacturers in the future to consolidate information from a variety of sources into a unified system, thereby improving data quality and accessibility. To gain comprehensive insights and make informed decisions about the future, you need to have a unified approach to managing your data.

There is a need for robust development strategies if we are to close the skills gap in the workforce. The manufacturers should invest in training programs that would assist workers in expanding their knowledge of artificial intelligence, data analytics, and other relevant technologies. Through collaboration with educational institutions, internships and apprenticeships can be provided to the workforce to ensure that they will be able to utilize new technologies in a timely manner and can help the company build a pipeline of skilled workers.

The importance of cyber security in the era of digital transformation cannot be overstated. An integrated approach to security is recommended for manufacturers, which includes network security as well as endpoint protection as well as continuous monitoring. The best practices to ensure cyber threats are controlled and sensitive data is protected are regular security audits, employee training, and incident response plans.

Figure 3. Overcoming challenges and making use of digital technolgies in the manufacturing industry

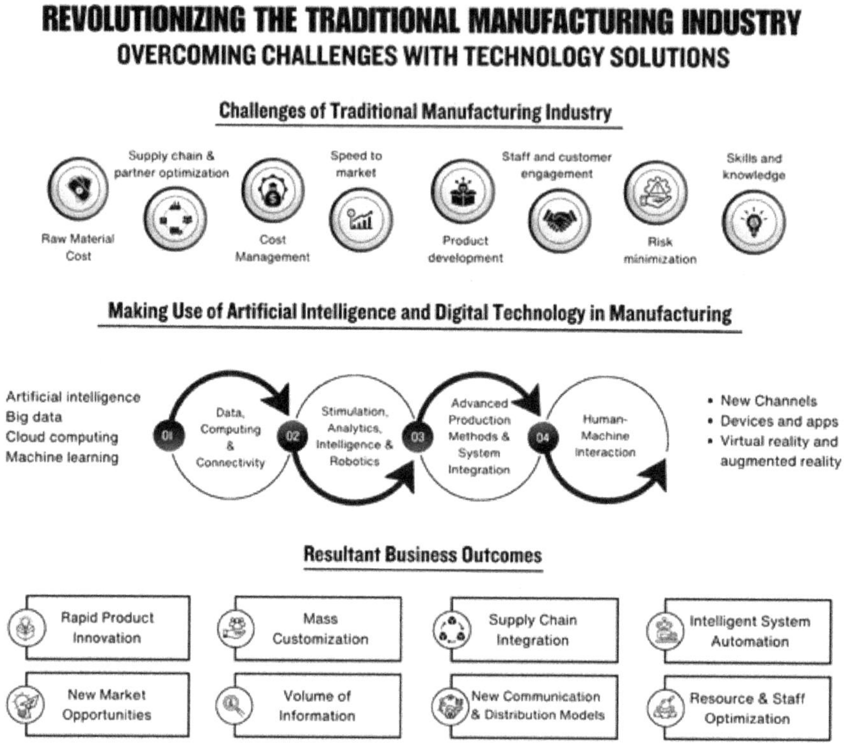

(Authors, 2025)

For securing the necessary funding, it is crucial to document a convincing business case for the use of AI and digital technologies. The manufacturers should conduct a thorough analysis of their return on investment (ROI) to figure out where they can save money, increase productivity, or gain a competitive advantage. Providing tangible proof of the benefits of your investment can be as simple as showcasing successful case studies and pilot projects, making it easier for you to justify yourself.

By analyzing data from sensors and machines to predict equipment failures before they happen, manufacturers can maximize the benefits of artificial intelligence and digital technologies when it comes to predictive maintenance through the use of data analytics and predictive maintenance tools. The proactive approach reduces downtime, extends the life of the equipment, and reduces the maintenance costs by reducing downtime. The optimization of production processes through real-time analysis of data and the use of advanced algorithms can enhance efficiency, improve product quality, and minimise waste in the manufacturing process.

Digital technologies also have an important role to play in supply chain optimization as well. A number of reasons why artificial intelligence can be used to make your supply chain more responsive and resilient, including predicting demand, optimizing inventory levels, and improving logistics. Using artificial intelligence-driven analytics in product development, it is possible to gather insights on customer preferences and market trends that guide the development of products that meet the needs of customers and reduce the time it takes to develop them.

Using artificial intelligence (AI) and digital technologies to optimize energy use and reduce waste can greatly contribute to sustainability efforts. By monitoring energy consumption with smart sensors and analytics, we can identify and identify areas for improvement, significantly reducing operating costs and assisting in the achievement of environmental goals. Manufacturers can effectively overcome challenges and harness the full potential of artificial intelligence as well as digital technologies by adopting these strategies, ensuring that they gain a competitive edge in the evolving market by using AI and digital technologies effectively.

The understanding of the benefits of digital integration in a manufacturing setting provides a more nuanced view of its advantages, emphasizing that digital technologies can have a positive impact on overall operations when they are properly integrated.

TOWARD AN INTEGRATED DIGITAL TECHNOLOGY FRAMEWORK FOR REVOLUTIONIZING MANUFACTURING

Developing an integrated framework for integrating digital technologies into manufacturing is essential for achieving a balance between cost-efficiency and flexibility in terms of manufacturing. Using this framework, the manufacturing operation can be implemented using a structured approach to integrating digital tools and processes with the goal of optimizing all aspects of the manufacturing process. An integrated system for IoT devices, artificial intelligence algorithms, and automated systems is included in the document, which facilitates seamless communication and coordination across the entire production line. It is through the implementation of this framework that manufacturers can systematically address inefficiencies, enhance productivity, and maintain their ability to adapt to changes in the market. The framework is also designed to support continuous improvement initiatives, thereby encouraging manufacturers to continuously evaluate and refine their processes on a regular basis. This integrated approach leads both to the utilization of digital technologies to their full potential, ensuring that manufacturing is able to take advantage of both efficiency as well as flexibility from digital technologies.

Streamlining processes and minimizing waste are some of the ways that digital technologies are transforming manufacturing. Using advanced sensors and automation technologies, manufacturers can keep a close eye on their production lines to identify inefficiencies and reduce material wastage by incorporating advanced sensors and automation techniques. The integration of automation systems into a company's production process allows precision and consistency in the operation. As a result, each product is guaranteed to meet high quality standards while minimizing defects and rework. The result of this is that production costs will be lower and profit margins will be higher. The use of digital technologies also facilitates real-time data collection and analysis that can be adapted immediately to optimize resource utilization and reduce operational costs, since they allow for real-time data collection and analysis. Digital technologies have taken the manufacturing industry to a new level of cost-competitiveness with the capability of responding quickly to the market expectations due to the integration of digital technologies.

Figure 4. Integrated framework encompassing digital technologies and manufacturing industry

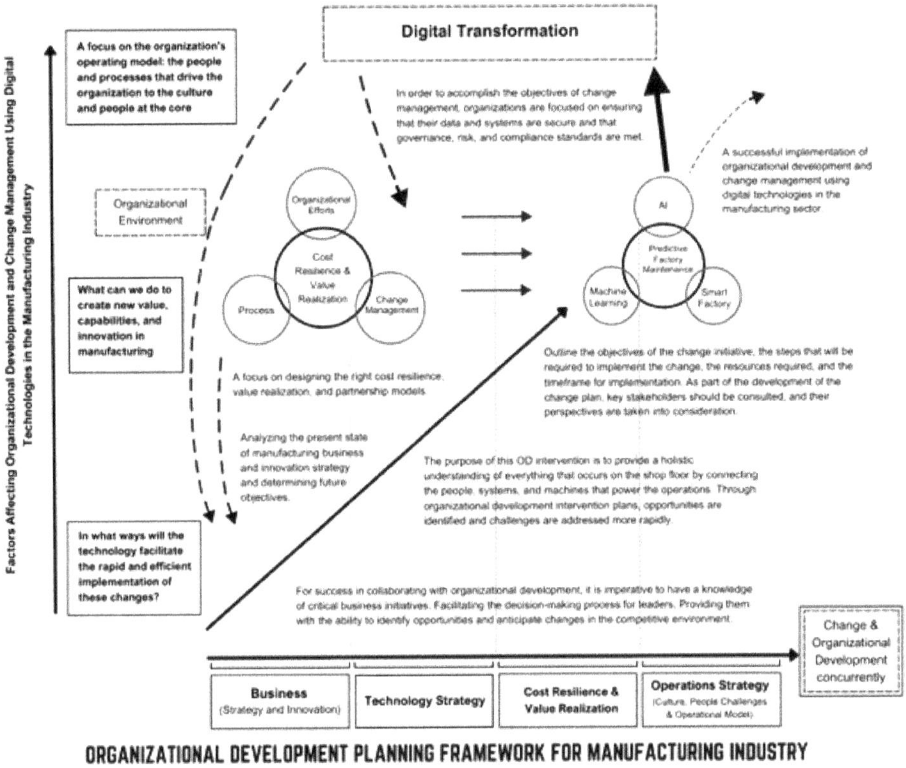

(Authors, 2024)

Below, the authors have listed the salient features of shifting towards an integrated framework and harnessing the benefits of digital transformation which will revolutionize manufacturing.

There is a growing trend in manufacturing today for new digital technologies to bridge the traditional gap between cost-efficiency and flexibility. Robotics and advanced automation are reducing labor costs and increasing production speed by automating repetitive tasks and automating repetitive tasks. The use of technologies such as 3D printing and flexible manufacturing systems allows for quick changes to be made in production lines so that different products can be accommodated or customized on the fly. A manufacturer can maintain high efficiency while at the same time being able to adapt to specific customer requirements or changes in the market as a result of this dual capability. It is also important to note that the integration of techniques and algorithms used in machine learning algorithms will allow for the optimization of production processes, thus ensuring the efficient use of resources without compromising the flexibility of the process. This combination of technologies results in a highly flexible and cost-effective manufacturing process that can be tailored to suit the needs of every customer.

Digital technologies have played a significant role in connecting and optimizing manufacturing systems over the last few years. A manufacturer is able to achieve a seamless flow of information between different departments and processes by integrating a variety of digital tools and platforms. With this type of connectivity, it is possible to increase coordination and efficiency, thereby reducing bottlenecks and improving overall productivity. The review also found that digital technologies contribute to the flexibility of manufacturers by allowing them to quickly reconfigure their production lines and make adjustments to them as needed. With the dual benefits of cost-efficiency and flexibility, manufacturers have the ability to respond to market demands while keeping operational costs low as well as responding to market demands quickly. Digital technologies are capable of transformative effects in manufacturing environments that are more integrated and responsive because of their transformative potential.

During agile manufacturing processes, digital technologies are used to foster close collaboration between human operators and machines, thereby improving the overall performance of the system. The use of advanced interfaces and collaborative robots (cobots) has allowed humans to interact seamlessly with automated systems, resulting in production processes that are more responsive and adaptable. As part of agile manufacturing, collaboration is crucial to achieve key attributes such as interoperability, configurability, programmability, and protectability, which are fundamental to the success of the process. Machines allow operators to react quickly to changing product designs or changes in production requirements, by reprogramming them within a short period of time, and each time the adjustments are fine-tuned by the operator in order to ensure the quality and efficiency of the process. It is also pertinent to mention that the protectability aspect plays an important role in ensuring that the systems remain resilient to disruptions and can continue to operate even in the face of difficulties. As a result of the increasing use of digital technologies, manufacturers have been facilitated in maintaining high levels of adaptability and agility.

Integrated with advanced sensors, IoT devices, and artificial intelligence algorithms, traditional manufacturing setups have been transformed into highly adaptive and responsive production environments as a result of the integration of these technologies. Machines are equipped with sensors that collect real-time data on a variety of parameters like temperature, pressure, and the performance of the equipment connected to them. Through IoT networks, the collected data is sent to a centralized system by AI algorithms that analyze the data to identify patterns and predict potential problems that might arise in the near future. The use of predictive maintenance powered by AI can, for example, prevent equipment failures from occurring before they occur, which allows for timely interventions and thereby reducing downtime during maintenance procedures. As a result, AI-driven analytics enable investors to gain actionable insights on their production processes, improve the quality of their products, and enhance the efficiency of their operations. Manufacturing is enabled by this interconnected ecosystem in such a way that it can react swiftly to changes and maintain optimal production conditions at all times.

Using digital technologies to facilitate real-time monitoring and analysis of manufacturing processes, manufacturing resources can be utilized to the fullest extent possible. With the help of Internet of Things sensors, businesses are monitoring the use and performance of their machines, while software solutions are monitoring the labor productivity and material consumption. By analyzing the data in real time, we can identify inefficiencies and optimize resource allocations according to the results. Using predictive analytics to determine whether a particular machine is underperforming, for instance, can suggest maintenance or adjustments to increase the efficiency of that machine. The workforce management system can also allocate labor based on skill sets and production needs. This will make sure that the right people are sent to the right places at the right time, and that the right tasks are performed. Reducing waste and lowering costs can be achieved by optimizing the use of materials. In order to maintain a high level of productivity and cost-effectiveness, it is essential for manufacturers to monitor and adjust resource utilization continuously.

With the help of artificial intelligence and machine learning, manufacturing control systems can be significantly enhanced by providing predictive analytics and real-time insights into data in real-time. During the production process, huge amounts of data are generated and these technologies can analyze it to identify trends, anomalies, and potential problems. Using predictive analytics to forecast demand fluctuations, equipment failures, and disruptions to the supply chain can enable manufacturers to take proactive measures in order to reduce those risks. With real-time data insights, production schedules, resource allocations, quality control processes, and resource allocations can be changed immediately in order to meet production goals. A dynamic adaptability of manufacturing operations ensures that they are efficient and responsive to changes as a result of this dynamic capability. A manufacturing environment that is able to predict and mitigate risks adds to overall operational stability and reduces downtime, resulting in a more resilient manufacturing environment that can deliver more consistent quality and reliability.

A significant improvement in the collection and conversion of data is achieved through digital technologies, which help streamline operations and improve decision-making. With the development of advanced data management tools, it is possible to seamlessly integrate data from diverse sources, such as sensors, ERP systems, and supply chain databases, in a seamless manner. Integrated data from this system can provide a holistic view of manufacturing operations, which can be used to better analyze and make better decisions about manufacturing operations. To optimize the production schedule and the allocation of resources, real-time data can be collected and analyzed in order to analyze production performance, inventory levels, and customer demand. With data conversion tools, we ensure that information is standardized and easily accessible across a wide variety of platforms, which means it will take less time and effort to interpret and act on the data. The collection and conversion of data can be improved to enable manufacturers to make more informed decisions, improve operating efficiency, and drive continuous improvement by improving data collection and conversion.

As a result of the use of digital technologies, human-machine collaboration has been fostered, making it easier to make changes to designs during the development process. The combination of advanced interfaces and collaborative robots (cobots) makes it easier for human operators to work closely with automated systems, making it much easier for modifications to be made to the design. A digital twin, for example, can be used by engineers to build a digital model of a new design to simulate how it will perform when it is used in a production environment, and thus improve the design. A rapid prototyping and testing process enables rapid adjustments and improvements based on real-time feedback from customers, enabling rapid prototyping and testing. It is also possible to employ AI-driven design tools to optimize the features of products and identify potential issues early on in the product development process, thus reducing costs. The seamless collaboration between humans and machines accelerates the development cycle, reduces the time it takes for products to reach the market, and ensures the quality of the finished products.

3D printing and virtual prototyping are two of the most innovative processes used to develop new products that significantly reduce the time to market. In order to improve designs, engineers can use 3D printing to create prototypes quickly, allowing them to test and refine their designs at a much faster rate. This iterative process has been proven to help identify and resolve design flaws early on in the development cycle, which minimizes the need for expensive revisions in the later stages of the development cycle. This technique refers to the creation of virtual models of products in order to simulate their performance under different conditions by using digital models. Simulating the behavior of a product and optimizing design before it is put into physical production can provide valuable insights into product behavior and help improving the product. These technologies are able to accelerate the prototyping and testing phases of the development process, which allows manufacturers to bring new products to market faster, thus giving them an edge over their competitors.

Manufacturing companies and suppliers with a well-established system and process for implementing new digital technologies are better positioned to achieve success when implementing them. These organizations were built upon the foundation of efficient practices and robust infrastructure that makes integrating new digital tools easier than in companies that do not have these approaches. The adoption of digital technologies is particularly well suited for companies that have an agile culture and are capable of continuously reconfiguring their labor and material resources on the fly. They have a high level of flexibility that allows them to quickly adapt to new systems and processes, ensuring a smooth transition to new processes and systems. To stay competitive in the long run, these companies are more likely to invest in continuous improvement initiatives, such as evaluating and refining their operations on a regular basis. Using their existing systems and agile capabilities, these manufacturers will be able to effectively adopt digital technologies and realize significant operational benefits by leveraging their existing systems and agility.

Analyzing real time data and using predictive modeling are two powerful tools that manufacturers can use to make better predictions about market trends, adjust their production schedules, and manage their inventory levels more effectively. AI algorithms are capable of predicting future demand with a high degree of accuracy after analyzing data from various sources, such as sales trends, customer behavior, and market conditions. A manufacturer can use insights such as these to align their production schedules with anticipated market needs, ensuring that they are producing the correct quantity of products at the right time as a result of these insights. As well as optimizing inventory levels, predictive modeling can also help identify patterns in stock usage and replenishment cycles to optimize inventory levels. With this proactive approach, overproductions or stockouts can be reduced, holding costs can be minimized, and overall supply chain efficiency can be improved as a result. Manufacturing operations are able to enhance their competitiveness and agility by anticipating and responding to market trends in a way that enhances their competitiveness and agility.

There are a variety of tools available to manufacturers through the use of digital technologies to optimize manufacturing processes and enhance operational efficiency. The automation of repetitive tasks reduces the cost of labor and speeds up the production process by streamlining repetitive tasks. With advanced analytics, you can gain insights into production performance, identify bottlenecks and inefficiencies, and address them in order to improve throughput by addressing them. Devices that are connected to the Internet of Things enable real-time monitoring of equipment and processes in order to maintain optimal performance by making adjustments as soon as possible. With the help of machine learning algorithms, it is possible to predict and prevent equipment failures based on historical data, reducing the amount of downtime and maintenance costs. It is in the interest of manufacturers and industry leaders to develop technologies that improve the efficiency, accuracy, and reliability of manufacturing operations, helping them to become more productive and efficient. Taking advantage of these digital tools in the manufacturing process can result in significant efficiency and competitiveness gains for manufacturers.

Among the latest technologies used in product development, such as 3D printing, are revolutionizing the way in which bespoke products are created, significantly reducing the time taken to develop them. By developing 3D printers, designers are able to produce custom parts and prototypes in a short amount of time, allowing them to test and refine their concepts quickly and efficiently. As a result, this capability is particularly valuable when it comes to designing custom-made products that are specifically tailored to the needs of each customer. The additive manufacturing process of 3D printing allows us to reduce setup times and costs by eliminating the need for traditional tools and molds. Manufacturing companies are able to produce small batches or one-of-a-kind items with this kind of flexibility. 3D printing can play a very important role in many areas of the product development process, such as accelerating the development cycle, reducing time to market, and offering customers more personalized products if it is integrated into the product development process.

With the advent of big data innovations, manufacturers are able to offer their customers more tailored products and solutions by gaining deeper insights into customer preferences and market trends through big data analytics. Manufacturers are able to identify specific customer needs by analyzing vast amounts of data from different sources and by designing products that precisely meet the needs of those customers. With the application of data-driven approaches, highly personalized products can be developed for customers, enhancing the level of customer satisfaction and loyalty. The use of big data analytics also provides manufacturers with the opportunity to optimize their supply chains and production processes, ensuring that they are able to deliver customized products as efficiently and as affordably as possible. Manufacturing companies are able to differentiate themselves from their competitors by offering tailored solutions that enable them not only to capture new market opportunities and drive growth, but also to differentiate themselves from their competitors.

A company that has developed robust operational processes is better positioned to utilize digital technologies effectively, resulting in significant improvements in the performance of the company. There is a culture of continuous improvement, strong controls over the process, and a commitment to quality that are common features of these organizations. The inclusion of such an environment is conducive to the successful adoption of digital technologies, as it provides a framework that is conducive to systematically evaluating and refining new tools and processes as they are developed. With the integration of digital technologies into these well-structured practices, these well-organized processes become more efficient, waste is reduced, and quality of the products is enhanced. These companies possess the experience and discipline that enable them to overcome implementation challenges in an efficient manner, ensuring that digital transformation initiatives deliver the value that they intended to.

The Kaizen continuous improvement methodologies as well as Just-In-Time (JIT) production techniques create an environment that is conducive to the adoption of digital technologies, as they are proven to result in a culture of efficiency that promotes the use of digital technologies in the workplace. Throughout the organization, a great deal of emphasis is placed on removing waste, optimizing processes, and seeking out new and better ways to improve performance as a result of these practices. The Kaizen culture at the organization is characterized by its focus on employee identification and implementation of improvements, thus encouraging all employees to take a proactive approach when it comes to problem-solving as a result. The JIT approach to production emphasizes the importance of maximizing inventory efficiency at a minimal cost while reducing waste and reducing inventory levels. These principles are perfectly aligned with the use of digital technology since it provides real-time data, predictive analytics, and automation tools which support continuous improvement and promote efficient production. There is no doubt that companies that embrace Kaizen and JIT are well placed to take advantage of digital technologies for the sake of further enhancing their operational efficiency.

A significant enhancement in cost-competitiveness in performance can be realized by incorporating digital technologies into existing manufacturing practices. Manufacturers can increase productivity levels as well as achieve cost savings by combining traditional efficiency methods with the use of digital tools that have advanced in recent years. The use of automation systems, for example, can allow repetitive tasks to be handled at a faster and more precise pace than by human workers, thereby reducing labor costs and enhancing output. Analyzing data can provide manufacturers with insights into their operational performance, which can be used to identify and eliminate inefficiencies as soon as possible. By anticipating equipment problems before they arise, predictive maintenance minimizes the amount of downtime and repair costs that are incurred during the operation. This combination of technologies and practices generates a synergistic effect, which in turn reduces production costs and enables businesses to become more competitive overall. The manufacturers who successfully integrate digital technologies into their operations will be better positioned to compete on the global market if they successfully integrate them into their operations.

CONCLUSION

Through digital transformation, the flexibility of the organization and the responsiveness of the organization are also significantly improved. The rigid structure of traditional manufacturing systems makes it difficult for them to adapt to the rapid change in market demand as a result of their rigid structures. Digital technologies, on the other hand, allow for a more agile approach to be taken. The use of AI-driven demand forecasting, for example, allows manufacturers to accurately forecast market trends with high levels of accuracy, allowing them to adjust production schedules and inventory levels accordingly. In this way, manufacturers are able to meet customer demands without incurring excessive costs due to this agility.

A significant transformation is taking place in the manufacturing industry as a result of the integration of digital technologies, which are on the cusp of transforming the industry. It is anticipated that this shift will have a revolutionary effect on traditional manufacturing processes by improving efficiency, flexibility, and enhancing overall competitiveness. A framework for digital transformation is essential in order to guide this evolution and ensure that manufacturers are able to leverage advanced technologies effectively in order to optimize their operations and respond dynamically to their customers' requirements. An integrated approach to digital transformation has a number of primary benefits, one of which is an improvement in operational efficiency, which is one of the main benefits. A manufacturer can monitor production lines in real time with the help of IoT sensors and advanced analytics, helping them to identify inefficiencies and bottlenecks in real time as well. The ability to see what is happening in real time allows for immediate adjustments to be made, reducing waste and increasing productivity overall. AI-powered predictive maintenance is a particularly useful tool for optimizing operations because it predicts equipment failures before they occur, thus reducing downtime and extending the lifespan of machines.

It is essential to have an integrated digital transformation framework in place in order to revolutionize the manufacturing industry. It has been proven that manufacturers can maximize operational efficiency, improve flexibility, foster innovation, optimize supply chain management, ensure quality, and support sustainability by leveraging advanced technologies. A comprehensive approach to digital transformation positions manufacturers to thrive in a market where there is an increasing level of competition and dynamic changes.

REFERENCES

Abdallah, A. (2020). The effect of corona pandemic on workers' psychology and performance in the manufacturing sector. *Systematic Reviews in Pharmacy, 11*(12), 1593–1599.

Baily, M. N., Gersbach, H., Scherer, F. M., & Lichtenberg, F. R. (1995). Efficiency in manufacturing and the need for global competition. *Brookings Papers on Economic Activity: Microeconomics, 1995*, 307–358. DOI: 10.2307/2534776

Beeson, P. E., & Husted, S. (1989). Patterns and determinants of productive efficiency in state manufacturing. *Journal of Regional Science, 29*(1), 15–28. DOI: 10.1111/j.1467-9787.1989.tb01219.x

Chafe, W. H. (1992). *The paradox of change: American women in the 20th century*. Oxford University Press. DOI: 10.1093/oso/9780195044195.001.0001

Fang, H., Jiang, C., Hussain, T., Zhang, X., & Huo, Q. (2022). Input digitization of the manufacturing industry and carbon emission intensity based on testing the world and developing countries. *International Journal of Environmental Research and Public Health, 19*(19), 12855. DOI: 10.3390/ijerph191912855 PMID: 36232155

Fratocchi, L., Nassimbeni, G., Sartor, M., Ancarani, A., Di Mauro, C., Zanoni, A., ... & Vignoli, M. (2013). Manufacturing Back-shoring and the Global Fragmentation of Production: What it is Changing after the Financial Crisis?. 40th Acdemy of International Business UK & Ireland.

Frosch, R. A., & Gallopoulos, N. E. (1989). Strategies for manufacturing. *Scientific American, 261*(3), 144–153. DOI: 10.1038/scientificamerican0989-144

Heshmati, A. (2003). Productivity growth, efficiency and outsourcing in manufacturing and service industries. *Journal of Economic Surveys, 17*(1), 79–112. DOI: 10.1111/1467-6419.00189

Holmström, J., & Partanen, J. (2014). Digital manufacturing-driven transformations of service supply chains for complex products. *Supply Chain Management, 19*(4), 421–430. DOI: 10.1108/SCM-10-2013-0387

Kapoor, K., Bigdeli, A. Z., Dwivedi, Y. K., & Raman, R. (2024). How is COVID-19 altering the manufacturing landscape? A literature review of imminent challenges and management interventions. *Annals of Operations Research, 335*(3), 1567–1599. DOI: 10.1007/s10479-021-04397-2 PMID: 34803204

Lyu, Y., Zhang, L., & Wang, D. (2023). The impact of digital transformation on low-carbon development of manufacturing. *Frontiers in Environmental Science, 11*, 1134882. DOI: 10.3389/fenvs.2023.1134882

Makinen, G. (2002). The economic effects of 9/11: A retrospective assessment. United States: Congress: Congressional Research Service (CRS).

Okano, M. T., Antunes, S. N., & Fernandes, M. E. (2021). Digital transformation in the manufacturing industry under the optics of digital platforms and ecosystems. *Independent Journal of Management & Production, 12*(4), 1139–1159. DOI: 10.14807/ijmp.v12i4.1375

Okorie, O., Subramoniam, R., Charnley, F., Patsavellas, J., Widdifield, D., & Salonitis, K. (2020). Manufacturing in the time of COVID-19: An assessment of barriers and enablers. *IEEE Engineering Management Review, 48*(3), 167–175. DOI: 10.1109/EMR.2020.3012112

Savastano, M., Amendola, C., & D'Ascenzo, F. (2018). How digital transformation is reshaping the manufacturing industry value chain: The new digital manufacturing ecosystem applied to a case study from the food industry. In *Network, Smart and Open: Three Keywords for Information Systems Innovation* (pp. 127–142). Springer International Publishing. DOI: 10.1007/978-3-319-62636-9_9

Westkämper, E. (2007). Digital Manufacturing in the global Era. In *Digital enterprise technology: Perspectives and future challenges* (pp. 3–14). Springer US. DOI: 10.1007/978-0-387-49864-5_1

Chapter 10
The Aerospace Industry:
A Country's Developmental Framework

Eliseo A. Aurellado
Independent Researcher, Philippines

ABSTRACT

Man has always been fascinated with flight. Since the Greek mythology character Icarus flew towards the sun in his crafted wings and fell to his death, early men dared to conquer the skies with often disastrous results. The dawn of aviation came when the Wright brothers flew the first heavier-than-air machine along the dunes of Kitty Hawk, North Carolina, in 1903. Soon, newly invented flying machines began to be used in sports and air shows. When the Great War broke out in 1914, machine guns were mounted on primitive aircraft, and airplanes became weapons of mass destruction. Advances in aircraft design and development led to the invention of more sophisticated and deadly military planes, which became a potent force in warfare in World War II. When the first satellite, the Soviet Sputnik, flew in 1957, the world entered the space age. A race between America and Russia began during the Cold War, culminating with Neil Armstrong landing on the moon's surface in 1969. Man has gone beyond the Earth's atmosphere and into the far regions of outer space.

INTRODUCTION

This paper provides an overview of the global aerospace industry, with a particular focus on its implications for the manufacturing sector. It also presents an integrative framework that the Philippines follows to develop its aerospace capability, while striving to keep pace with its ASEAN neighbors. In its simplest form, the story of aerospace is a story about humanity's desire to achieve possibility at the core of

DOI: 10.4018/979-8-3373-1082-4.ch010

the human spirit. The aerospace industry today stands at a critical juncture. We are witnessing a recovery in passenger traffic around the world; private launch providers are facilitating access to space; and governments are redefining security strategies in a world of shifting alliances. Despite the noise of turbines and the plumes of rockets, there is a quieter engine of change that drives the economy: manufacturing. Aircraft manufacturers now have to develop lighter airframes, fault-tolerant avionics, and propulsion systems that sip rather than guzzle fuel, and these factors determine which countries capture value and which countries merely purchase it. The manufacturing of aerospace products has never been an easy task. There are millions of individual parts used in modern aircraft that are sourced from a constellation of suppliers that span multiple time zones, tiers, and regulatory regimes that are spread across the globe.

Artificial intelligence is revolutionizing aviation through applications such as predictive maintenance and passenger experience enhancements (Addepto, n.d.). The initial aerospace manufacturing road map outlined foundational strategies for industry growth in the Philippines (Aerospace Industries Association of the Philippines, 2013). A decade later, the updated road map set new targets for export capacity and workforce development (Aerospace Industries Association of the Philippines, 2023). Aviation contributes significantly to global economic growth, supporting jobs and GDP worldwide (Air Transport Action Group, n.d.). Emerging urban air mobility solutions require integrated simulation frameworks to ensure safety and efficiency (Altair Engineering, n.d.). The Philippine aircraft maintenance market is expected to expand as post-pandemic travel demand rebounds (Balinbin, 2022). Participation in global value chains has positioned the Philippines as a competitive hub for certain aerospace components (Bamber et al., 2016). Tiered supply-chain structures help manage complex aerospace production networks (Bescast, n.d.). Government policy envisions the Philippines as an attractive location for aerospace investors (Department of Trade and Industry, n.d.).

Industrial park incentives support cluster formation for aerospace firms (First Philippine Industrial Park, 2023a). Diversification into electronics manufacturing complements aerospace capabilities (First Philippine Industrial Park, 2023b). Connected aircraft use IoT sensors to optimize flight operations and maintenance (Imaginovation, n.d.). IoT-enabled analytics can reduce turnaround times in airport operations (Joshi, n.d.). Industry leaders anticipate new growth drivers in the defense sector (KPMG International, 2016). Digital twins are transforming MRO processes into predictive, data-driven practices (Lahane, n.d.). Advanced composites and smart materials enhance aircraft performance and reduce weight (Mehra, 2023). NASA's research on electric-aircraft propulsion aims to lower emissions in future fleets (National Aeronautics and Space Administration, n.d.). The government has pledged full support for domestically produced satellites such as MULA (Presidential

Communications Office, n.d.). Special economic zones created under Republic Act No. 7916 provide fiscal incentives to aerospace investors (Republic Act No. 7916, 1995). Global aerospace manufacturing includes OEMs as well as specialized drone producers (RevFine, n.d.).

The Philippines stands at an exciting juncture, with a wealth of human capital, a geographical advantage, and an ethos of collaboration conducive to rapid learning, which makes it an ideal location for rapid development. The Philippines as a nation aims to contribute to the writing of aerospace's next chapter not just through manufacturing excellence but by integrating forward-looking governance into a deliberate, integrated framework. As the sector moves towards quieter engines, smarter factories, and cleaner skies, nations that align their industrial gears now will find themselves in a cruising altitude in the future when new technologies take flight.

A widening competitive landscape is also taking place in the industry. The global aerospace industry is both a frontier of extreme complexity and a canvas of extraordinary possibilities at the same time. There are numerous forces that are reshaping the industry. Digital transformation, geopolitical flux, environmental urgency demand nimble responses from incumbents and aspirants alike.

THE GLOBAL AEROSPACE INDUSTRY

The global aerospace industry has been extensively studied in the context of technological innovation, geopolitical influences, and supply chain disruptions (KPMG, 2016; Duke University, 2016). Scholars have emphasized the impact of digitization, AI integration, and sustainability mandates (NASA, 2022). However, much of this literature focuses on dominant markets like the U.S., Europe, or China, leaving Southeast Asian contexts, particularly the Philippines, underexplored. Most academic contributions overlook the adaptive strategies small economies use to develop aerospace capacity under financial, infrastructural, and regulatory constraints. This chapter aims to fill that lacuna by focusing on the Philippine experience.

Aerospace is defined as referring to the atmosphere and outer space. It relates to the science or technology of flight. The term aviation is often used interchangeably with aerospace; the subtle difference is that aerospace extends to outer space while aviation is confined within the Earth's atmosphere.

Aerospace activity is diverse, with commercial, industrial, and military applications. Some typical applications are in aeronautical engineering, satellite communications, guidance and tracking systems, destructive and non-destructive materials testing, propulsion and engine testing, surveillance, and electronic warfare. The aerospace sector is one of the largest and most powerful industries in the United

States, and supplies five markets: military aircraft, missiles, space, commercial airliners, and general aviation.

The global aerospace industry contributes to worldwide economic prosperity by boosting global trade and tourism. Despite a decline in global aerospace exports in 2023 in the wake of the COVID-19 pandemic, the global supply is expected to continue to increase at 2.1% annually, according to a published international report. By 2038, it is forecasted to contribute over US$6 trillion to the global economy. It is also projected to provide over 140 million jobs by the same year. The Defense segment of this industry has grown steadily in previous years due to the need to modernize aging military fleets to counter political and territorial conflicts, such as the Russia-Ukraine war in 2022 and the increasing tensions in the South China Sea. Global defense spending averages around 2.5% of a country's GDP. The commercial aircraft segment also drives the industry with growth in passenger travel demand and the need for more fuel-efficient aircraft.

The aerospace industry is highly regulated because of safety concerns. It requires a high level of technological skill and expertise, and a high cost of capital and human resources, resulting in a high barrier to entry.

EMERGING TECHNOLOGIES IN AEROSPACE

According to certain studies, the development of autonomous flight systems, 3D printing, artificial intelligence, the Internet of Things, new materials such as composites, advances in manufacturing technology, electric propulsion, and supersonic flight are among the emerging trends in the global aerospace market. These factors present opportunities for countries that can respond positively.

Autonomous flight systems, which are increasingly being developed, enable aircraft to operate without human intervention. These systems are used by unmanned aerial vehicles or drones. They depend on computer systems, sensors, and algorithms to operate the aircraft.

3D printing, or additive manufacturing, is a process of layering materials under computer control to manufacture a three-dimensional object from a digital model. This can be used in manufacturing airplane parts that would otherwise be difficult to fabricate.

Aside from 3D printing and the use of composite materials, other manufacturing technologies used are laser welding and robotics, smart machine tools (SMT) or intelligent machine tools (IMT) that use artificial intelligence in smart drilling, which is important in cutting operations for airplane parts.

Artificial intelligence enables machines to simulate human intelligence. IATA reports that AI is used in aviation for non-critical tasks like customer service and safety-critical operations such as trajectory prediction, conflict detection between aircraft, traffic advisory systems, and autonomous flight capabilities.

Internet of Things (IoT), refers to a network of connected devices that facilitates communication between them and the digital cloud. Some of the applications in the aerospace industry are automated check-ins, baggage tracking, maintenance operations, and in-flight entertainment.

Materials such as carbon fiber, fiberglass, or aramid fibers, and epoxy resin are combined to produce composites, which are lighter, stronger, and more durable than traditional materials such as steel and aluminum. These have better strength-to-weight ratios, resulting in fuel efficiency and higher performance. They also contribute to a cleaner environment.

Electric propulsion in the aerospace industry uses electricity to create thrust for a more efficient flight. This emerging technology is being explored for drones and larger commercial aircraft. NASA's Glenn Research Center is developing Electrified Aircraft Propulsion for more efficient commercial air transportation.

Supersonic flight means that an aircraft can travel through the air at speeds greater than the speed of sound (Mach 1). The first aircraft to pierce the sound barrier was the Bell X-1 experimental plane flown in 1947; the first supersonic transport (SST) was flown by the Soviets in 1969; the first passenger-carrying supersonic plane was the Concorde, which flew at Mach 2 in 1973. The Concorde's flight was discontinued in 2000 when it crashed upon take-off, resulting in numerous fatalities. Due to high costs of production, SSTs for commercial flights were discontinued, but the development of hypersonic planes is continuing for the military.

These advances have created new opportunities for aerospace manufacturing. It has also produced more sophisticated aircraft like the sixth-generation war planes featuring stealth technology, and drones that have many applications.

SUPPLY CHAIN

The complexity of aircraft production has created a network of suppliers and manufacturers making up the industry's supply chain. Studies indicate that supply chains are increasingly outsourced offshore to maintain profitability because of increased competition and globalization.

The global commercial aerospace market comprises a few prime contractors that manufacture aircraft and engines, and numerous small and medium-sized firms that supply components and subsystems. The aerospace value chain is characterized by engine manufacturers and system supplier firms that supply exclusively to aircraft manufacturers.

A typical aerospace supply chain consists of OEM or Primes, Tier 1, Tier 2, and Tier 3 Suppliers. The OEMs are tasked with the design, manufacturing, and assembly functions of an aircraft, considered the most critical component of the value chain. Boeing and Airbus are common players at this level. Tier 1 Suppliers support the Primes as they provide equipment and systems such as engines, flight control, fuel, electrical power, propulsion, and pneumatic systems. Examples of Tier 1 suppliers are General Dynamics, Rolls-Royce, Lockheed Martin, Northrop Grumman, and Raytheon. Tier 2 Suppliers manufacture and develop parts as specified by Primes and Tier 1 suppliers. Hydraulic pumps, motors, and controls are usually supplied from this level. Some Tier 2 suppliers are Moog Inc, Honeywell, and Eaton. Tier 3 Suppliers supply basic products and components such as pistons, cylinders, connectors, and O-rings to vendors.

Strict quality standards are imposed on companies wanting to participate in the aerospace supply chain. They must comply with the AS9100 quality management system, and other regulations to ensure reliability and safety.

With globalization, system manufacturers are outsourcing more to shorten development time and focus on higher value-added systems. It is estimated that 80% of the airplane is outsourced. Product designing is also becoming a globally collaborative activity because of the need to reduce the time to market the product. Specialization also seems to be increasing because complex airplanes require expertise that no one company can meet. US companies have also resorted to outsourcing aircraft maintenance as a cost-cutting measure. It is estimated that almost $140 billion of aerospace manufacturing was outsourced in 2008. Technical expertise is considered the most important criterion for selecting an outsourcing partner. Sharpening the technical expertise through education, transfer technology, and training is deemed crucial to the success of a local company wanting to become part of the supply chain.

MANUFACTURING SERVICES IN THE INDUSTRY

Two important areas in manufacturing and service, which aid in the continuous growth of the industry, are OEM and MRO.

The OEM Segment

The aerospace industry includes aerospace parts manufacturing, which is a multi-billion-dollar business. Aircraft parts manufacturing is characterized by a complicated process involving the production of various components with varying technology requirements. High standards of quality and inspection are required for aircraft parts to be useful and approved. Manufacturers of these parts are called Original Equipment Manufacturers, or OEMs. OEM is defined as an organization that makes devices from parts bought from other organizations.

An aircraft has numerous parts and systems such as engine, avionics, aerostructures, cabin interiors, insulation components and equipment, and support systems. However, the major components of an aircraft in terms of value (65% of production cost) are the airframe and engine. Systems and avionics account for another 25 percent.

A diagram of the typical external parts of an aircraft is shown in Figure 1.

Figure 1. Exterior parts of the aircraft

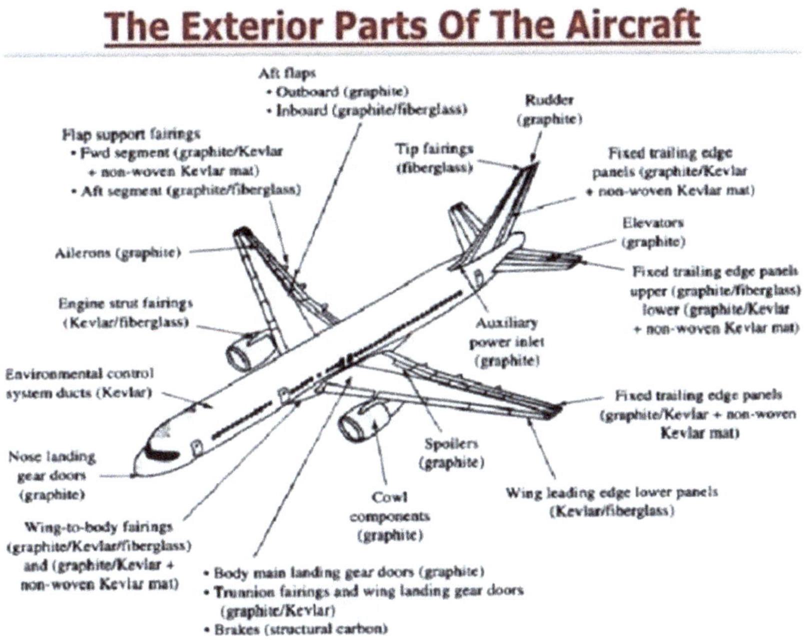

(ResearchGate)

With the advent of new technologies, it becomes imperative for the aircraft industry to replace its obsolete and aging fleet with more advanced aircraft. Planes have to be replaced within a set period, normally after 3 years, since it becomes more expensive to maintain them. New components that are produced when an aircraft is replaced or refurbished contribute significantly to the aerospace parts manufacturing market. Statistics show that more than 42,000 aircraft valued at $5.8 trillion are expected to be delivered from 2023 to 2042 by major manufacturers such as Boeing and Airbus.

The MRO Segment

MRO, short for Maintenance, Repair, and Operations, refers to activities and resources needed to maintain equipment and facilities in optimal working condition. This includes tasks like routine maintenance, repair, and various operational activities essential for businesses to run smoothly.

MRO in the aerospace industry consists primarily of airframe maintenance, engine and component work, and line maintenance. Engine maintenance accounted for around 48 percent of the global civil air transport MRO market in 2020. It is projected to reach $64.2 billion by 2030. Shown in Figure 2 is a composition of the MRO market and an indication of its size over a four-year period ending 2020.

Figure 2. Global MRO components and market size

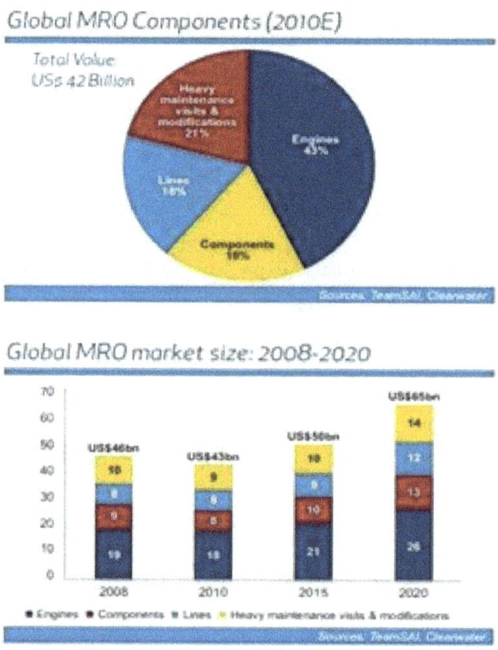

(Aerospace Global Report 2011: A Clearwater Industrials Team Report)

The leading regions in the aircraft MRO market are the United States, Canada and Mexico in North America; Germany, UK, France, Italy, Russia and Turkey in Europe; and China, Japan, Korea, India, Australia, Singapore, Indonesia, Thailand, Philippines, Malaysia and Vietnam in the Asia-Pacific.

A comprehensive report on the aircraft MRO market indicates that the global market size was valued at $80.6 billion in 2022 and is expected to expand to $130 billion by 2030, for a CAGR of 5.2%. The compound annual growth rate of the market in the Asia-Pacific region is expected to grow by 1.8 percent during the period 2023 to 2033.

The Asia-Pacific region is expected to drive the demand and production of aerospace parts because of increasing air passenger traffic and the rapid development in emerging economies in the area.

The ASEAN-5

The aerospace industries of five (5) ASEAN countries, Singapore, Malaysia, Indonesia, Thailand, and Vietnam, are growing and are strongly supported by their governments through aggressive policies and business incentives. Several international aerospace companies have located their manufacturing facilities in these countries.

Singapore is the number 1 MRO hub in Asia, contributing 10% to global MRO output. It has adopted multiple strategies to build value chain clusters such as creating an aerospace park to attract investments from major players, assistance to SMEs to compete through business alliances, development of QMS and capability systems, establishment of a strong research and development, supplier conferences and air shows, and development of engineering talent through education. Singapore has one of the largest and most diverse concentrations of aerospace companies in Asia. The industry's exports reached an output of S$13.3 billion for 2022.

Malaysia is the second largest aerospace hub in ASEAN. The Malaysian government considers aerospace a high-value industry and has implemented a blueprint to drive industry growth. Its total revenue generated in 2019 amounted to RM16.2 billion.

Indonesia is the second-largest country in Southeast Asia with a complex aircraft manufacturing capability. Despite aerial accidents attributed to poor maintenance and lack of pilot training, It has created its own national aircraft. Expanding its aviation industry, it signed a Memorandum of Agreement with Boeing to set up business operations in Indonesia to boost the national economy. A total MRO value of $65.3 million was forecast for 2020.

Thailand has a growing aviation industry with full government support and maximum investment incentives. It has a capable and efficient workforce, cost-effective inputs such as stainless steel and aluminum alloy, fiberglass application technology, polymer supply, and available and developed support industries.

Vietnam is the world's seventh fastest-growing aviation market with an expected 17.4 percent average growth rate compared to the 6.1 percent in ASEAN. The major drivers of this development include a robust tourism industry with growing airport infrastructure, the emergence of low-cost carriers, and strong economic growth. The Vietnam MRO market is still in its early development stage. As of 2018, it was worth US$542 million.

THE PHILIPPINES AS AN AEROSPACE INDUSTRY PLAYER

The Philippines is one of the pioneers in the ASEAN aerospace industry. It draws upon a rich tradition in aviation. The Philippine Air Lines (PAL) is the first commercial airline in Asia to fly in 1941. The FEATI University, in collaboration with the country's Air Force, designed and flew in 1954 its light aircraft, the *Tagak*, a single-engine, twin-boom, high-wing monoplane made of reinforced woven bamboo materials.

In 1973, the country began to recognize the importance of a viable and reliable aerospace industry with the passage of a presidential decree creating a government agency, the Philippine Aerospace Development Corporation (PADC), to rationalize and spearhead aerospace development. The agency's charter included the "assembly and manufacture of all forms of aircraft, device, equipment or contraption, and the maintenance, repair/overhaul, and modification of aerospace and associated flight and ground equipment and components to provide technical services and overhaul support to the Philippine Air Force, the national airline, foreign airline companies, foreign air forces and to the aviation industry in general."

In 1981, PADC embarked on a program to develop a high-performance trainer plane for the use of the Philippine Air Force. The program was launched as an innovative step to build an aircraft of local design, utilizing indigenous materials. A single, piston-engine trainer aircraft was conceived, designed, built, and flown within three years. The Defiant 300 prototype flew twice, the first on February 22, 1988. Weak government support eventually ended this pioneering effort.

The aerospace agency's major accomplishments since its inception include the assembly of helicopters; and aircraft under a licensing agreement with a German and a British company. It also included the assembly of turboprop trainer aircraft and jet trainers for the Philippine Air Force. It also exported helicopter components, i.e., helicopter tail booms and fiberglass components. Later, the aerospace agency's thrust was in MRO activities. The experience of this agency presents a workable platform to develop the industry further.

In 1997, the Philippines launched its first satellite *Agila-2,* into orbit. This was part of the country's space program, which was initiated in the 60s. Since then, the country has launched several satellites, relying on other nations' orbital launch capability. These satellites are designed to provide useful data and information concerning the status of natural resources through transmitted images to improve the lives of Filipinos. They are also used to provide connectivity to remote areas in the country. While the initial satellites were sourced externally, all present active satellites are built and operated by the country's Department of Science and Technology (DOST) and its child agencies. The DOST is behind Diwata-1, the first satellite built and designed by Filipinos, which was launched into space in 2016. Following this was

Maya-1, the first nano-satellite owned by the Philippines, which was launched in 2018. Additional Maya satellites were developed and launched in cooperation with Japan's Kyushu Institute of Technology. The government administration is presently supporting the development of a 130-kg multi-spectral Unit for Land Assessment (MULA) satellite that will be developed and built by an all-Filipino team of engineers to be launched in 2025. Coordinating all these space efforts is the Philippine Space Agency, which has been the lead government organization of the country's space program since 2019.

In 2014, a light sport aircraft, the RP-S512, was built by a local machining company. It was a 2-seater sports aircraft with a range of 700 kilometers and a speed of 25 kilometers per hour that used unleaded gasoline. The plane has yet to be cleared by the Civil Aviation Authority

Given these modest and encouraging achievements, and the right government support and private initiatives, the Philippines expects to present an aerospace capability that can be competitive within the ASEAN region.

AN INTEGRATED FRAMEWORK

A country needs to have an aerospace industry for economic, political, and national security reasons. Further, the demand for fast and safe air transportation requires a robust and reliable aerospace industry.

The Philippine aerospace manufacturing industry envisions the country to be the hub of original equipment manufacturing (OEM) and maintenance, repair, and overhaul (MRO) of aircraft and parts in ASEAN. Based on trade data in 2022, the country's aerospace exports registered $426.2 million. It projects a $1.9 billion export in 2033 and expects to contribute $3.9 billion to the country's GDP in the same year.

A study conducted by Duke University in 2016 indicated that the Philippine aerospace industry's modest participation in the global value chain was in the manufacture and assembly of parts and electronic and mechanical components, subassemblies in the interiors and flight control systems, and some post-sales services such as MRO activities. It also indicated certain challenges that would affect its trajectory, such as supply chain gaps, high energy costs, weak access to loans, and a lack of availability of experienced human resources.

Although the Philippines trails its ASEAN neighbors, Singapore, Malaysia, Indonesia, Thailand, and, lately, Vietnam, in aerospace exports and imports, it has the potential to service the requirements of major aircraft manufacturers such as Boeing and Airbus. It has several positive attributes, such as a strategic location, the quality of its human resources, who are not only technically skillful but also proficient in English, and the presence of three (3) Tier 1 and several Tier 2 suppliers that can create a niche in the global aerospace supply chain. It can also design and manufacture light aircraft prototypes and satellites, and localize certain import-dependent services in aerospace.

In most developed countries, the aerospace industry is a cooperation of public and private industries. For example, in the United States, there is NASA and the Department of Defense around which private enterprises revolve to create a network of supply chains that fulfill aerospace requirements.

The Philippines has developed a unique system for growing its aerospace manufacturing industry. It relies on a network of players involving government, manufacturers, suppliers, the academe, and financing institutions. Driving this initiative is the industry's aerospace association, which seeks to collaborate with stakeholders in the industry, as depicted in Figure 3 below.

Figure 3. Integrative framework

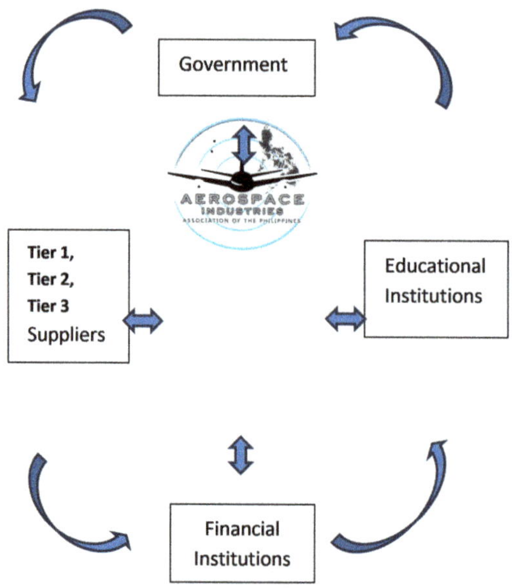

(Author, based on AIAP strategy model, 2023)

The government provides major assistance to strengthen the aerospace industry by attracting and locating players in free export trade zones, where they enjoy incentives such as support facilities and tax holidays. These special locations are expected to include additional Tier 1 and Tier 2 suppliers. Through the Department of Trade and Industry, the government assists aerospace companies in marketing their services globally. Through the Board of Investments, it provides incentives to companies with pioneering ventures to hasten the industry's growth. The government also provides skills and trade certifications through its manpower training and science-based technology centers.

The private sector membership in the association is composed largely of manufacturers, with 3 Tier 1 suppliers leading the charge. These are Moog, Collins, and JAMC. Lufthansa, a major MRO company, is also a member. AIAP helps elevate production and manpower standards through collaborative alliances among manufacturers. educational institutions, and government agencies. AIAP also initiates policy reforms through dialogues with the government's transportation and communication agency and its instrumentalities, such as civil aviation consultative councils and the PADC, the primary agency tasked with the development of Philippine aerospace.

In 2011, the Association commissioned a study to craft a roadmap for the aerospace industry, which was submitted to the government. It was intended to bring to the government's attention the need for a robust and exceptional aerospace industry as one of the pillars for sustainable economic development. It presented the global and regional realities prevailing in the industry against which the country has to compete and assert its presence. It outlined the opportunities and threats posed by the environment, identified the industry's strengths and weaknesses in terms of resources, capabilities, and support systems. It put together a set of prerogatives and programs designed to close the gaps in infrastructure and logistics.

During a decade of existence, AIAP has encouraged its members to gain accreditation and improve their OEM and MRO capabilities. Through conferences and workshops, it has captured members' sentiments regarding the need for government assistance. It has forged a strong alliance between the government and the private sector to grow an industry that will generate revenues and employment, enrich Filipino lives, and contribute to GDP.

The AIAP has grown in strength over the past ten (10) years with the addition of Tier 2 members. It has also enlisted the participation of several government agencies such as the Metals Industry Research Development Corporation, which provides solutions for metal fabrication, the Civil Aviation Authority of the Philippines, which regulates air transportation, the Philippine Aerospace Development Corporation and the Philippine International Trading Corporation, which engages in exports and trade services as assistance to local companies. They are admitted as honorary members of the association for better liaison and coordination of aerospace activities. AIAP also expanded its membership to include aviation schools, MRO/ Approved Maintenance Organization, General Aviation, and companies that issue Air Operator certifications. It facilitated the AS 9100 training of member firms through funds provided by government agencies, resulting in the certification of nine (9) local companies. Likewise, two (2) member firms were certified under the National Aerospace and Defense Contractors Accreditation Program (NADCAP), which is an organization composed of technical experts from prime contractors, suppliers, and government representatives who work together to ensure conformity with 'special processes'. The association has also hosted several forums to discuss concerns about the industry.

Another study was commissioned by the AIAP in 2023 to prepare a road map for the industry in the succeeding ten years. An environmental scan and internal assessment featured in these studies, which were undertaken in 2011 and 2023, confirm certain opportunities arising from strong growth and demand for products in the aerospace sector, such as interiors and flight controls, electrical systems, improved manufacturing processes, and engineering services. One of the identified threats is the instability of government policies, which produce changes with each new government administration. Another external challenge is the strong competition presented by Singapore and Malaysia, which are backed by government subsidies.

The industry's major strengths cited are affordable engineering talent who are English-speaking, an efficient and competitive Export Processing Zone regime, where incentives are granted to locators, good industrial relations, and experience in the automotive and electronics industries, which are helpful to avionics.

The Philippines features some of the best flight schools, and more and more foreigners are enrolling. The schools are relatively inexpensive compared to other countries' schools. Typically, aviation training courses require at least 250 hours of pilot training to qualify for a pilot license, but in the Philippines, aviation schools require over 1,500 hours. Courses offered by the top-flight schools include bachelor's degrees in Aeronautical Engineering, Air Transportation, Aircraft Maintenance Technology, Avionics Technology, and Airline Business Administration.

Special economic zones were created by legislative enactment in 1995, administered by the Philippine Economic Zone Authority (PEZA) to encourage private enterprise; and provide incentives to needed investments. Some of the objectives and purposes of this economic law are to transform selected areas in the country into highly developed agro-industrial, industrial, commercial, tourist, banking, investment, and financial centers, where highly trained workers and efficient services are made available to commercial enterprises; promote the flow of foreign and local investors into these special economic zones to generate employment opportunities and establish linkages among industries in and around the economic zones; provide attractive climate and incentives for business activity; and promote financial and industrial cooperation between the Philippines and industrialized countries through technology-intensive industries that will modernize the country's industrial sector and improve productivity levels by utilizing new technological and managerial know-how. PEZA currently gives the following fiscal incentives for export manufacturing enterprises: income tax holiday (ITH), i.e., 100% exemption from corporate income tax, 4 years ITH for non-pioneer projects, and 6 years ITH for pioneer projects.

Within these special economic zones, industrial peace and harmonious relations are promoted so that locators are not disturbed by labor disruptions.

The Philippines has a rich history of working with electronics technology. The electronics industry in the country started in the 70s with the production of semiconductor components, followed by advanced components like microprocessors, CD-ROMs, and computers. According to the Semiconductor & Electronics Industries in the Philippines Foundation Inc. (SEIPI), "the Philippine Electronics Industry is classified into 73% Semiconductor Manufacturing Services (SMS) and 27% Electronics Manufacturing Services (EMS). Electronic companies in the country practice the best known methods in manufacturing with capabilities ranging from integrated circuit packaging, printed circuit board assembly and full product assembly." The Philippines hosts seven of the world's top 20 chipmakers, which comprise 10% of the global semiconductor manufacturing and supply. The country also supplies 50% of the world demand for 2.5" hard disk drive (HDD) and 10% of the world demand for 3.5" HDD. Hitachi, Fujitsu, and Toshiba, three of the largest HDD producers have their plants located in the Philippines. The country's semiconductors and electronics industry specializes in manufacturing assembly, testing, packaging, and distribution. In 2021, Philippine electronics exports amounted to US$ 29.99 billion. On the other hand, the automotive industry in the Philippines is robust, with over 440 thousand vehicles sold in 2023. The industry started in 1975 with the country's progressive car manufacturing program. The automobile industry has two sectors: motor vehicle assembly and vehicle parts and components manufacturing. The country is a major exporter of wiring harnesses, which is a key part of the electronics industry, alongside semiconductors and other electronic components. With workers adept at electronics and automotive technology, their skill can easily be adapted to aerospace manufacturing, especially in avionics.

The aerospace industry's major challenges are supply chain gaps in raw materials availability and key processes, inadequate infrastructure and related services, high energy costs, and the absence of bilateral safety agreements such as the Wassenaar Arrangement.

The identified challenges faced by the industry are certain inadequacies in production and process capability, such as the use of high-end machine tools and metrology equipment, fine machining, heat treat and surface treatment process, gear manufacturing, non-destructive inspection (NDI), and general functional testing for compliance with aerospace requirements. There is a high capitalization requirement for machine tools and metrology equipment. There is a lack of supply chain integration such as in raw material supply/distribution, transport and import/export turnaround, complete process capability, and tooling and chemical supply distribution. Inadequate training and education in CNC machining/programming and advanced metrology, surface treatment and testing technology, heat treat, gear manufacturing and metrology, AS9100 and ADCAP accreditation are also major concerns. Insufficient availability of and improvement of government programs and incentives for potential investors in process capabilities is also a handicap.

The studies also recommend short-term, medium-term, and long-term strategies to enhance the industry's growth. The strategies consist of process upgrades such as multi-axis machining, chemical and heat treatment, and upgrading into higher value products in two existing aircraft systems: interiors and flight controls, and wire harnesses. Manufacturing suppliers are also recommended to be certified with AS9100 (i.e., an international quality management system (QMS) standard specifically for the space, aviation, and defense industries) and NADCAP to facilitate their entry into the global value chain.

It is recommended that the Philippines continue to develop as a competitive alternative in the ASEAN region in MRO operations. There is awareness within AIAP that the industry's internal capability should address three critical elements: quality, cost, and delivery. Since the aerospace industry puts a premium on these three factors, the industry must respond effectively through an efficient and well-equipped supply chain that optimizes the talent of its major and minor players. Infrastructure constraints, such as airport inadequacies, are also recommended to be addressed by the government. The studies point out the potential of the lucrative after-market product segment to supply and service spare parts for aircraft.

THE PHILIPPINE AEROSPACE CAPABILITIES

Commercial Aircraft Production

The capability of the Philippine aerospace industry is bolstered by the presence of Tier 1 suppliers that have relocated to export processing zones that offer tax incentives. These suppliers are Moog Philippines, Collins Aerospace, and JAMCO.

MOOG Philippines supplies standard and customized products across a large number of categories. They design, manufacture, and integrate precision motion control components and systems. About 80% of the above flight control actuation systems are currently being produced in the Philippines.

Collins Aerospace provides "nose-to-tail" solutions for commercial and business jet aircraft. They provide all classes of seating products from super first-class environments to business and main cabin class seats, and VIP and business jet seating. Their oxygen systems are available on Airbus and Boeing aircraft, regional jets, and on most business jets and military aircraft. In addition, Collins manufactures gourmet-catering solutions such as Endura beverage makers, Selectra Gourmet mode selectable chillers, and the 3000 series ovens. Their lighting systems, crew rests, monuments, and aerospace fasteners and consumables offer unique solutions for virtually any aircraft platform. The local unit of Collins Aerospace has invested over P1 billion to produce aircraft galleys in its existing facility. Aircraft seats have been the primary target, with the country having capable manufacturers of plastics and rubbers.

JAMCO Philippines specializes in aircraft interiors and components such as plane galleys and first-class seats. It is a major supplier of Boeing. Its manufacturing operations include machining, panel assembly and anodic coating.

PRODUCT VALUE CHAIN IN THE PHILIPPINES

Industry challenges limit full capabilities for the Philippines to participate aggressively in the global value chain. Activities are largely confined to machining, assembly, production, and delivery. Tier 1 major suppliers are mainly involved in these processes. Some of the machining processes are still done in neighboring Asian countries. The United States, Canada, United Kingdom, Singapore and some European countries are the main sources of raw materials. This raw material gap in value chain goes with the country's inadequacy in terms of special processing, chemical/ metallurgical testing and fine machining. Though some of the Tier 1 players have the capabilities on some of those processes, the industry still resorts to outsourcing from Singaporean companies.

Among AIAP's programs for the next ten years, as outlined in its industry roadmap, are improvements in systems and processes. This includes the certification of all active Tier 2 and 3 suppliers with AS 9100 and NADCAP for special process suppliers; strengthening of supply chain capabilities in CNC machining and metrology, surface and heat treatment, non-destructive testing, and composites manufacturing. Linkages will also be established with government and private institutions to continually improve education and training in the above manufacturing areas. An annual assessment of skilled workforce in the field of aeronautical engineering, CNC machining and programming operations, aviation maintenance, surface treatment (heat treatment and plating), and additive manufacturing is also recommended to be conducted. In addition to building a strong manufacturing base, with skilled manpower and focus on science and engineering, the AIAP aims also to establish a comprehensive range of nose-to-tail MRO capabilities, development of supply chain integration, establishment of an R & D hub in the country, development of talents to meet the needs of industry through training, and exploration of the feasibility of establishing a light aircraft manufacturing capability.

POLICY REFORMS AND PRIORITY PROJECTS

The aerospace manufacturing industry's sustainable development relies on government actions in terms of policy reforms, which will clear administrative bottlenecks; and implement priority projects. A major policy reform that is recommended by AIAP is the creation of a Philippine Aerospace Manufacturing Council. While the Philippines has several government agencies that are involved in aerospace activities, there is no central agency that takes a holistic approach to aerospace development. A case in point is the recently created Philippine Space Agency, which is concentrated mainly on the development of satellites. An aerospace council staffed by public and private entities can spearhead the development and rationalization of the industry. This council will represent the interests of aerospace manufacturers and stakeholders in leading educational programs and industry research among its members and be a neutral platform for effective coordination of the industry.

Another policy reform clamor is the inclusion of aerospace manufacturing as a pioneering industry in the country's investment priority plans to include extension of tax holidays from 4-6 years to 6-8 years.

The government is also requested to fill in supply chain gaps through investments in process capabilities such as surface finishing, special testing, and NDI or non-destructive inspection, which evaluates the condition of aircraft parts and components without causing damage. The product value chain analysis will identify the status of the Philippines in terms of raw material procurement, packaging, and delivery processes. Current capabilities are limited.

The government should also support training programs and certifications related to aerospace manufacturing to develop the diverse skills required to function effectively in a modern manufacturing environment. These skills and inputs of talented workers are required in the design and manufacture of aerospace products. Training priorities should be made in technology areas, especially in CNC machining and manufacturing, surface treatment, heat treatment, gear manufacturing, AS 9100, and NADCAP.

There is also a major concern about administrative bottlenecks that should be cleared by invoking the law on ease of doing business. Some of these initiatives are streamlining import/export lead times and procedures for faster transaction processing, and promoting supply chain integration.

This set of goals, objectives, strategies, and work plans is expected to usher the fledgling industry on the road to achieving its full potential to becoming a force in the region. This holistic strategy involves public-private partnership that counts among its participants all concerned sectors in the industry. If the concerned government agencies and participating players implement the initiatives recommended by AIAP to provide or sustain infrastructure, investment incentives, and credit facilities to

Philippine aerospace enterprises, the country may be able to achieve its vision for the industry to become a major MRO/OEM hub in ASEAN.

A macro view of the aerospace industry, coupled with a micro perspective featuring an Asian country's developing experience, offers useful insights into the industry's future direction. The challenges are enormous, especially from the point of view of geopolitics, yet solvable, with the advent of new technologies injecting fresh ideas and requiring innovative strategies. The value of collaboration, communication, and coordination among the players in the industry cannot be overemphasized. These mindsets must pervade the supply chain for resources to be moved with efficiency. Likewise, corporate initiatives must be effectively integrated into the government's policies for greater synergy. With the myriad factors influencing the industry, adaptability and resilience are important attributes that stakeholders in the supply and value chain must possess in good measure. Operational issues need to be resolved with care and precision because of their safety implications. Human capital must be factored into the equation to ensure a vibrant and progressive industry.

The concluding section looks at future strategies that may be adopted to improve and strengthen the supply chain and enhance the value of services offered. Captains of industry and managers can consider these strategies seriously to mitigate risks attendant to aerospace activities and reduce vulnerabilities.

FUTURE STRATEGIES AND CONCEPTS

Vision 2040: Digital Horizons for Philippine Aerospace

Looking further ahead, the Philippine aerospace community envisions an ecosystem that is not merely reactive to external market forces but proactively shapes regional standards for safety, sustainability, and advanced manufacturing. By 2040, the industry aspires to integrate real-time data loops that connect aircraft in operation, maintenance crews on the tarmac, engineers at design consoles, and regulators overseeing flightworthiness. The ambition is to move from periodic upgrade cycles toward a living-digital-twin paradigm in which airframe performance, passenger comfort, and environmental metrics are continuously optimized.

In this scenario, original equipment manufacturers in Cavite and Batangas no longer ship components halfway around the world for final assembly; instead, they collaborate across a distributed lattice of micro-factories that specialize in additive manufacturing, composite lay-up, and smart machining. Each facility is equipped with modular production cells designed to be reconfigured at a moment's notice. When demand spikes for a particular component, capacity can be shifted across the network without the lengthy tooling delays that once hamstrung expansion plans.

The workforce of 2040 is equally dynamic. Technical colleges partner with flight schools and AI laboratories to offer stackable micro-credentials that allow machinists to evolve into data analysts and avionics specialists to transition into cyber-resilience architects. Apprentices rotate through virtual and physical stations, mastering tasks in immersive environments before interacting with cutting-edge machinery. As a result, learning curves flatten, and the talent pool remains agile in the face of shifting technological frontiers.

Sustainability targets are no longer side notes but the central scoreboard by which success is measured. Every plant publishes a live carbon dashboard, visible to investors and the general public, tracking emissions down to individual production lots. Electric ground vehicles, hydrogen fuel cells powering test stands, and rooftop solar arrays feeding micro-grids become standard features rather than experimental showcases. Water-intensive processes, once a persistent pain-point in plating and surface treatment, are replaced by closed-loop chemistries that recover and recycle nearly all fluids.

On the commercial front, Philippine airlines leverage the nation's geostrategic position as an archipelagic bridge between Northeast and Southeast Asia. New point-to-point routes emerge, served by medium-range hybrid-electric aircraft optimized for island hopping. This expansion stimulates local demand for maintenance, repair, and overhaul services, creating a virtuous cycle in which operational data feeds back to design teams who iterate on airframe geometry and power-management algorithms.

Meanwhile, the country's nascent space program matures into a constellation of small satellites tasked with maritime domain awareness, disaster-response imaging, and precision agriculture. These spacecraft are no longer assembled exclusively overseas; instead, final integration occurs in a purpose-built cleanroom complex north of Manila, staffed by graduates of the National Space Engineering Institute. Launch partnerships with regional providers reduce cost and foster a culture of cooperative risk sharing.

Regulation keeps pace through adaptive governance. Rather than static rulebooks, authorities deploy machine-readable policies that automatically cross-check design files for compliance at the moment of submission. Cybersecurity protocols follow zero-trust architectures, allowing suppliers to exchange sensitive build data without exposing proprietary algorithms or customer records.

In this envisioned 2040 landscape, the Philippine aerospace industry operates less like a traditional hierarchy and more like an extensible operating system. Stakeholders—from raw-material producers to end-of-life recyclers—plug into a common data backbone that enforces standards, notarizes transactions, and rewards innovation through tokenized intellectual-property exchanges. Financial institutions underwrite production expansion based on real-time ESG scores, while international investors view the country as a bellwether for inclusive, technology-driven industrialization.

To reach this horizon, strategic priorities for the next decade include scaling additive-manufacturing capacity, nurturing a national composites center, and institutionalizing a dual-ladder career path that values both technical mastery and digital fluency. By aligning infrastructure upgrades with skills development and research incentives, the Philippines positions itself not as a follower but as a first-mover among emerging aerospace nations.

Risk oversight frameworks also evolve. Traditional quarterly reviews give way to continuous assurance models driven by AI-enabled anomaly detection. If a vibration sensor in a machining spindle exceeds its engineered signature, the deviation is flagged, correlated with operator logs, and dispatched to a national predictive-maintenance cloud. Insurance premiums are dynamically adjusted based on demonstrated reliability, incentivizing firms to invest in resilient tooling and robust data pipelines.

Financing mechanisms diversify beyond conventional bank loans. Sovereign green bonds, indexed to measurable CO2 reductions, funnel capital toward renewable-powered hangars and cold-soak test chambers. Crowdfunding platforms reserved for accredited local investors allow diaspora Filipinos to take fractional stakes in export-oriented aerospace start-ups, aligning national pride with economic returns.

Crucially, the sector cultivates a vibrant research culture. A tri-city corridor stretching from Quezon City to Clark Global City and Batangas Science Park hosts living laboratories where students, engineers, and entrepreneurs co-create prototypes with on-site 5-axis mills and low-vacuum autoclaves. Annual hackathons challenge multi-disciplinary teams to redesign a legacy component for 50 percent weight reduction, with winners receiving seed grants and access to test rigs at the Aerospace Industries Association of the Philippines' innovation center.

These collective efforts converge toward a single north-star metric: technology sovereignty balanced by collaborative openness. By deliberately cultivating local capabilities in critical domains—powder metallurgy, battery management systems, and flight-control firmware—the Philippines safeguards supply-chain continuity while remaining an enthusiastic participant in cross-border knowledge networks.

The journey is ambitious, but each incremental milestone—be it certifying a new composite winglet, achieving AS9145 compliance for advanced product quality planning, or commissioning a bio-derived jet-fuel pilot plant—cements the Philippines' reputation as a credible, future-ready contributor to global aviation. More importantly, it demonstrates that small and medium economies can leapfrog structural disadvantages by weaving digital infrastructure, human capital, and environmental stewardship into a coherent industrial vision.

Ultimately, the Philippine aerospace success story of 2040 embodies resilience, inclusivity, and technological audacity—proof that purposeful collaboration can bend even the loftiest ambitions into achievable flight paths.

The author of this paper recognizes a need to strengthen operational resilience and adapt strategy in the aerospace industry since global disruptions have occurred recently. Industry leaders must reconsider their approaches to managing these complex systems when vulnerabilities in aerospace supply chains escalate into production delays and significant financial losses. To sustain competitive advantage, aerospace companies must also work towards sustainability goals, workforce enhancement, and innovative business models. The author in this book chapter has explored strategic methods for enhancing aerospace supply chain resilience, including dynamic supplier reconfiguration, predictive resilience modeling, artificial intelligence (AI), and resilient-by-design manufacturing. (Refer to Figure 4). The book chapter connects these operational strategies to the broader corporate strategies for sustainability, human capital development, and adaptive business diversification.

Figure 4. Strengthening supply chain in aerospace industry strategies and concepts

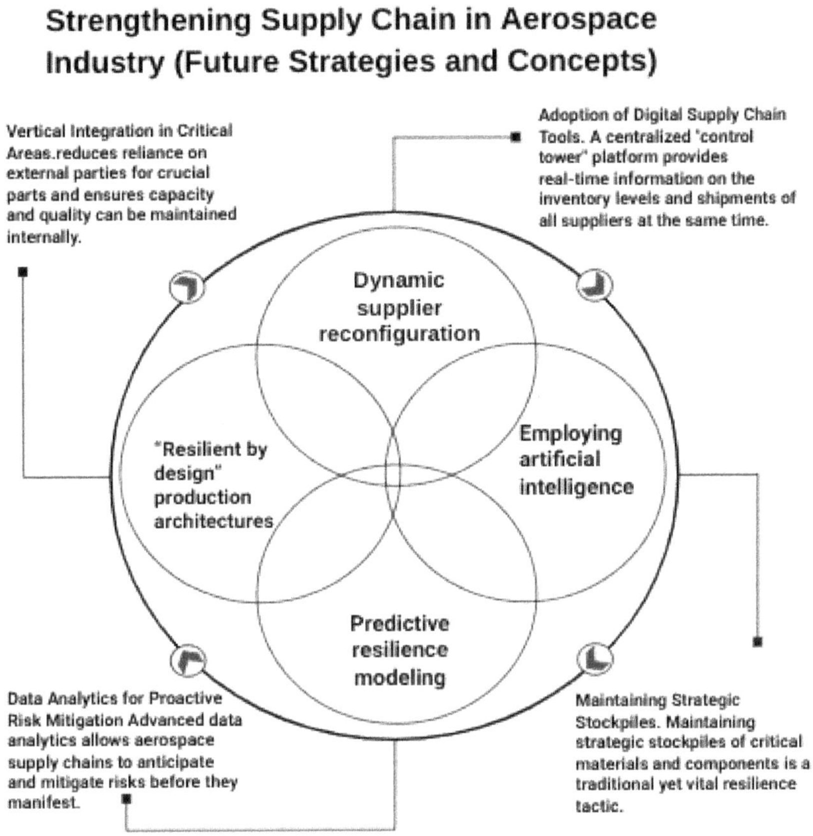

(Author's synthesis based on current aerospace SCM literature)

Predictive resilience modeling utilizes advanced analytics and simulation to identify supply chain risks early. By leveraging data-driven analytics, companies are proactively mitigating risks, and conducting scenario planning and forecasting to identify potential disruptions. Strategic stockpiles of important materials are integrated into this strategy, providing necessary buffers against unexpected events. The use of predictive models can be used to identify geopolitical risks, material shortages, or other vulnerabilities, thus enabling preemptive measures such as stockpiling or sourcing alternatives, so that downtime can be minimized.

The strategic integration of artificial intelligence and digital technologies has significantly enhanced the resilience of the supply chain. As a result of the deployment of digital "control tower" platforms, centralized, real-time visibility of inventory and logistics across the entire supply chain network is provided across the entire enterprise. With the use of artificial intelligence, predictive analytics can be powered, anomalies can be detected, and real-time decisions can be made, significantly improving efficiency, responsiveness, and accuracy. Using artificial intelligence to forecast demand, optimize inventory and automate processes enables proactive rather than reactive management, ensuring agility during disruptions to your business processes.

Production architectures that are "resilient by design" entail embedding this feature into the core design of aerospace manufacturing systems in order to increase their adaptability to changes that occur during the production process. An approach based on this production architecture advocates the use of flexibility, redundancy, and robustness as fundamental principles of the design of systems. An integral part of resilient-by-design strategies is modular production processes, standardized interfaces for components and geographically distributed manufacturing facilities, as well as additive manufacturing (3D printing). The use of such practices enhances the ability of the organization to adapt to disruptions and ensure the continuity of production during such disruptions.

The use of an interconnected set of operational strategies featuring dynamic supplier reconfigurations, predictive modeling, AI integrations, and resilient-by-design practices that reinforce each other will guarantee the enhancement of the value chain. Similarly, supply chain activities that are intelligently managed using predictive analytics, modular architectures facilitate supplier adjustments, while AI integration supports intelligent decision-making. These factors contribute to a cohesive framework that enables aerospace supply chains to manage risk and disruption effectively.

The aerospace industry has increasingly integrated sustainability into its strategic planning in addition to operational resilience. Increasingly, aerospace companies are making environmental sustainability a key element of their future growth strategies. It is very important for companies to set clear objectives in order to reduce carbon emissions, achieve carbon neutrality, and comply with the strictest environmental standards.

CONCLUSION

The aerospace industry will continue to be a big contributor to the global economy as it upscales itself using digital technology. The aircraft as well as its variants will rely continuously on an improved value and supply chain that is well integrated. The developed economies as well as the emerging ones can contribute to the sustainability of the aerospace industry with its emphasis on safety and concern for the environment. While defense is a major factor in the growth of the aerospace industry, peaceful and constructive applications are equally important to a well-connected and inter-related world order. This paper has demonstrated the role of a Southeast Asian country in the global aerospace framework as it navigates its way to a more robust and vibrant participative role in the global value chain.

REFERENCES

Addepto. (n.d.). *Top 10 practical applications and use cases of AI in the aviation industry.* https://addepto.com/blog/top-10-practical-applications-and-use-cases-of-ai-in-the-aviation-industry/

Aerospace Industries Association of the Philippines. (2013). *Aerospace manufacturing: An industry road map.*

Aerospace Industries Association of the Philippines. (2023). *Aerospace manufacturing: An industry road map.*

Air Transport Action Group. (n.d.). *Economic growth.* https://aviationbenefits.org/economic-growth/

Altair Engineering. (n.d.). *Guide to urban air mobility* [White paper].

Balinbin, A. L. (2022, March 31). PHL aircraft maintenance market seen growing as travel industry recovers. *BusinessWorld.* https://www.bworldonline.com/economy/2022/03/31/439548/phl-aircraft-maintenance-market-seen-growing-as-travel-industry-recovers/

Bamber, P., Frederick, S., & Gereffi, G. (2016). *The Philippines in the aerospace global value chain.* Duke University, Center on Globalization, Governance & Competitiveness. https://hdl.handle.net/10161/12483

Bescast. (n.d.). *Aerospace supply chain tiers explained.* https://www.bescast.com/blog/aerospace-supply-chain-tiers/

Department of Trade and Industry. (n.d.). *Aerospace: Securing the future.* https://industry.gov.ph/industry/aerospace/

First Philippine Industrial Park. (2023a, October). *Aerospace manufacturing industry* [Brochure]. https://fpip.com/sites/default/files/2023-10/FPIP%20-%20Aerospace%20Manufacturing%20Industry%20brochure.pdf

First Philippine Industrial Park. (2023b, October). *Electronics industry* [Brochure]. https://fpip.com/fpip-official/sites/default/files/2023-10/FPIP%20-%20Electronics%20Industry%20brochure.pdf

Imaginovation. (n.d.). *IoT in aviation: How connected aircraft re-define flying.* https://imaginovation.net/blog/iot-in-aviation/

Joshi, N. (n.d.). *Revolutionizing aviation with IoT.* Allerin Blog. https://www.allerin.com/blog/revolutionizing-aviation-with-iot

KPMG International. (2016). *Global aerospace and defense outlook: The dawn of a new day.*

Lahane, R. (n.d.). *Smart MRO: The digital revolution in aviation, aerospace & defense.* Ramco Blog. https://www.ramco.com/blog/aviation/tech-revolution-in-aviation-and-defense-mro

Mehra, M. (2023). Smart materials and their applications in aerospace engineering. *Journal of Air Transport Management, 114*, 102437. Advance online publication. DOI: 10.1016/j.jairtraman.2023.102437

National Aeronautics and Space Administration. (n.d.). *EAP mission overview.* https://www.nasa.gov/mission/eap/

Presidential Communications Office. (n.d.). *PBBM gives all-out support for RP-made MULA satellite.* https://pco.gov.ph/news_releases/pbbm-gives-all-out-support-for-rp-made-mula-satellite/

Republic Act No. 7916. (1995). *Special Economic Zone Act of 1995 (Philippines).*

RevFine. (n.d.). *Aerospace companies: List of aircraft and spacecraft manufacturers.* https://www.revfine.com/aviation-industry/

Chapter 11
Machine Learning and Artificial Intelligence (AI) in Manufacturing

Binastya Anggara Sekti
https://orcid.org/0000-0001-5489-4888
Universitas Esa Unggul, Indonesia

ABSTRACT

The integration of machine learning (ML) and artificial intelligence (AI) into manufacturing has transformed traditional processes, creating smarter, more efficient, and adaptable production systems in line with Industry 4.0 principles. Advanced AI applications can enhance automated quality control by detecting defects in real time, thus improving product reliability and reducing waste. Technologies like digital twins and generative design allow manufacturers to simulate, optimize, and innovate production processes before physical implementation, cutting down on development time and costs. In supply chain optimization, ML algorithms enhance inventory management, demand forecasting, and logistics efficiency. The benefits are substantial are challenges such as big data management, integration with legacy systems, workforce reskilling, and ethical concerns. Future trends include the emergence of autonomous manufacturing systems, sustainable AI applications, and improved interoperability through the Internet of Things (IoT) and edge computing in the manufacturing sector.

DOI: 10.4018/979-8-3373-1082-4.ch011

INTRODUCTION TO MACHINE LEARNING AND AI IN MANUFACTURING

Manufacturing has long been a cornerstone of economic growth and technological advancement, driving industrial revolutions and shaping global economics. Over the years, this sector has evolved from manual production to mechanization, automation, and digitalization. The advent of Industry 4.0 marks the latest evolution, blending traditional manufacturing processes with cutting-edge technologies to create intelligent, interconnected systems. In developments in the manufacturing industry, sustainable development activities in the manufacturing industry are always integrated with Industry 4.0. (Hussain, Akhunzada, Iqbal, Bibi, & Gani, 2021; Patalas-Maliszewska & Łosyk, 2022).

The rise of machine learning (ML) and artificial intelligence (AI) has catalyzed a monumental shift in the manufacturing sector. These technologies are revolutionizing how manufacturers operate by unlocking data-driven insights and enabling unprecedented levels of efficiency, innovation, and adaptability. Industry 4.0, characterized by the convergence of digital, physical, and cyber systems, has been a pivotal force in this transformation. Key technologies such as big data analytics, Industrial Internet of Things (IIoT), and machine learning are increasingly adopted to enhance productivity, optimize processes, and improve decision-making capabilities (Antony et al., 2024; Nacchia, Fruggiero, Lambiase, & Bruton, 2021). Figure 1 shows the scope of machine learning in manufacturing.

Figure 1. Machine learning in manufacturing

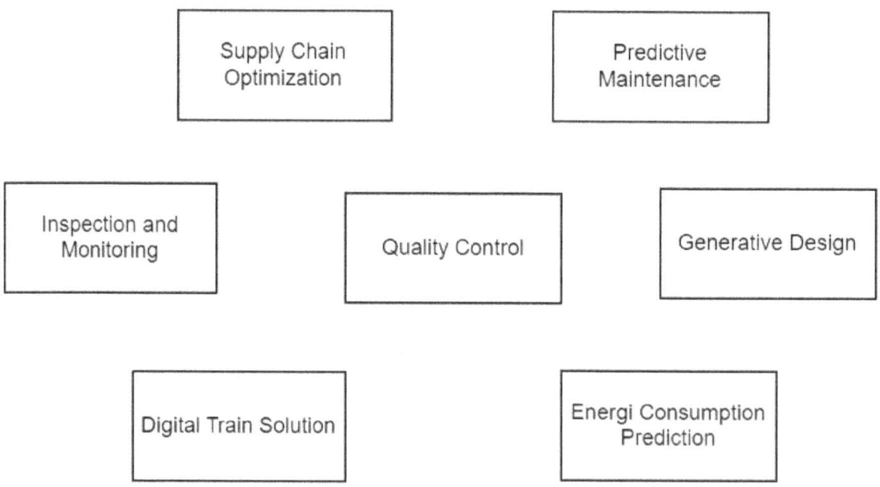

As global competition intensifies and consumer demands grow more dynamic, manufacturers face increasing pressure to innovate while maintaining cost-effectiveness and sustainability. ML and AI provide solutions to these challenges by enabling smart factories, predictive systems, and autonomous decision-making capabilities. We will introduce the transformative role of ML and AI in manufacturing, delving into their applications, benefits, and implications across various domains. From predictive maintenance and quality control to sustainable practices and advanced robotics, ML and AI are integral to reshaping the industry for the demands of a dynamic and competitive global market.

The Role of ML and AI in Industry 4.0

The emergence of Industry 4.0 has redefined the landscape of manufacturing, blending advanced technologies with traditional processes to create intelligent, interconnected, and highly efficient ecosystems. Machine learning (ML) and artificial intelligence (AI) play pivotal roles in this transformation, empowering manufacturers to transition from reactive, labor-intensive workflows to agile, predictive, and autonomous operations. These technologies leverage vast amounts of data generated by IoT devices, sensors, and interconnected machinery to provide actionable insights, optimize performance, and drive innovation. Significant benefits of Industry 4.0 for sustainable manufacturing and identifies tools and elements of Industry 4.0 for developing environmental sustainability (Dogan & Birant, 2021; Dubey et al., 2020; Javaid, Haleem, Singh, Suman, & Gonzalez, 2022; Zonta et al., 2020).

Key Contributions of AI and ML in Manufacturing

Production Flexibility. AI-driven systems enable manufacturers to achieve remarkable adaptability in the face of changing market demands. By analyzing consumer trends, production data, and external factors, ML models help manufacturers quickly adjust production lines, facilitate rapid customization, and reduce time-to-market. For instance, in industries such as automotive and electronics, AI-powered tools support mass personalization, allowing companies to meet the rising demand for tailored products without compromising efficiency.

Predictive Analytics. Several ML applications in the industry have been tried, predictive analyst is one of the most transformative ML applications on Industry 4.0. By processing real-time data from industrial equipment, ML models can forecast potential failures, identify bottlenecks, and predict maintenance needs well in advance. This minimizes unplanned downtime, reduces operational costs, and extends the lifespan of machinery. For example, predictive maintenance systems are widely used in sectors like aerospace and heavy manufacturing, where even minor disruptions can lead to substantial financial losses.

Optimization Across Processes. AI technologies excel at optimizing complex and interconnected processes across the manufacturing value chain. From inventory management to production scheduling, ML algorithms streamline operations by aligning production with real-time market demands, reducing material waste, and improving overall efficiency. For instance, in supply chain management, AI-powered systems analyze supplier data, transportation networks, and demand forecasts to recommend cost-effective and sustainable solutions.

Automation and Safety. AI-powered robotics and automation systems are transforming factory floors by taking over repetitive, labor-intensive, or hazardous tasks. These systems not only enhance worker safety but also reduce human errors, increase throughput, and improve product quality. In addition, collaborative robots (cobots) equipped with AI capabilities are being deployed to work alongside human operators, further enhancing productivity while maintaining flexibility.

Emerging Technologies Supporting Industry 4.0

In the context of AI and ML integration, several technologies have emerged as foundational tools for manufacturers seeking to modernize their operations:

Digital Twins. Digital twins—virtual replicas of physical assets—are revolutionizing manufacturing processes by providing real-time simulations and predictive insights. These systems use AI to model machinery performance, simulate scenarios, and optimize workflows. Manufacturers can use digital twins to test modifications, forecast the impact of production changes, and identify inefficiencies before implementing them on the factory floor.

RealTime Analytics. The capability to analyze data in real time has revolutionized manufacturing operations. By integrating big data analytics with AI, manufacturers can oversee production processes as they occur, ensuring that any deviations or anomalies are swiftly addressed. Real-time analytics also offer vital insights into energy usage, resource allocation, and production quality, thereby enabling more informed decision-making.

AI-Driven Robotics. AI-enhanced robotics are becoming more advanced, able to perform intricate tasks with great accuracy and flexibility. These robots not only boost automation but also foster intelligent human-machine collaboration. For example, AI-driven robots can modify their actions based on feedback from operators, making them particularly useful in industries where precision and adaptability are essential.

The Benefits of Integrating ML and AI in Industry 4.0

The incorporation of ML and AI into Industry 4.0 frameworks offers numerous advantages, such as heightened productivity, enhanced sustainability, and increased competitiveness. These technologies enable real-time decision-making and operational improvement, helping manufacturers stay adaptable in a rapidly changing global market. Furthermore, as sustainability becomes increasingly important, ML and AI support resource-efficient manufacturing by reducing waste, lowering energy consumption, and promoting circular economy practices.

AI and ML as Drivers of Digital Transformation

Artificial intelligence (AI) and machine learning (ML) are at the forefront of digital transformation in manufacturing, fostering data-centric ecosystems that redefine operational efficiency, innovation, and competitiveness. As key enablers of Industry 4.0, AI and ML allow manufacturers to move beyond conventional methods, harnessing the power of big data and intelligent algorithms to drive insights and actionable outcomes. From predictive analytics and process optimization to real-time decision-making, these technologies underpin a profound shift in how businesses operate in a hyper-connected and rapidly evolving industrial environment.

Transformative Applications of AI and ML in Manufacturing

AI-driven analytics platforms are central to this transformation, enabling manufacturers to forecast market trends, optimize resource allocation, and streamline supply chain operations. For example, advanced AI tools can analyze global economic data, supply chain disruptions, and consumer behavior to predict demand fluctuations, ensuring production systems remain aligned with market needs (Sharma & Villányi, 2022). Similarly, ML algorithms utilize real-time data from interconnected devices and sensors to optimize operations, improve quality control, and enhance predictive maintenance, resulting in a more agile and resilient production environment (Rai, Tiwari, Ivanov, & Dolgui, 2021).

A Holistic Impact on Manufacturing

AI and ML significantly influence manufacturing, offering a wide range of benefits across various areas:

Improved Productivity

Automating routine and repetitive tasks through AI enables faster, more reliable production cycles. For instance, robotic process automation (RPA) combined with ML allows manufacturers to reduce downtime, improve workflow efficiency, and enhance overall throughput. Studies indicate that manufacturers adopting AI-driven automation have reported productivity gains of up to 30% (Sharma & Villányi, 2022).

Cost Efficiency

By minimizing waste and energy consumption, AI and ML contribute to substantial cost reductions. Predictive analytics, for instance, identifies inefficiencies in equipment performance and energy usage, allowing for timely interventions. A case study in the automotive sector showed that predictive maintenance powered by ML algorithms reduced equipment-related costs by 25% (Bertolini, Mezzogori, Neroni, & Zammori, 2021; Kegyes, Süle, & Abonyi, 2021; Penumuru, Muthuswamy, & Karumbu, 2020; Usuga Cadavid, Lamouri, Grabot, Pellerin, & Fortin, 2020).

Sustainability

AI supports environmentally sustainable practices by optimizing energy use, reducing material waste, and facilitating circular economy models. For example, AI-powered monitoring systems help factories minimize their carbon footprint by identifying areas for energy savings and waste reduction (Jamwal et al., 2022). This aligns with the growing demand for sustainable manufacturing processes in a carbon-conscious global market.

Workforce Empowerment

Although automation replaces some manual tasks, AI empowers employees by enhancing their capabilities. It allows workers to concentrate on higher-value activities such as strategic planning and innovation, rather than routine or repetitive tasks. Additionally, AI-driven tools like augmented reality (AR) and virtual reality (VR) offer training and support, helping workers improve their skills and adapt to the evolving requirements of smart factories (Makridis & Han, 2021).

AI and ML: A Strategic Necessity

The implementation of AI and ML is now essential for organizations striving to stay competitive in the constantly evolving industrial sector. Companies that do not adopt these technologies risk lagging behind their competitors in terms of efficiency, adaptability, and sustainability. Additionally, AI and ML equip manufacturers with the flexibility required to manage complex global supply chains and respond to unexpected challenges. For instance, during the COVID-19 pandemic, AI-driven tools played a crucial role in mitigating disruptions and restoring operational stability (Modgil, Singh, & Hannibal, 2022).

Transforming Traditional Manufacturing

The transformation of traditional manufacturing through machine learning (ML) and artificial intelligence (AI) provides a new paradigm in the way factories operate. These advanced technologies enable manufacturers to move beyond outdated, reactive methods toward proactive and predictive approaches that drive efficiency, innovation, and adaptability. By incorporating intelligent, data-driven systems, manufacturers are unlocking new levels of productivity while addressing challenges related to cost, quality, and market demand.

The Rise of Smart Factories

One of the most prominent examples of this transformation is the advent of smart factories, where IoT and AI devices work together to create an interconnected, automated production environment. At every stage of production, real-time data can be captured from sensors embedded in machines, production lines, and supply chains to dynamically optimize operations. For example, AI can analyze machine performance in real time to identify inefficiencies or potential breakdowns, allowing adjustments to be made instantaneously. Smart factories embody the principles of Industry 4.0, offering unparalleled levels of adaptability, precision, and efficiency (Ortiz-Ruiz, Bermejo, Sicilia, & Bermejo, 2024). The application of smart manufacturing can encourage managers and decision-makers to implement a wider range of enabling technologies in the various stages of the value chain (Büchi, Cugno, & Castagnoli, 2020). Some of the components of the smart industry are shown in Figure 2.

Figure 2. Smart industry

Just-in-Time Inventory Management

AI has revolutionized inventory management through the implementation of just-in-time (JIT) principles, which ensure that materials and components are available precisely when needed. By analyzing historical data, market trends, and real-time supply chain information, the AI system makes the inventory system optimal by preventing overstocking while reducing production disruptions due to out-of-stock. This not only lowers storage costs but also enhances cash flow and operational agility. For instance, manufacturers in the automotive and electronics industries have adopted AI-powered JIT systems to synchronize production with fluctuating customer demands, achieving greater cost efficiency and responsiveness (Dogru & Keskin, 2020).

AI-Powered Robotic and Automation

Robots equipped with AI capabilities are transforming factory floors by taking on complex tasks that were previously beyond the scope of traditional automation. These AI-driven robots excel in tasks such as precision assembly, quality inspection, and packaging, where accuracy and consistency are critical. In addition to improving productivity, these robots can learn and adapt to new processes through reinforcement learning, reducing the need for extensive reprogramming during production line changes. Collaborative robots (cobots) designed to be able to work alongside humans, will further improve efficiency by combining human intuition with robotic precision, creating a hybrid workforce optimized for complex production scenarios (Dogru & Keskin, 2020).

Lean Manufacturing Supported by AI

The principles of lean manufacturing, which emphasize the judicious use of resources, waste reduction, and continuous process improvement, have been further strengthened by the adoption of AI. By providing data-driven insights, AI enables manufacturers to pinpoint inefficiencies in energy usage, workflows, and material handling. In example, AI can analyze production data to identify bottlenecks, suggesting process improvements that align with lean principles. Furthermore, AI supports zero-waste initiatives by optimizing raw material usage and enabling recycling efforts, helping manufacturers align with environmental sustainability goals (Fatimah, Govindan, Murniningsih, & Setiawan, 2020).

The Holistic Impact on Traditional Manufacturing

AI and ML are not only transforming operational processes but are also reshaping the strategic outlook of manufacturers. Traditional systems relying on siloed data and reactive decision-making are being replaced by intelligent systems promoting transparency, collaboration, and agility. This positions manufacturers to tackle globalization, supply chain disruptions, and evolving customer expectations. By embracing ML and AI, traditional manufacturing evolves into a forward-thinking, sustainable, and competitive industry. The integration of AI and ML marks a milestone in manufacturing's evolution. From smart factories and JIT inventory management to AI-driven robotics and lean practices, these technologies drive a new era in manufacturing, enhancing efficiency, sustainability, and resilience in a global market.

Sustainability and Innovation through AI

Artificial intelligence (AI) is reshaping manufacturing by driving sustainability and innovation, addressing both the need for eco-friendly practices and the pursuit of efficiency. In an era of heightened environmental awareness, manufacturers must adapt to regulatory and societal pressures for greener operations. AI technologies provide powerful solutions to optimize energy consumption, reduce waste, and improve resource efficiency, aligning manufacturing processes with sustainability objectives while fostering innovation.

Optimizing Energy Consumption

One of the most impactful contributions of AI is its ability to manage and optimize energy consumption in manufacturing. AI-powered predictive energy management systems use machine learning (ML) algorithms to forecast energy needs based on historical data, production schedules, and external factors such as weather conditions. These systems enable manufacturers to better utilize renewable energy sources and reduce reliance on non-renewable energy, resulting in lower operational costs and a reduced carbon footprint (Bai et al., 2022). For example, in energy-intensive industries such as steel and cement, AI-based tools dynamically adjust machinery operations to minimize energy waste.

Reducing Waste Through Generative Design

Generative design, a cutting-edge application of AI, allows manufacturers to create lightweight, material-efficient designs that reduce production waste. Using advanced algorithms, generative design tools explore thousands of design possibilities based on specified constraints, such as material usage, weight, and performance requirements. These AI-driven designs not only minimize material waste during production but also lead to more sustainable end products. For instance, the aerospace and automotive industries leverage generative design to produce lightweight components, resulting in significant energy savings during transportation and operation (Choi et al., 2023).

Enhancing Recycling and Material Recovery

AI is revolutionizing recycling and material recovery processes by improving the accuracy and efficiency of sorting operations. Traditional recycling systems often face challenges in identifying and separating materials, leading to inefficiencies and contamination. AI-powered systems equipped with computer vision and

ML algorithms can accurately identify materials based on visual and spectral data, ensuring proper sorting and repurposing. For example, AI-driven robotic arms in recycling facilities can distinguish between different types of plastics, metals, and glass, significantly improving recycling rates (Monostori et al., 2023). This innovation supports circular economy models, allowing manufacturers to reduce reliance on virgin materials and adopt sustainable production practices. Table 1 highlights key AI innovations that are driving sustainability initiatives.

Table 1. Key AI innovations driving sustainability

AI Innovation	Description	Example Applications
Generative Design	AI generates lightweight, material-efficient designs to reduce waste and optimize costs.	Aerospace (lightweight parts), Automotive (energy-efficient components)
Predictive Energy Management	ML models forecast energy needs, optimizing renewable energy utilization and reducing waste.	Steel production (energy savings), Cement manufacturing (energy load balancing)
Recycling and Material Recovery	AI enhances sorting and recycling processes, improving material recovery and reducing contamination.	Waste management facilities (AI-powered robotic sorting), Plastic recycling plants
Sustainable Process Optimization	AI monitors and optimizes processes to minimize resource use and emissions.	Chemical manufacturing (optimized resource use), Food production (reduced spoilage)

Competitive Edge Through Sustainability

AI-driven sustainability efforts not only reduce the environmental impact of manufacturing but also create competitive advantages in a marketplace increasingly influenced by eco-conscious consumers. A 2022 study by Ivanov & Dolgui highlights that manufacturers adopting AI-based sustainability practices experience increased customer loyalty, enhanced brand reputation, and access to new markets focused on green products. Moreover, regulatory compliance is simplified through AI's ability to monitor and report environmental metrics in real time, reducing the risk of fines or operational disruptions.

Overcoming Challenges in AI Adoption

While machine learning (ML) and artificial intelligence (AI) offer transformative opportunities in manufacturing, their implementation is not without challenges. Manufacturers face hurdles in data management, workforce readiness, financial constraints, and technical complexity, all of which can hinder the seamless integration of these technologies into existing systems. Overcoming these challenges is critical to unlocking the full potential of AI and ML in the manufacturing sector.

Data Quality and Integration

Data forms the backbone of AI and ML systems, but issues with data quality and integration remain a significant obstacle. Many manufacturers operate with legacy systems and siloed data repositories, resulting in inconsistencies and gaps that impede effective AI deployment. For instance, poorly labeled datasets, incompatible data formats, and missing information can reduce the accuracy and reliability of AI models. Additionally, integrating new AI technologies with existing infrastructure often requires significant time, resources, and expertise.

To address these challenges, manufacturers must invest in modern data infrastructure and adopt standardized protocols for data collection, storage, and sharing. Technologies like data lakes and cloud computing can facilitate seamless data integration, while advanced data cleaning and preprocessing tools can improve data quality. Furthermore, adopting industrial IoT (IIoT) devices and sensors that capture real-time, high-quality data can significantly enhance the effectiveness of AI systems (Bai et al., 2022).

Skill Gaps in the Workforce

A major barrier to AI adoption in manufacturing is the lack of skilled personnel capable of managing and optimizing these advanced systems. Engineers and technicians may be proficient in traditional manufacturing techniques but lack the specialized knowledge required for AI model development, data science, and system integration. This skills gap is especially pronounced in small and medium-sized enterprises (SMEs) that often lack the resources to hire or train AI specialists AI have significant impacts on the relationships between corporate digital entrepreneurship and its predictors (Chatterjee, Chaudhuri, Vrontis, & Basile, 2022; Hansen & Bøgh, 2021).

To bridge this gap, manufacturers must prioritize workforce development through targeted training programs and upskilling initiatives. Partnerships with academic institutions and online learning platforms can provide access to specialized AI and ML courses tailored to the needs of the manufacturing sector. Additionally, the use of low-code or no-code AI platforms can empower existing staff to implement AI solutions without requiring deep technical expertise (Ivanov & Dolgui, 2020).

Cost and Complexity of Implementation

The high upfront costs and complexity associated with AI implementation present another challenge, particularly for SMEs. From procuring advanced hardware and software to hiring specialized personnel, the financial burden can be significant. Moreover, integrating AI systems into existing manufacturing processes requires careful planning and coordination, which can further increase costs and extend implementation timelines.

To mitigate these barriers, governments and industry organizations can offer financial incentives such as grants, subsidies, and tax breaks to encourage AI adoption. Manufacturers can also adopt phased implementation strategies, starting with pilot projects that demonstrate clear value before scaling up. Open-source AI tools and pre-trained models offer cost-effective alternatives for businesses with limited budgets (Choi et al., 2023).

The Role of Collaboration in Accelerating Adoption

Collaborative efforts between industry, academia, and policymakers can play a vital role in overcoming these challenges. Industry partnerships can drive the development of standardized AI solutions tailored to manufacturing needs, while academic research can explore new methodologies for enhancing AI efficiency and reducing costs. Policymakers can establish supportive frameworks that promote AI adoption, such as regulatory guidelines, workforce development initiatives, and funding programs.

For example, initiatives like the EU's "Factories of the Future" program and the U.S. "Manufacturing USA" network are designed to foster innovation and AI adoption in the manufacturing sector. These programs provide funding for research, facilitate knowledge sharing, and support the creation of AI-enabled manufacturing ecosystems (Monostori et al., 2023).

A Path Forward

Addressing the challenges of AI adoption in manufacturing requires a multifaceted approach that combines technological investments, workforce development, and collaborative initiatives. By improving data infrastructure, closing skill gaps, and making AI adoption more accessible and cost-effective, manufacturers can fully realize the transformative potential of AI and ML. As industries continue to adapt to the demands of Industry 4.0, overcoming these barriers will be essential for creating competitive, sustainable, and future-ready manufacturing systems.

Conclusion

Machine learning and AI represent a paradigm shift in manufacturing, offering tools that go beyond automation to empower manufacturers with insights, precision, and adaptability. From predictive maintenance to supply chain optimization, these technologies are integral to the industry's future. While challenges persist, the ongoing evolution of AI and ML promises to unlock unprecedented opportunities for innovation, efficiency, and sustainability in manufacturing.

As this book progresses, we will explore the practical applications of AI and ML in greater detail, highlighting their transformative impact across various facets of manufacturing.

Machine learning (ML) and artificial intelligence (AI) have revolutionized manufacturing by enabling the sector to leverage data and computational intelligence to achieve unprecedented levels of efficiency, innovation, and adaptability. These technologies empower manufacturers to extract actionable insights from vast amounts of data generated by interconnected devices, such as sensors, machines, and IoT systems. By analyzing this data, ML and AI facilitate optimization of complex manufacturing processes, enabling predictive maintenance, quality control, and real-time decision-making. Di era Industri 4.0 beberapa teknologi seperti Internet of Things Industri, analitik data besar, dan pembelajaran mesin (ML) semakin luas didunakan untuk meningkatkan produktivitas dan efisiensi pada sistem dan proses manufaktur (Antony et al., 2024; Nacchia et al., 2021; Nuhu, Zeeshan, Safaei, & Shahzad, 2023).

As integral components of Industry 4.0, ML and AI serve as catalysts for digital transformation within the manufacturing ecosystem. They enhance production flexibility, allowing manufacturers to respond swiftly to market fluctuations and customized consumer demands. For instance, AI-driven analytics can identify emerging market trends, while ML models optimize supply chain operations, ensuring just-in-time inventory management and reducing waste. Moreover, ML and AI enable predictive analytics, offering foresight into equipment performance, potential bottlenecks, and product lifecycle management. This capability supports proactive decision-making, minimizing downtime and maximizing productivity. Automation powered by AI-driven robotics further streamlines workflows, reduces human error, and enhances workplace safety.

PREDICTIVE MAINTENANCE AND EQUIPMENT OPTIMIZATION

Predictive maintenance stands out as one of the most impactful applications of machine learning (ML) and artificial intelligence (AI) in manufacturing. This approach leverages data-driven insights to transition from traditional, reactive maintenance methods to proactive strategies, enabling manufacturers to maximize equipment efficiency, minimize downtime, and reduce operational costs.

By harnessing data from sensors, IoT devices, and other interconnected systems, predictive maintenance systems utilize advanced ML algorithms to monitor equipment performance continuously. These algorithms analyze variables such as temperature, vibration, pressure, and operational cycles to detect subtle anomalies that might indicate impending equipment failures. For instance, time-series analysis and supervised learning models are commonly employed to study historical data and recognize patterns associated with wear, tear, or malfunction.

One of the key advancements in predictive maintenance is the use of anomaly detection techniques. These methods enable real-time monitoring of equipment, identifying deviations from standard operating conditions. When irregularities are detected, automated alerts are generated, allowing maintenance teams to intervene before minor issues escalate into costly breakdowns.

Deep learning models further enhance predictive maintenance by processing large, complex datasets, such as audio signals from machinery or thermal imaging data. These models excel at identifying nuanced patterns that traditional methods might overlook, providing even greater accuracy in predicting failures. Additionally, reinforcement learning is being explored for its potential to optimize maintenance schedules dynamically, balancing operational demands with maintenance needs.

The benefits of predictive maintenance extend beyond minimizing downtime. By addressing issues proactively, manufacturers can extend the lifespan of machinery, improve operational safety, and reduce maintenance expenses. Furthermore, this approach supports sustainability goals by minimizing waste, conserving energy, and reducing the carbon footprint associated with equipment replacement and unplanned outages.

Examples of successful predictive maintenance implementations include smart factories where AI systems predict and prevent failures in assembly lines or power plants where ML models optimize turbine performance. These cases demonstrate how predictive maintenance not only safeguards critical infrastructure but also enhances overall productivity and operational excellence.

QUALITY CONTROL WITH AI-DRIVEN COMPUTER VISION

Automated quality control, powered by AI-driven computer vision, is transforming manufacturing processes by introducing precision, speed, and scalability in product inspection. Computer vision, a specialized field within artificial intelligence (AI), enables machines to "see" and analyze visual data through high-resolution cameras and sophisticated algorithms, ensuring that production lines maintain consistent quality and minimal waste.

One of the primary applications of computer vision in manufacturing is defect detection. High-speed cameras capture detailed images of products on assembly lines, while AI-powered image recognition algorithms analyze these images in real-time to identify defects such as surface scratches, dimensional inaccuracies, misalignments, or discoloration. Unlike traditional inspection methods, computer vision systems can process vast volumes of products rapidly and without fatigue, ensuring that even subtle anomalies are detected.

Advanced AI techniques, such as convolutional neural networks (CNNs), play a pivotal role in enhancing the capabilities of computer vision systems. These algorithms are trained on extensive datasets containing images of both defect-free and flawed products, enabling them to accurately distinguish between acceptable and unacceptable items. As the system continues to operate, it learns from new data, improving its ability to identify complex patterns and rare defects over time.

In addition to defect detection, computer vision systems are employed for tasks such as dimensional measurement, assembly verification, and packaging inspection. For instance, in the automotive industry, these systems ensure that components are assembled with precise alignment, while in the food and beverage sector, they verify that packaging labels are correctly applied and readable.

The benefits of AI-driven quality control extend beyond defect detection. By minimizing human error and enabling continuous monitoring, these systems reduce production downtime and improve operational efficiency. They also support sustainability initiatives by reducing material waste and energy consumption associated with rework and defective products. Furthermore, data collected by these systems can be analyzed to identify trends and root causes of defects, providing manufacturers with actionable insights to enhance production processes.

Emerging advancements in computer vision include the integration of hyperspectral imaging and 3D vision technologies. Hyperspectral imaging allows systems to analyze material composition and detect internal defects that are invisible to the human eye, while 3D vision enables precise spatial measurements, critical for industries requiring high precision, such as aerospace and electronics manufacturing.

DIGITAL TWINS IN MANUFACTURING

Integration of digital twins—virtual replicas of physical assets, systems, or processes—has revolutionized how manufacturers design, monitor, and optimize their operations. By creating a dynamic, data-driven representation of the physical world, digital twins allow manufacturers to simulate real-world scenarios, analyze workflows, and make informed decisions without interrupting actual production.

At the heart of a digital twin is its ability to gather real-time data from sensors, Internet of Things (IoT) devices, and connected machinery. This steady influx of data offers a detailed view of the asset's current status, performance metrics, and operating conditions. Machine learning (ML) and artificial intelligence (AI) enhance this capability by analyzing data trends, identifying anomalies, and forecasting potential issues.

One of the major advantages of digital twins is their capability to perform predictive and prescriptive simulations. Manufacturers can utilize digital twins to test various "what-if" scenarios, such as adjusting production schedules, upgrading equipment, or modifying supply chain logistics. These simulations help businesses identify inefficiencies, assess potential risks, and optimize processes before implementing changes in the real world.

In manufacturing, digital twins are widely applied to monitor the performance of complex machinery and systems. For example, in aerospace manufacturing, digital twins of jet engines monitor performance during test flights, providing engineers with critical insights into wear, tear, and efficiency. Similarly, automotive manufacturers use digital twins to optimize assembly line operations and ensure quality control.

AI plays a transformative role in enhancing the functionality of digital twins. Through adaptive learning and dynamic updates, AI ensures that digital twin models remain accurate as conditions evolve. Algorithms analyze incoming data to detect deviations from expected performance, enabling manufacturers to respond proactively to issues such as equipment failures or supply chain disruptions. This adaptability also facilitates continuous improvement, as AI-driven insights help refine operational strategies over time.

The integration of digital twins and AI also supports the shift toward mass customization in manufacturing. By leveraging data from digital twins, manufacturers can design and simulate customized products tailored to individual customer preferences. This approach minimizes waste and accelerates production timelines, meeting the growing demand for personalized goods.

Emerging technologies, such as edge computing and 5G connectivity, further enhance the capabilities of digital twins. Edge computing allows real-time data processing at the source, reducing latency and improving responsiveness. Meanwhile,

5G connectivity ensures seamless data transfer between physical assets and their digital counterparts, enabling precise and timely decision-making.

In addition to optimizing existing processes, digital twins are instrumental in prototyping and innovation. Virtual replicas allow engineers to experiment with new designs, materials, and production techniques, reducing the need for costly physical prototypes. In industries such as pharmaceuticals and renewable energy, digital twins accelerate research and development, bringing innovative products to market more quickly.

SUPPLY CHAIN AND INVENTORY OPTIMIZATION

Artificial intelligence (AI) and machine learning (ML) are transforming supply chain management and inventory optimization, enabling manufacturers to achieve higher efficiency, cost-effectiveness, and responsiveness in their operations. By leveraging predictive analytics, reinforcement learning, and real-time data processing, these technologies address the complexities of modern supply chains, from demand forecasting to logistics optimization.

Predictive Analytics for Demand Forecasting

Predictive analytics powered by ML algorithms play a pivotal role in accurately forecasting demand fluctuations. By analyzing historical sales data, market trends, and external factors such as seasonality, economic indicators, or geopolitical events, ML models generate precise demand predictions. This allows manufacturers to adjust production schedules, manage inventory levels, and minimize the risk of overstocking or stockouts. For example, retailers use AI-driven demand forecasting to prepare for peak shopping seasons, ensuring sufficient inventory while avoiding excess.

Dynamic Inventory Management

ML models enable dynamic inventory management by continuously monitoring stock levels, sales velocity, and lead times. Algorithms predict when inventory replenishment is needed, triggering automated orders to suppliers. Additionally, classification techniques identify high-priority or high-value items requiring special attention, enabling manufacturers to adopt targeted strategies like just-in-time (JIT) inventory practices. This approach minimizes holding costs, reduces waste, and ensures optimal inventory turnover.

Reinforcement Learning for Logistics Optimization

Logistics optimization is another area where AI and ML demonstrate their value. Reinforcement learning algorithms analyze complex logistical networks to identify the most efficient transportation routes. These models consider multiple variables, including traffic patterns, fuel costs, delivery deadlines, and environmental impact. By optimizing delivery routes, manufacturers can reduce transportation costs, shorten delivery times, and lower carbon emissions, contributing to more sustainable supply chain operations.

Real-Time Analytics for Supply Chain Visibility

Real-time analytics powered by AI enhance end-to-end supply chain visibility, enabling manufacturers to monitor every stage of the supply chain in real time. Sensors and IoT devices track the location, condition, and status of goods in transit, while ML algorithms analyze this data to provide actionable insights. For example, real-time visibility helps identify bottlenecks in the supply chain, enabling swift corrective actions to minimize delays. This capability is particularly crucial in industries such as pharmaceuticals and perishable goods, where timely delivery is critical.

Response to Disruptions and Market Shifts

AI-powered supply chain systems excel in responding to disruptions and market shifts. Whether dealing with sudden demand surges, supplier shortages, or global crises like pandemics, ML models can simulate scenarios and recommend adaptive strategies. By integrating data from diverse sources, such as weather reports, geopolitical events, and market analytics, AI systems ensure manufacturers remain agile and resilient in the face of uncertainty.

Enhancing Supplier Collaboration

AI tools facilitate better collaboration with suppliers by providing insights into supplier performance, lead times, and reliability. Predictive models evaluate supplier risk and suggest alternative options, reducing dependency on single suppliers and ensuring continuity of operations. Additionally, AI-driven negotiation tools enhance procurement efficiency by analyzing historical pricing trends and market conditions.

Blockchain Integration for Traceability

When combined with blockchain technology, AI further strengthens supply chain traceability and transparency. Blockchain provides a decentralized ledger for tracking transactions, while AI analyzes this data to detect anomalies, such as counterfeit products or unauthorized alterations. This integration builds trust among stakeholders, enhances compliance with regulations, and improves consumer confidence in product authenticity.

Autonomous Supply Chain Systems

Emerging trends in AI include the development of autonomous supply chain systems. These systems use advanced ML algorithms and robotics to automate tasks such as warehouse management, order picking, and last-mile delivery. For instance, AI-powered drones and autonomous vehicles are being deployed to expedite deliveries, especially in remote or densely populated areas.

Sustainability in Supply Chain Management

AI also contributes to sustainability by optimizing resource utilization and reducing waste. ML algorithms identify opportunities for greener supply chain practices, such as consolidating shipments to reduce emissions or optimizing energy usage in warehouses. These initiatives align with corporate social responsibility (CSR) goals and enhance the long-term viability of supply chain operations.

Conclusion

AI and ML are reshaping supply chain and inventory management, providing manufacturers with the tools needed to navigate an increasingly complex and competitive landscape. By leveraging these technologies, manufacturers can achieve greater efficiency, adaptability, and sustainability in their supply chain operations, ultimately driving business success and customer satisfaction. As AI and ML continue to evolve, their role in supply chain optimization will only become more integral to achieving industrial excellence.

ADVANCED ROBOTICS AND HUMAN-MACHINE COLLABORATION

The integration of advanced robotics and artificial intelligence (AI) in manufacturing has ushered in a new era of human-machine collaboration. Modern robotics, particularly collaborative robots (cobots), leverage machine learning (ML), computer vision, and natural language processing (NLP) to create systems that not only perform tasks with precision but also work seamlessly alongside human operators. This synergy enhances productivity, safety, and adaptability in manufacturing environments.

Collaborative Robots (Cobots)

Collaborative robots are designed to operate in shared spaces with human workers, unlike traditional industrial robots that require isolated zones for safety. Cobots are equipped with sensors, adaptive control systems, and ML algorithms that enable them to respond to real-time inputs, such as human gestures or changes in the workspace.

- Safety Features: Advanced safety mechanisms, including force-limited joints and real-time motion planning, ensure cobots can detect and avoid collisions, making them safe to work alongside humans.
- Flexibility: Cobots are easily programmable and adaptable, allowing manufacturers to redeploy them for various tasks without extensive downtime. For instance, a cobot might switch from assembly to packaging tasks with minimal reconfiguration.
- Cost Efficiency: Cobots are often more affordable than traditional industrial robots, making them accessible to small and medium-sized enterprises (SMEs) looking to automate processes.

Enhancing Repetitive and Hazardous Tasks

Cobots excel at handling repetitive or hazardous tasks that would otherwise pose ergonomic risks or safety concerns for human workers. Tasks such as welding, material handling, and quality inspections are automated with high accuracy and consistency.

- Repetitive Tasks: Cobots can execute repetitive assembly line tasks tirelessly, reducing worker fatigue and human error.
- Hazardous Environments: In environments involving extreme temperatures, toxic substances, or heavy lifting, cobots mitigate risks by taking over dan-

gerous operations. For example, in chemical manufacturing, cobots handle substances requiring precision and safety compliance.

Human-Machine Interaction

Advancements in natural language processing (NLP) and machine vision have significantly improved the interaction between robots and human workers. These technologies enable intuitive and efficient collaboration.

- Natural Language Communication: NLP allows workers to issue verbal commands or queries to robots, streamlining complex operations. For example, a worker might instruct a robot to "fetch the component from bin 3," reducing the need for specialized programming knowledge.
- Gesture Recognition: Machine vision systems enable robots to interpret human gestures, such as pointing or signaling, for seamless task delegation.
- Augmented Reality (AR) Integration: Coupling cobots with AR tools provides operators with real-time visual guidance, such as overlaying instructions for assembling components, further enhancing productivity.

Adaptive Learning and Self-Optimization

Machine learning enables cobots to adapt to changing conditions and learn from their interactions with the environment and human operators.

- Dynamic Task Adjustment: ML algorithms allow robots to recalibrate their actions in response to environmental changes, such as variations in component dimensions or production schedules.
- Continuous Improvement: Through reinforcement learning, robots optimize their efficiency over time, improving task performance with each iteration.
- Error Detection: Integrated sensors and ML models detect anomalies during operations, such as misaligned components, ensuring higher accuracy and quality.

Real-World Applications

- Automotive Manufacturing: Cobots assist in assembling intricate components, such as wiring harnesses and engine parts, while maintaining precision and speed.
- Electronics Industry: Robots handle delicate electronic components with high accuracy, minimizing the risk of damage during assembly.

- Pharmaceuticals: Cobots automate repetitive tasks like packaging and labeling, ensuring compliance with stringent regulatory standards.

Impact on Workforce and Skill Development

The adoption of advanced robotics requires a shift in workforce roles and skills. While robots handle repetitive tasks, human workers focus on higher-value activities like system monitoring, troubleshooting, and innovation.

- Reskilling Initiatives: Manufacturers invest in training programs to upskill employees in programming, maintenance, and robotics integration.
- Collaboration and Creativity: Workers collaborate with robots to optimize workflows and develop innovative manufacturing solutions, fostering a culture of creativity and problem-solving.

Challenges and Future Directions

While advanced robotics offers numerous benefits, its implementation comes with challenges:

- Integration Complexity: Seamlessly integrating robots into existing workflows requires careful planning and robust infrastructure.
- Cost Barriers: Initial investments in robotics technology can be prohibitive for some manufacturers.
- Ethical Considerations: Ensuring equitable workforce transitions and addressing concerns about job displacement are critical for sustainable adoption.

Future advancements in AI, robotics, and edge computing will further enhance the capabilities of cobots, making them smarter, faster, and more versatile. Autonomous robots capable of self-repair, intelligent decision-making, and collaborative problem-solving will shape the manufacturing landscapes of tomorrow.

AI IN SUSTAINABLE MANUFACTURING

As sustainability becomes a key priority across industries, manufacturing is leveraging the transformative power of artificial intelligence (AI) to achieve green production goals. AI is driving efficiency and innovation in sustainable practices, enabling manufacturers to minimize waste, reduce energy consumption, and lower carbon footprints. These advancements are reshaping traditional manufacturing processes, ensuring environmental responsibility while maintaining competitiveness in a rapidly evolving market.

Energy-Efficient Systems and Resource Optimization

AI-powered systems analyze vast datasets to optimize resource consumption across manufacturing operations.

- Energy Management: AI monitors and predicts energy usage, enabling manufacturers to shift loads during off-peak hours, balance demand, and integrate renewable energy sources into operations. This minimizes energy costs and supports grid stability.
- Smart Sensors and IoT: Internet of Things (IoT) devices, combined with AI, provide real-time data on machine performance, energy consumption, and process efficiency. These insights drive precise adjustments to reduce energy waste.
- Dynamic Energy Optimization: AI algorithms dynamically adjust production schedules and equipment settings to align with energy efficiency goals, reducing overall operational emissions.

Waste Reduction and Circular Economy

AI helps manufacturers embrace the principles of a circular economy, focusing on waste minimization and resource recovery.

- Predictive Waste Reduction: Machine learning models identify inefficiencies in production lines that lead to material wastage. By analyzing patterns, manufacturers can adjust processes to prevent overproduction or defective outputs.
- Recycling and Material Recovery: AI-driven sorting technologies enhance the recycling process by accurately identifying and segregating materials, ensuring higher quality and yield in recycled resources.
- Production Planning: AI streamlines production planning to minimize leftover inventory and scrap materials, aligning manufacturing outputs more closely with demand forecasts.

Generative Design and Material Efficiency

Generative design, an innovative AI application, revolutionizes product development by creating designs that use fewer resources without compromising functionality or quality.

- Optimized Product Design: AI generates multiple design iterations based on specified constraints, such as weight, strength, and material usage. This results in lightweight components that meet performance standards while reducing material consumption.
- Sustainable Material Selection: AI systems recommend alternative materials that are more sustainable, durable, or recyclable, helping manufacturers transition to eco-friendly options.
- Lifecycle Analysis: Generative design incorporates lifecycle impact assessments to ensure that products are not only efficient to produce but also environmentally responsible throughout their use and disposal phases.

AI-Enhanced Sustainable Supply Chains

A sustainable manufacturing process requires a green supply chain, and AI plays a critical role in its optimization.

- Demand Forecasting: AI accurately predicts demand patterns, preventing overproduction and reducing waste across the supply chain.
- Logistics Optimization: Machine learning algorithms optimize delivery routes and schedules, reducing transportation emissions and costs.
- Supply Chain Transparency: AI systems track and analyze supply chain data, enabling manufacturers to assess the sustainability of their suppliers and ensure compliance with environmental standards.

AI in Carbon Footprint Reduction

Manufacturers are leveraging AI to measure and reduce their carbon emissions.

- Carbon Accounting: AI automates the tracking of carbon emissions at each stage of production, providing actionable insights for reduction strategies.
- Energy Transition Support: AI supports the integration of renewable energy sources by forecasting energy availability and optimizing their use within production processes.

- Emission Prediction Models: AI models predict emissions under various scenarios, helping manufacturers plan for greener operations and achieve regulatory compliance.

Real-World Applications

- Automotive Industry: AI-driven energy management systems optimize power usage in electric vehicle manufacturing, reducing production emissions.
- Textile Manufacturing: AI predicts fabric demand and optimizes dyeing processes, significantly reducing water and chemical usage.
- Electronics Sector: Generative design creates lighter, more efficient circuit boards, reducing the consumption of raw materials and energy.
- Food and Beverage Industry: AI monitors water and energy usage during processing, enabling efficient resource allocation while minimizing waste.

Challenges and Future Directions

Despite its potential, AI-driven sustainable manufacturing faces challenges such as high initial implementation costs, data integration complexities, and a need for skilled talent to manage advanced technologies. Overcoming these hurdles will require collaboration among stakeholders, investments in workforce development, and the establishment of clear sustainability metrics.

Future advancements in AI are expected to amplify its impact on sustainability. Innovations in quantum computing, federated learning, and edge AI will enable even more precise optimizations, while integration with blockchain technology can ensure transparency and accountability in sustainable practices.

CYBERSECURITY AND RISK MITIGATION

As Industry 4.0 progresses, the integration of interconnected systems, IoT devices, and cloud-based platforms has greatly enhanced operational efficiency and innovation. However, this digital transformation has also broadened the cybersecurity threat landscape, exposing manufacturing systems to a wider range of risks. Protecting critical infrastructure, sensitive data, and intellectual property has become a top priority. AI and machine learning (ML) technologies offer revolutionary solutions to strengthen cybersecurity and mitigate risks in manufacturing environments.

AI-Driven Threat Detection and Response

AI-powered cybersecurity solutions provide a proactive approach to identifying and responding to threats.

- Anomaly Detection: Machine learning models assess extensive amounts of network traffic data to pinpoint unusual patterns or irregularities that might suggest potential threats. For instance, real-time alerts can be generated for unexpected surges in data transfer or unauthorized access attempts.
- Behavioral Monitoring: AI observes user activities to establish standard behaviors and identify deviations that could signal potential account compromises or insider threats.
- Automated Incident Response: AI systems can take immediate action against detected threats by isolating impacted systems, blocking suspicious IP addresses, or providing security teams with actionable information.

Vulnerability Management

AI enhances vulnerability management by automating the identification and prioritization of security gaps.

- Predictive Vulnerability Scanning: AI predicts potential vulnerabilities in systems and software by analyzing historical data and known exploits, allowing manufacturers to address issues before they are exploited.
- Patch Management: AI streamlines the deployment of security patches by identifying critical systems at risk and automating patch installations during low-activity periods.
- Threat Intelligence Integration: AI integrates global threat intelligence feeds to provide manufacturers with up-to-date information on emerging risks and vulnerabilities, ensuring timely protective measures.

Securing IoT Devices

The widespread use of IoT and edge devices in manufacturing has significantly increased vulnerabilities to cyberattacks. AI provides specialized solutions to mitigate these risks.

- Device Authentication: AI ensures secure access to IoT devices through advanced authentication mechanisms such as biometric verification and anomaly-based authorization.

- Edge Computing Security: AI-powered solutions deployed at the edge monitor and secure device-to-device communication, reducing latency and preventing unauthorized data access.
- Dynamic Risk Assessment: ML models assess risks in real time by analyzing device behavior, connectivity patterns, and software updates, ensuring a secure IoT ecosystem.

Protecting Intellectual Property

Intellectual property (IP) is a critical asset for manufacturers, and its protection requires robust security measures.

- Data Encryption and Access Control: AI-driven tools ensure sensitive IP data is encrypted and accessible only to authorized personnel.
- AI-Powered Intrusion Detection: AI systems monitor access logs and detect suspicious activities that may indicate unauthorized attempts to access proprietary information.
- Digital Watermarking: AI embeds digital watermarks into files containing sensitive data, enabling traceability and accountability in case of data leaks.

Cybersecurity for Supply Chain Integrity

The interconnected nature of modern supply chains makes them vulnerable to cyber threats. AI strengthens supply chain security by enhancing visibility and resilience.

- Supply Chain Monitoring: AI continuously monitors supply chain networks for anomalies, ensuring timely detection of potential cyber threats.
- Third-Party Risk Assessment: AI evaluates the security practices of third-party vendors, identifying weak links that could compromise the supply chain.
- Blockchain Integration: Combining AI with blockchain technology ensures secure and transparent tracking of goods and transactions throughout the supply chain.

Real-World Applications

- Pharmaceutical Manufacturing: AI detects attempts to counterfeit medications by monitoring production processes and tracking distribution channels.

- Automotive Industry: AI identifies vulnerabilities in connected car manufacturing systems, preventing potential cyberattacks on autonomous vehicle networks.
- Aerospace Sector: AI enhances cybersecurity in aerospace manufacturing by protecting sensitive designs and ensuring the integrity of critical communication systems.
- Food and Beverage: AI safeguards IoT-enabled supply chains, preventing disruptions or contamination risks caused by cyber incidents.

Challenges and Future Directions

Despite its potential, AI-driven cybersecurity in manufacturing faces several challenges:

- High Implementation Costs: Advanced AI solutions require significant financial investment, which can be a barrier for smaller manufacturers.
- Skill Gaps: Implementing and managing AI-based security tools necessitates skilled professionals, and there is a shortage of cybersecurity expertise in many organizations.
- Complexity of Integration: Integrating AI solutions with existing legacy systems can be challenging and may require extensive customization.

Future advancements in AI and ML, such as federated learning and explainable AI, are expected to further strengthen cybersecurity in manufacturing. These technologies will provide enhanced collaboration across organizations, improved transparency, and more robust protection against evolving threats.

CHALLENGES IN IMPLEMENTING ML AND AI IN MANUFACTURING

The integration of machine learning (ML) and artificial intelligence (AI) into manufacturing has the potential to revolutionize the industry, yet its adoption is not without challenges. These obstacles span technological, operational, and human dimensions, posing significant barriers to widespread implementation.

Data Quality and Availability

One of the foremost challenges in implementing AI and ML in manufacturing is the quality and availability of data.

Data Silos and Fragmentation

Many manufacturing facilities operate with data stored in isolated systems, making it challenging to consolidate and utilize for AI and ML applications. These silos limit the ability to create holistic models that require comprehensive datasets.

Inconsistent Data Quality

Data inconsistencies, inaccuracies, and gaps hinder the training of effective ML models. For instance, differences in data formats across devices and systems, as well as missing entries, can lead to unreliable insights.

Real-Time Data Challenges

For AI to drive real-time decision-making, continuous data streams are essential. However, not all manufacturing environments are equipped with IoT sensors or infrastructure capable of providing timely and accurate data.

Solution: Manufacturers must invest in robust data integration frameworks and adopt advanced data cleaning and preprocessing techniques to improve data usability.

Integration with Legacy Systems

Integrating AI solutions into existing infrastructure presents significant challenges due to the prevalence of legacy systems in manufacturing.

Compatibility Issues

Legacy systems were not designed to support modern AI technologies, making integration difficult. Often, these systems require costly upgrades or replacements.

High Implementation Costs

Retrofitting AI capabilities into existing infrastructure can be prohibitively expensive, particularly for small and medium-sized enterprises (SMEs) with limited budgets.

Downtime Risks

Integrating AI systems may necessitate temporary production halts, which can disrupt operations and result in significant costs.

Solution: Hybrid approaches, such as phased implementation or using middleware to bridge legacy systems with AI platforms, can help reduce costs and minimize disruption.

Workforce Upskilling

The adoption of AI and ML demands a skilled workforce, yet many manufacturers face challenges in this area.

- Skills Gap. AI and ML technologies require expertise in data science, software engineering, and system integration. However, a shortage of skilled personnel poses a significant barrier to effective deployment.
- Resistance to Change. Workers may resist AI adoption due to fears of job displacement or skepticism about the technology's reliability.
- Training Challenges. Comprehensive training programs are required to upskill employees, but these initiatives can be resource-intensive and time-consuming.

Solution: Organizations should implement continuous learning programs and emphasize the role of AI as a tool to augment, rather than replace, human capabilities.

Ethical and Regulatory Concerns

AI adoption in manufacturing raises important ethical and regulatory issues.

- Data Privacy Issues. Manufacturers must ensure that data collected for AI applications complies with regulations such as GDPR, HIPAA, or local data protection laws. Mismanagement of sensitive information could lead to legal repercussions and reputational damage.
- Ethical AI Usage. Deployment must align with ethical guidelines to prevent biased decision-making, ensure transparency, and maintain accountability.
- Regulatory Complexity. Navigating varying regulations across regions and industries can complicate AI implementation, especially for global manufacturers.

Solution: Establishing clear ethical policies and investing in explainable AI technologies can help organizations build trust and ensure compliance with regulations.

Organizational and Cultural Challenges

Organizational and cultural factors often impede AI adoption.

- Alignment of Stakeholders. Successful implementation requires buy-in from executives, engineers, and production staff. Misalignment among stakeholders can slow progress.
- Organizational Readiness. Many organizations lack the maturity needed to adopt AI effectively. Insufficient infrastructure, unclear workflows, and a lack of strategic vision hinder progress.
- Change Management. Transitioning to AI-driven operations necessitates cultural shifts and adjustments to established practices, which can be difficult to achieve.

Solution: Strong leadership, clear communication, and an iterative approach to adoption can help mitigate these challenges and drive cultural transformation.

Cybersecurity Risks

The reliance on interconnected devices and cloud platforms for AI systems introduces new cybersecurity risks.

- Vulnerability to Attacks. AI systems become prime targets for cyberattacks, which could disrupt operations or compromise sensitive data.
- Dependence on External Vendors. Manufacturers often rely on third-party providers for AI tools, creating additional vulnerabilities if security protocols are not stringent.

Solution: Incorporating AI-driven cybersecurity measures and adopting a zero-trust framework can enhance system resilience against threats.

FUTURE DIRECTIONS AND TRENDS

The future of machine learning (ML) and artificial intelligence (AI) in manufacturing is rich with transformative potential, marked by groundbreaking trends that will redefine how industries operate. These advancements aim to enhance efficiency, innovation, and sustainability while addressing the evolving demands of a global market. Below are key directions that will shape the integration of AI and ML in manufacturing in the coming years.

Autonomous Systems

One of the most promising advancements is the development of fully autonomous systems, often referred to as "lights-out" manufacturing. These factories leverage AI-powered robotics, predictive maintenance, and intelligent production planning to minimize human intervention. Autonomous systems can operate 24/7 with high precision, reducing operational costs while maximizing efficiency and output. For example, AI-driven autonomous assembly lines can adapt to varying production demands and self-correct errors in real time, ensuring consistent quality (Ivanov & Dolgui, 2020).

Edge AI for Real-Time Decision-Making

Edge AI is becoming a transformative trend, allowing for localized data processing at the point of data collection. By shifting AI computations from the cloud to the edge, manufacturers can make real-time decisions with minimal delay. This is especially beneficial in scenarios like predictive maintenance, where quick responses are essential to prevent expensive equipment breakdowns. Additionally, Edge AI improves data security by reducing the need to transmit data to centralized servers, making it a suitable choice for industries with strict data privacy regulations (Choi et al., 2023).

AI-Driven Innovation

Generative AI is poised to revolutionize product design and manufacturing by exploring novel concepts and techniques. Using advanced algorithms, generative AI can create optimized designs based on predefined parameters such as weight, material, and structural integrity. This not only accelerates the design process but also enables the discovery of innovative solutions that might not have been achievable through traditional methods. For example, aerospace manufacturers are already using generative AI to design lightweight components that improve fuel efficiency while maintaining structural strength (Bai et al., 2022).

Interoperability in Industry 4.0

The seamless integration of AI with other technologies such as IoT, blockchain, and augmented reality (AR) is a crucial future direction. Interoperability across these technologies will enhance collaboration and transparency throughout the manufacturing value chain. For instance, combining AI with IoT sensors enables real-time monitoring of production lines, while blockchain ensures secure and traceable supply chain transactions. This convergence will result in smarter, more connected factories capable of adapting to rapidly changing market conditions (Monostori et al., 2023). Table 2 presents key future trends in AI and ML for manufacturing, highlighting emerging innovations and their potential impact on the industry

Table 2. Key future trends in AI and ML for manufacturing

Trend	Description	Benefits	Example Applications
Autonomous Systems	Fully automated factories with minimal human intervention.	Increased efficiency, reduced labor costs, continuous operation.	Autonomous assembly lines, automated material handling.
Edge AI	AI processing performed locally at the edge of the network.	Real-time decision-making, low latency, enhanced data privacy.	Predictive maintenance, quality control at production sites.
AI-Driven Innovation	Generative AI creates novel product designs and optimizes manufacturing methods.	Faster innovation, lightweight and efficient designs, cost savings.	Aerospace components, 3D printing design optimization.
Interoperability in Industry 4.0	Integration of AI with IoT, blockchain, and other advanced technologies.	Improved collaboration, supply chain transparency, adaptive manufacturing.	IoT-enabled smart factories, blockchain for supply chain tracking.

Implications for the Industry

These trends highlight a shift toward smarter, more adaptive manufacturing ecosystems capable of addressing the complexities of a dynamic global market. Autonomous systems and edge AI will empower manufacturers to operate at unprecedented speeds and efficiencies. AI-driven innovation will facilitate rapid product development cycles, while interoperability will enable seamless collaboration across stakeholders, fostering resilience and adaptability.

CONCLUSION

Machine learning (ML) and artificial intelligence (AI) have become essential in driving the digital transformation within the manufacturing sector. These technologies enable predictive maintenance, improve quality control, streamline supply chains, and promote sustainable practices. They equip manufacturers with the necessary tools to stay agile, competitive, and responsive in an ever-changing and complex global environment.

Although there are challenges related to data integration, workforce readiness, and implementation costs, the swift advancements in AI and ML technologies, coupled with strategic planning and investments, are unlocking new possibilities for innovation and operational excellence. These developments are not only transforming traditional manufacturing processes but are also redefining the industry's approach to efficiency, adaptability, and sustainability.

As manufacturers adopt AI and ML technologies, the industry is set to reach new heights of productivity and innovation. By overcoming current obstacles through teamwork, employee training, and robust infrastructure, the manufacturing sector can fully exploit the transformative power of these technologies. This will enable manufacturers to successfully meet the challenges of Industry 4.0 and retain a competitive advantage in the global market.

REFERENCES

Antony, J., Jalušić, D., Bergweiler, S., Hajnal, Á., Žlabravec, V., Emődi, M., Strbad, D., Legler, T., & Marosi, A. C. (2024). Adapting to Changes: A Novel Framework for Continual Machine Learning in Industrial Applications. *Journal of Grid Computing*, *22*(4), 71. Advance online publication. DOI: 10.1007/s10723-024-09785-z

Bertolini, M., Mezzogori, D., Neroni, M., & Zammori, F. (2021). Machine Learning for industrial applications: A comprehensive literature review. *Expert Systems with Applications*, *175*, 114820. DOI: 10.1016/j.eswa.2021.114820

Büchi, G., Cugno, M., & Castagnoli, R. (2020). Smart factory performance and Industry 4.0. *Technological Forecasting and Social Change*, *150*, 119790. Advance online publication. DOI: 10.1016/j.techfore.2019.119790

Chatterjee, S., Chaudhuri, R., Vrontis, D., & Basile, G. (2022). Digital transformation and entrepreneurship process in SMEs of India: A moderating role of adoption of AI-CRM capability and strategic planning. *Journal of Strategy and Management*, *15*(3), 416–433. DOI: 10.1108/JSMA-02-2021-0049

Dogan, A., & Birant, D. (2021). Machine learning and data mining in manufacturing. *Expert Systems with Applications*, *166*, 114060. Advance online publication. DOI: 10.1016/j.eswa.2020.114060

Dogru, A. K., & Keskin, B. B. (2020). Artificial intelligence in operations management: Applications, challenges and opportunities. *Journal of Data, Information and Management. DOI, 10.*

Dubey, R., Gunasekaran, A., Childe, S. J., Bryde, D. J., Giannakis, M., Foropon, C., Roubaud, D., & Hazen, B. T. (2020). Big data analytics and artificial intelligence pathway to operational performance under the effects of entrepreneurial orientation and environmental dynamism: A study of manufacturing organisations. *International Journal of Production Economics*, *226*, 107599. Advance online publication. DOI: 10.1016/j.ijpe.2019.107599

Fatimah, Y. A., Govindan, K., Murniningsih, R., & Setiawan, A. (2020). Industry 4.0 based sustainable circular economy approach for smart waste management system to achieve sustainable development goals: A case study of Indonesia. *Journal of Cleaner Production*, *269*, 122263. DOI: 10.1016/j.jclepro.2020.122263

Hansen, E. B., & Bøgh, S. (2021). Artificial intelligence and internet of things in small and medium-sized enterprises: A survey. *Journal of Manufacturing Systems*, *58*, 362–372. DOI: 10.1016/j.jmsy.2020.08.009

Hussain, Z., Akhunzada, A., Iqbal, J., Bibi, I., & Gani, A. (2021). Secure IIoT-enabled industry 4.0. *Sustainability (Basel), 13*(22), 12384. Advance online publication. DOI: 10.3390/su132212384

Jamwal, A., Agrawal, R., Sharma, M., Kumar, A., Kumar, V., & Garza-Reyes, J. A. A. (2022). Machine learning applications for sustainable manufacturing: A bibliometric-based review for future research. *Journal of Enterprise Information Management, 35*(2), 566–596. DOI: 10.1108/JEIM-09-2020-0361

Javaid, M., Haleem, A., Singh, R. P., Suman, R., & Gonzalez, E. S. (2022). Understanding the adoption of Industry 4.0 technologies in improving environmental sustainability. *Sustainable Operations and Computers, 3*, 203–217. DOI: 10.1016/j.susoc.2022.01.008

Kegyes, T., Süle, Z., & Abonyi, J. (2021). The applicability of reinforcement learning methods in the development of industry 4.0 applications. *Complexity, 2021*(1), 7179374. DOI: 10.1155/2021/7179374

Makridis, C. A., & Han, J. H. (2021). Future of work and employee empowerment and satisfaction: Evidence from a decade of technological change. *Technological Forecasting and Social Change, 173*, 121162. DOI: 10.1016/j.techfore.2021.121162

Modgil, S., Singh, R. K., & Hannibal, C. (2022). Artificial intelligence for supply chain resilience: Learning from Covid-19. *International Journal of Logistics Management, 33*(4), 1246–1268. DOI: 10.1108/IJLM-02-2021-0094

Nacchia, M., Fruggiero, F., Lambiase, A., & Bruton, K. (2021). A systematic mapping of the advancing use of machine learning techniques for predictive maintenance in the manufacturing sector. *Applied Sciences (Basel, Switzerland), 11*(6), 2546. Advance online publication. DOI: 10.3390/app11062546

Nuhu, A. A., Zeeshan, Q., Safaei, B., & Shahzad, M. A. (2023). Machine learning-based techniques for fault diagnosis in the semiconductor manufacturing process: A comparative study. *The Journal of Supercomputing, 79*(2), 2031–2081. DOI: 10.1007/s11227-022-04730-x

Ortiz-Ruiz, E., Bermejo, J. R., Sicilia, J. A., & Bermejo, J. (2024). Machine Learning Techniques for Cyberattack Prevention in IoT Systems: A Comparative Perspective of Cybersecurity and Cyberdefense in Colombia. *Electronics (Basel), 13*(5), 824. Advance online publication. DOI: 10.3390/electronics13050824

Patalas-Maliszewska, J., & Łosyk, H. (2022). An approach to maintenance sustainability level assessment integrated with Industry 4.0 technologies using Fuzzy-TOPSIS: A real case study. *Advances in Production Engineering & Management, 17*(4), 455–468. DOI: 10.14743/apem2022.4.448

Penumuru, D. P., Muthuswamy, S., & Karumbu, P. (2020). Identification and classification of materials using machine vision and machine learning in the context of industry 4.0. *Journal of Intelligent Manufacturing, 31*(5), 1229–1241. DOI: 10.1007/s10845-019-01508-6

Rai, R., Tiwari, M. K., Ivanov, D., & Dolgui, A. (2021). Machine learning in manufacturing and industry 4.0 applications. *International Journal of Production Research*. Taylor & Francis.

Sharma, R., & Villányi, B. (2022). Evaluation of corporate requirements for smart manufacturing systems using predictive analytics. *Internet of Things : Engineering Cyber Physical Human Systems, 19*, 100554. DOI: 10.1016/j.iot.2022.100554

Usuga Cadavid, J. P., Lamouri, S., Grabot, B., Pellerin, R., & Fortin, A. (2020). Machine learning applied in production planning and control: A state-of-the-art in the era of industry 4.0. *Journal of Intelligent Manufacturing, 31*(6), 1531–1558. DOI: 10.1007/s10845-019-01531-7

Zonta, T., da Costa, C. A., da Rosa Righi, R., de Lima, M. J., da Trindade, E. S., & Li, G. P. (2020). Predictive maintenance in the Industry 4.0: A systematic literature review. *Computers & Industrial Engineering, 150*, 106889. Advance online publication. DOI: 10.1016/j.cie.2020.106889

Chapter 12

Augmented Reality as an Efficient Tool During the Product Design Process in the Manufacturing Industry

Gerardo Reyes Ruiz
https://orcid.org/0000-0003-0212-2952
Center for Advanced Naval Studies, Mexico

ABSTRACT

Manufacturing processes are divided into stages: the initial phase is the design of the product, then continues with the cutting of the fabric and concludes, roughly speaking, with the assembly of the previously elaborated pieces. All these stages are carried out to achieve the desired product, however, the assembly stage could be considered as the most important in the development or manufacture of a garment. To make manufacturing processes more efficient, it is important to make use of new technologies, such as augmented reality, which show the physical reality of the product virtually. In this work, interviews and observation of maquiladora companies in the apparel industry were used and, derived from the results obtained, it is possible to propose a system based on augmented reality that shows a different way of designing patterns (which will later be called prototypes), as well as highlighting the possible benefits that could be obtained with the use of this technology in some manufacturing processes.

DOI: 10.4018/979-8-3373-1082-4.ch012

INTRODUCTION

Manufacturing areas are key elements in the manufacture of products of any kind, which are of fundamental importance and require, in turn, new methods to help optimize all their processes (Paszkiewicz et al., 2020); consequently, cutting-edge methods are emerging, which help to complement and improve production processes, such is the case of additive manufacturing or also called 3D printing (Jandyal et al., 2022). Undoubtedly, 3D printing is revolutionizing the manufacturing industry, due, among other things, to the fact that 3D printing serves to create large-scale parts (Thang and Kim, 2025; Roschli et al., 2019). Other developments that include smart technology of various types and come together as important parts of the production process are computer-based models, manufacturing systems and sensor technologies, virtual/augmented reality (Sakr and Abdullah, 2024), the Internet of Things (IoT) (Hassan et al., 2020; Ahmad and Zulkifli, 2022), and technology trends that correspond to global patterns, which reveal a wide range of sectoral applications (Turovets et al., 2019). These tools must be contextualized in research that focuses them towards points of help to build intelligent and avant-garde factories, by the technological era in which we live and, above all, capable of improving and facilitating all processes related to manufacturing so that, as a result, useful products for people are more accessible and of better quality (Sun, Shahzad, Ali and Razzaq, 2025).

The production phase begins with the design of a product, which will serve as a prototype and must meet certain parameters related to measurements, materials, and specifications. The pattern of a product is geometrically elaborated following the indications of the model created by a designer, who is usually a person fully qualified for this profession. Once the product has been designed, the model (prototype or sample) is translated into assembly terms by the manufacturing area personnel, who give their construction interpretation, especially in the elementary details of the garment; this interpretation may sometimes be erroneous or different from the original proposal. If this interpretation is correct, then a quality product will be obtained, otherwise, a product with defects may be obtained, and consequently, it may not be sold or its value may be diminished. These problems show the need to introduce state-of-the-art technological tools that help optimize manufacturing processes through the visualization of patterns, which can be coupled/adapted to each person who must interact with the model. Undoubtedly, on this type of technology adjacent to the factories, research should be conducted that is oriented towards the advancement of manufacturing processes in a digital world, which is present in everyday life and is essential for any type of activity.

A tool that lately converges towards the requirements of production processes is Augmented Reality (AR) (Schumann et al., 2021), which integrates multiple technological tools that are oriented to different construction purposes, among which is the design of objects (Bellalouna, 2021). The elements that make up AR (, such as videos, 3D images, and animations, open a new world of opportunities with a full perspective of the supply chain (Plakas et al., 2020); these technologies not only allow for increasing the efficiency and flexibility of manufacturing and distribution processes but also allow modifying the existing relationship between the different scenarios involved in the supply chain, above all, with a special emphasis towards the benefit of consumers (Ho et al., 2022; de Souza Cardoso, Martins Queiroz Mariano and Zorzal, 2020).

The new technology called AR goes beyond video games, entertainment, and technology, which is why its benefits in the field of production should be highlighted mainly in places where technologies are not used to force users to perform new activities but to persuade and motivate them to perform their same daily activities with other alternatives but with better work experiences (Siriborvornratanakul, 2018; Parekh et al., 2020). With this premise, probably, we are probably already empowered to use these techniques and new tools so that, in turn, we can provide different benefits than what people are used to. In other words, new technologies - such as AR-, have facilitated new construction and production techniques, which have been applied in different ways and in different contexts (Reljić et al., 2021), which depend on the type of manufacturing that is desired, but have undoubtedly led to and facilitated new production systems based on AR (Egger and Masood, 2020).

In this scenario, AR serves to prioritize the design of a product (in our context it would be a garment), not only as a "sample" of what an object could be or how that "sample" product would be visualized at a given time. That is, through RA, a "product" can be observed as a useful way to more deeply represent objects to be mass-produced and, in this way, understand its implications in other areas, such as, for example, the maintenance of a production area (Shyr, Tsai, Lin and Liau, 2022). To exemplify this case, Khalek et al (2019), investigated the design of a building where the decisions taken at the beginning of that phase -the maintenance of the building- can have an important impact, measured by different visualization means, which allow people without previous experience to identify maintenance problems in a design model, i.e., this maintenance could be achieved, in a more efficient way, with the implementation of a system with AR. Continuing with this theme, we can mention the existence of tutorials that have helped to complete the construction processes where the technology used does not only use 2D images (Qiu, Yu, Zhao et al., 2024; Chen, Li, Tian and Zhang, 2023), that is, there are also tutorials based on more innovative technologies; such is the case, of a software that helps the construction of wooden structure frames, which uses a software based on AR with a real scale visualization and audio (Arena, Collotta, Pau and Termine, 2022). With this technology, builders can be helped to follow directions and measurements more accurately and, consequently, facilitate the adaptation of those frames in wooden houses (for this case AR works because the frames are of standard type), generally oriented to the construction market of the United States of America (Xu, Feng, Jelodar and Guo, 2024; Adebowale and Agumba, 2024).

Another industry that is engaged in the development and construction of machines for the manufacture of high-precision tools and tooling equipment for the machining of parts in industrial enterprises is the company Emuge-Franken (https://www.emuge-franken-group.com/mx/mx). This German company has virtual technical manuals for tool-related tasks in thread cutting and milling technology, which form the basis of the business. Also, this German company has adopted the animated models of the HF20 threaded shank chuck assembly in some of its production processes, which is based on AR. The experience of this company shows that this implementation could be necessary and helpful for companies in the Russian market, mainly because this process deals with metal processing by drilling and thread generation (Xiong, Hsiang, He et al., 2021; Gren et al., 2018). As a consequence of these dynamics, these technologies have been considered one of the key tools for companies to improve their competitiveness and develop their relationships with the interrelated nodes in their supply chain, both upstream and downstream. Therefore, the application of these technologies for supply chain management has become a fruitful area of research due to, among other results, their clear applications in manufacturing and their strong implications in the management of their processes (Liang, Shiu and Liu, 2025).

Thus, AR has already been valued for inclusion in future factories to help workers multitask more efficiently (Molina Vargas, Vijayan and Mork, 2020). This dynamic is intended to facilitate the possible shift from mass production towards personalized production, however, it is not clear how these promises could be fulfilled in an industrial scenario, since its specific conceptualization, oriented to the development of AR-based tutorials to produce objects -wearables-, has been visualized and detected a higher performance in tasks with a higher degree of difficulty (Voinea, Gîrbacia, Duguleană, Boboc and Gheorghe, 2023). This background, undoubtedly, is the basis of future smart factories, which are waiting to get the best techniques that help manufacturing processes (Adel, 2022). It is thus that this research work is built on envisioning the formation of those manufacturing companies focused on processes and not only in the forms of design but rather, in the discernment of the abstract to subsequently, superimpose it with AR (Santi, Ceruti, Liverani and Osti, 2021).

This research focuses on the area of the garment industry, where the design must have an adaptation that leads to the quality of the garment, which is achieved by following exactly the geometric characteristics of a prototype that functions as a guide for the transformation of the raw material into garments, which has been created before starting the manufacture of the garment (Pereira, Alves and Arezes, 2023). In this context, the workshops, generally, are the spaces where the garments are manufactured, so that, for each workshop, a physical sample is destined and the design or preparation must be memorized by the people who will be in charge of its "assembly" (Roberto, Newnes and Shokrani, 2024). This "assembly" of the garment includes both the hidden details as well as those that are not hidden but are necessary for the garment to be visualized and function as planned from the original model. A difficulty in this "assembly" stage is that, on many occasions, these ergonometric or functional details are difficult to memorize or to specify, and, not having the garment in sight that serves as the "design pattern", the assembly can be carried out erroneously. As a consequence of this dynamic, the manufacturing company loses material resources, assembly time, and, of course, work performed. In addition, when errors are corrected in the garments already produced, then delivery times are delayed and, consequently, the allocation and obtaining of profits, which in turn implies a significant economic deficit for the company.

It is at this precise moment in the process when AR takes on great importance because, through this new technology, excellent results have been obtained for factories and, specifically, in some manufacturing sectors (Morales Méndez and del Cerro Velázquez, 2024). For example, through machinery repair tutorials, to make inventories that require space optimization, and through instructions to build tools or other types of objects, to mention just a few applications of AR. Therefore, this proposal allows for to complement of the manufacturing processes of some garments by allowing from assembly to quality control (Devagiri et al., 2022). Undoubtedly, the improvement of all these manufacturing processes will facilitate the obtaining and elaboration of better products that converge quickly for the optimization of the cost-benefit equation. Thus, this chapter considers the following:

GENERAL OBJECTIVE

Improve the process associated with garment manufacturing through technological tools that help optimize delivery times, resources used, and garment quality.

PARTICULAR OBJECTIVES

- Realize an AR-based system with resources designed to assist workers when assembling products.
- Implement the AR system in a garment manufacturing workshop.

- Analyze the performance of the AR system to generate recommendations.

METHODOLOGY

Problem

For the development of this work, the applied methodology was used, whose main characteristic is precisely that of immediately applying the knowledge resulting from the confrontation of theory with real problems. This dynamic was carried out to acquire the concrete and real solution to those problems and direct those adjacent results toward the improvement of the activities of the individuals involved in the manufacturing process, and considered in this research (Aguinis, 2024). In this way, the work method was divided into 3 faces:

A. Study of the problem.
B. System based on AR.
C. Results.

Study of the Problem

Manufacturing processes are developed in stages, which at first glance may seem easy and autonomous (Kubik et al., 2025). However, although manufacturing can be considered a procedure where people assemble the parts without any complication, there are also several levels of difficulty associated with the assembly where it is essential to take into account several aspects: the union of parts that are different, the details that are hidden and the combination of different colors. The problem arises when the assembly is not done properly, which can result in production errors and, consequently, in losses, which can be considered in time, money, and, particularly, in workers' effort. The phases of the manufacturing process of a garment can be listed consecutively, as shown below:

- At the parent factory.

    1) The garment is designed,
    2) Cut the pieces indicated in the design,
    3) The pieces are labeled and packaged,
    4) They are sent to the maquiladoras together with a garment that will serve as a pattern.

- In the maquiladora.

    1) Parts are unpacked,
    2) The parts are distributed to the workers,
    3) The workers assemble them in a "chain" (sequentially),
    4) When the assembly is finished, they are transferred to the sections of:

- Finished,
- Ironing and,
- Quality control and packaging.

Of the phases mentioned above, the one that concerns this research is the assembly phase which is described in the following actions:

- When the packages of cut fabric arrive at the maquiladora, the person who delivers them indicates and shows the garment called "sample" or "pattern" to the assemblers, who then observe it in detail and save as much information as they can in their memory to then proceed to elaborate or start the assembly process.
- The parts of a garment are made in series, that is, there is a team of people who do certain work in detail, such as the series of assemblies of collars, sleeves, backs, "fronts", etc. Subsequently, these pieces are transferred to the people who do the finishing, which includes the placement of buttons, buttonholes, zippers, and/or snaps and even the shredding. Finally, these almost-finished pieces are transferred to the ironing area to conclude in the packaging section.
- Quality control is carried out at the packaging stage. At this stage, where the garments are folded to be packaged, the people in charge of folding them check or notice any defect in the assembly. Some errors may occur in several garments, since, as mentioned above, they are made in series and, therefore, the error may also be repeated sequentially.
- As the quality of the garment is checked during its packaging or, even worse, when the garments are delivered to the factory, it is at that precise moment when the personnel becomes aware of the errors in that batch of garments. Consequently, the defect in the garments is corrected and it is essential to correct the assembly before the final garments are accepted by the parent factory.
- In the assembly industry, there are processes where workers have to deal with the various parts of the product, i.e., they must analyze complex components on multiple occasions, which have been captured in the "patterns" of the garment (original designs) and rarely consulted by workers dedicated to the

assembly of these parts. Of course, and as it is logical, on many occasions the details of the garments are stored in their short-term memory, which can cause these workers to forget important details at the time of assembly.

The difficulties frequently encountered in assembly can be of various types, although everything depends on the design of the garment. The following are some examples:

- Side seams are not aligned correctly.
- Stitching is the wrong size.
- Some flaps are not done correctly.
- Buttons are placed on the wrong side.
- The color scheme is reversed.
- Pouches are not in the correct place.
- The number of buttons is not correct.
- The hems do not correspond to what is required.
- Collars can be simple and make them with "collar foot" or vice versa.
- Bags are not aligned.
- There may be color confusion.
- Embroidery is in the wrong place.

As mentioned above, there are many errors in the assembly stage. Of course, some of these errors will depend on the garment itself. However, the most common problem is the assembly error in several samples, because as mentioned above, as the components are assembled in series, then an error detected in one piece can result in a defect for hundreds of garments. The types of manufacturing errors result in considerable monetary losses, i.e., when garments are unpicked to correct the assembly, other types of errors may arise, some of them considered extreme. For example, when working with a type of fabric that is fragile, then the garment can be damaged. In addition, lead times are extended, which can result in delayed payments, credit delays, production losses, and, of course, reduced profits.

Based on the problem posed, this paper poses the following question:

What factors affect the improvement of a manufacturing system for a factory employing AR?

As a consequence of the previous question, the following hypothesis Ho, is constructed in this work:

AR is a technological tool capable of improving the process associated with manufacturing in a garment factory.

The Importance and Significance of AR

The technology known as AR supports the development of applications that take into account the diversity of clients visiting stores; these clients will have heterogeneous and changing characteristics in terms of age, schooling, occupation, and interests, among others (Lin and Huang, 2024). Many companies seek to provide an easy, friendly, and adequate sales service to their multiple clients, who will consequently refer the business to their contacts and justify their preference. It is also clear that the world has been invaded by mobile devices, a fundamental reason to add new marketing elements to products offered in stores (Hoffmann and Mai, 2022). For that reason, innovative and avant-garde technological resources must be created so that clients can appreciate and use them (Tan, Chandukala and Reddy, 2022; Mekonnen, 2024). The hardware and software of these technological resources must be widely accepted and used on an everyday basis, such as a mobile device or smartphone (Luceri, Bijmolt, Bellini and Aiolfi, 2022; Alves and Machado, 2024).

For their part, information systems use accurately compiled codes to support people in making their decisions conveniently, easily, and quickly. These decisions refer to different areas, from knowledge acquisition, medicine, and tourism to applications focused on the sale and purchase of products and the offering and presentation of products to consumers, that is, regular marketing practices (Ramos, Rita and Vong, 2024). To create innovative marketing actions, it is necessary to increase the avant-garde technological elements used in sales systems (Alt, 2020; Agnihotri, Chaker, Dugan, Galvan and Nowlin, 2023). Therefore, it is necessary to build inventive information systems; as Wallace (2021) argues, our capacity to choose will be improved by optimizing information resources, an essential part of the invention and improvement of information and communication technologies. These types of systems accelerated their evolution when they were combined with features such as mobility, big data for business applications, and cloud computing (Laudon and Laudon, 2020; Gupta and Goyal, 2020).

As a consequence, new technologies must use the internet and mobile devices to meet the foremost requirements of users, for example, potential customers. Al-Ansi et al (2023) have defined AR as a technology that uses the visualization of a real environment and adds computer-generated visual elements to a device. This definition of AR entails that the customer is not guided through a fantasy but presented with virtual objects superposed on their physical reality (Riar et al., 2023). Despite the availability of technological resources using RA, the use of these tools in the field of marketing is still underexploited (Massa and Ladhari, 2023). For that purpose, these new systems are proposed as a tool to help sellers display a product to their potential customers in the way that they intend; the essentials of the food product must be established to determine what to show to the customer, and the product is then presented using multimedia elements (Schifferstein, Lemke and de Boer, 2022). The AR-based design is shown below.

Design of an Information System based on AR

In the design phase, the behavior of the population under study is considered, and a solution is proposed; the latter must consider improvements to the two pre-established systems in this document. These improvements should address aspects such as speed of response, economy of resources, efficiency, effectiveness, and visually attractive design. The design of these systems based on RA must follow certain criteria, but in general, all of them must have similar benefits, such as the following:

a. Their design should be visually attractive.
b. They must contain multimedia elements so that users can understand and retain the food product's features and facilitate their purchase decisions.
c. The information provided must be easy to process.

The full development and reliable operation of these tools require many basic technologies, but for simplicity, and according to Pressman and Maxim (2020), they could be developed in the following phases:

1. Analysis of the requirements of the RA-based system.
2. Formulate the abstract and physical design so that software coding and creation can in time take place.
3. Preparation of files, identification, and creation of test data.
4. Finally, software testing and integration.

Figure 1. AR-based system

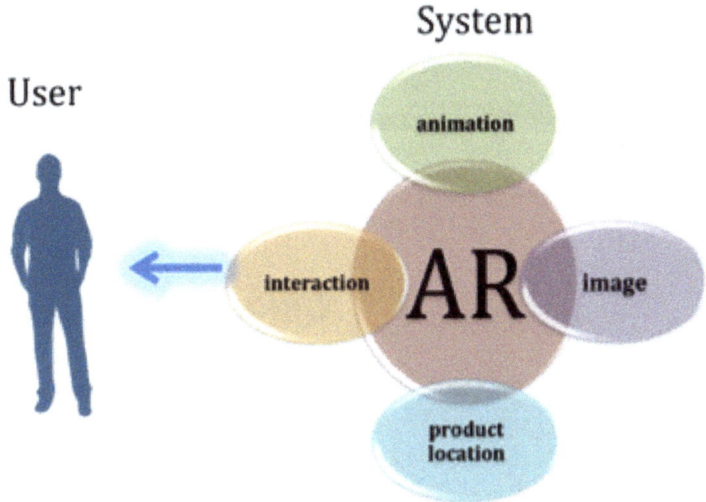

(Author's own elaboration)

Additionally, the necessary elements involved in both systems are:

Hardware: 1) Camera. Most mobile devices have a built-in camera that can be used to trigger the AR software. 2) Screen. It showcases the image of the product, generally in 3D, using the device with which the AR-based system interacts (Li et al., 2025). Thus, the screen superimposes the AR elements using 3D images and presents a virtual tour; this environment is returned as a query result. The screen of a mobile device is normally used for this purpose. 3) Audio. This accessory is also included in mobile devices and computers, which are used to complement the AR with sound (in most cases using a Podcast).

Bookmarks: The data associated with these 2D images, also called reference images, is easily extracted due to their predefined shape and color features (Ghasemi et al., 2022). The AR is triggered by these elements when the mobile device detects them via its camera.

3D images: These elements have a realistic appearance, and their X, Y, and Z coordinates produce models that overlap the AR to simulate physical reality.

Virtual presentation: This specialized software application creates image-based presentations (usually panoramic) for the user to view the garment through a virtual environment by simply pressing the arrow keys (Krajčovič, Matys, Gabajová and Komačka, 2024).

Software: The software used to create the AR should be aimed at solving these types of problems. This is because AR-based applications work by association; in the two systems dealt with in this article, the AR will be associated with the bookmark to show the 3D images and reproduce the simultaneous audio recording (Feist, Jacques de Sousa, Sanhudo and Poças Martins, 2024).

Figure 2. AR-based system

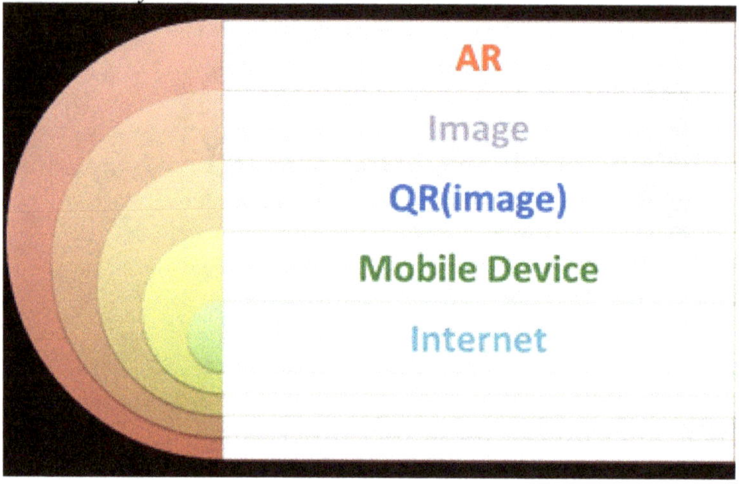

(Author's own elaboration)

RESULTS AND DISCUSSION

The proposed solution is the realization of an AR-based system, which would be sent to the maquiladoras along with the manufacturing materials, and these, in turn, can be distributed among their production team. With this system, the assembly team members could consult it whenever necessary, using their own mobile device, to visualize the model superimposed on the fabric of the part to be built, including instructions in audio or video format. As it is already well known, AR is an emerging technology that is defined as digital take or computer-generated information, either with images, audio, video, or tactile sensations and the superimposition of them in real-time, which can be the mixture between real and the synthetic world (Vilar, Filgueiras and Rebelo, 2022). AR is already particularly suitable for the maintenance area in the industry, being easily implemented in processes that can improve the view of users in different scenarios and that also include visual animations, sounds, and written or static instructions (Aquino, Rapaccini, Adrodegari and Pezzotta, 2023).

The production line of this technology is on the rise, since several applications have been found, among which the following stand out: military use, medicine, design, engineering, robotics, manufacturing applications, maintenance and repair, teaching and learning, entertainment, psychological treatments, just to mention a few (Cannavacciuolo et al., 2023). This research is proposed as a support to help garment manufacturing be done more simply, so that anyone who has this technology in their hands will be able to perform their activities more efficiently and reliably. Virtual manufacturing is defined as an integrated synthetic environment, which is exercised to improve all levels of decision-making and control. Its classification can be focused on design, production, and control (Eswaran and Raju Bahubalendruni, 2023). Thus, based on this type of definition, digital elements that combine physical and virtual reality are used, making use of the wide and varied design tools available for their construction. The potential of the AR-based manufacturing context supports faster product and process designs as well as control and maintenance activities (Morales Méndez and del Cerro Velázquez, 2024). Under this scheme, AR has good construction tools, although its design should be done simply to avoid the user feeling confused rather than helped. Thus, taking into account these principles for the realization of the proposed AR system, it was summarized in the following aspects:

The designer of the garment, in addition to making it in 2D image patterns, with measurements and assembly instructions, must also draw it in 3D images to facilitate the visual display of the details of the garment virtually (Song, Koeck and Luo, 2021). These images will also gather very strict details of the model such as measurements, colors, number of components, and instructions issued by audio or text, because by fulfilling all these steps the company will meet the primary objective of each company, which is, among other things, to manufacture products that meet the needs of customers (Carthy, Cormican and Sampaio, 2021). Of course, use is also made of videos that clearly explain and show the procedure, making use of images, audio, and animation, which means more detailed teaching of the assembly process of each piece of the product, specifically for those garments that are more complex. For the explanation of the AR system, an example taken from a maquiladora will be taken into consideration, using a specific garment, shown in Figure 3. Although the system can work with all the garments that are manufactured, the following phases are considered for the elaboration of the AR-based manufacturing system:

1. The garment is modeled and its general details are added, which are necessary for the AR deployment. These characteristics give specificity to the garment and are denoted as: name, color, sizes, and general conditions, as shown below:

   Name: Shirt-Bona.
   Color: orange.
   Sizes: 28-42.
   Details: the combination of colors and decorative ribbons.

Figure 3. Prototype (sample) of a Bona-type shirt

(Author's own elaboration)

2. The construction of a garment is made up of several pieces, which have been designed, cut, and integrated into packages for the construction of several garments. Each package of parts contains all the necessary elements, for example, a Bona-type shirt consists of sleeves, collar, back, and front; each package must match its size and prototype. In the example in Figure 4, you can see the parts of a sleeve for this type of garment.

Figure 4. Assembly parts for a sleeve

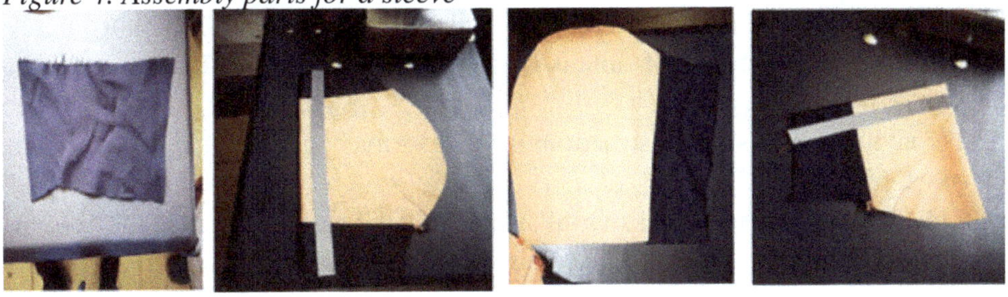

(Author's own elaboration)

The back of the shirt is shown in the example in Figure 5.

Figure 5. Assembly of the back of a Bona-type liner

(Author's own elaboration)

3. For the development of the AR-based system, the following components must be in place:
    - 3D images,
    - Markers or images that will function as AR deployment markers,
    - Videos,
    - Audio,
    - Cloud storage,
    - AR management application.

In this context, the 3D images will be used to model the garments in detail and are formed from 2D images (Chen, Li, Tian and Zhang, 2023), which can be photos or drawings.

The markers are images that will serve to "trigger" the AR when the camera of the mobile device is focused on them.

The videos can be videos of the interpretation of the most complex processes, filmed separately, and trying to keep them short. This is to resolve the doubt quickly and try that the investment of time is translated to the benefit of improvement in the work.

The audios are files with instructions in voice format that can guide users through the assembly process.

Cloud storage is used to make available the digital material used in the system. This system can be implicit in the existing AR managers and can be freely accessible. On the other hand, if there are few components of the AR used, then there is the possibility of hiring a paid manager, of course, depending on the convenience of the company.

The digital component manager application, which is used to design, build, and implement the AR, contemplates all the elements in the aforementioned formats, which will be used to build the AR-based manufacturing system. Currently, several AR managers have similar functions but differ in their development environments, "IDE" and also have different features that translate into advantages for the AR creator user. Some of these managers are detailed below:

Aurasma. It is an AR editor with which you can display 2D images, videos, and animations created in Adobe Flash Player flash video. (https://www.auganix.org/hud/aurasma/).

Vuforia (https://developer.vuforia.com/home). Software that works with Unity (https://unity.com/), and is quite effective for AR development. It is a program with excellent capabilities without any doubt, but it is important to mention that its degree of difficulty of use is higher compared to its analogs. This is because it requires a series of long and complex steps to achieve a product that, with other software, would be done in a short time and easily.

ArToolKit (http://www.artoolkitx.org/) is a freely available program whose features are orientation tracking and pattern recognition with markers to create ARs, which require difficult programming skills, although it is important to mention that there are variants that make this process easier, such as ARTag, ATOMIC Authoring Tool, FRARTooKit, NyARToolKit, and ARDesktop.

Layar (https://www.blippar.com/) is an AR content manager that contains digital multimedia elements, which include images, text, videos, audio, social networks, and buttons. It is worth mentioning that in addition to the aforementioned content, any type of AR software that Layar contains also serves to project 3D images, and that its content is classified by categories to help the user realize digital AR themes, which can be adapted to the purpose of their requirements.

Aumentaty (http://www.aumentaty.com/) is a software that was designed to create AR content. it is free to access and easy to use. This software has the functionality to handle digital elements in the form of video, 2D and 3D images, and points of interest with GPS; its complement is the App Scoppe (https://scopeapp.io/), with which AR can be observed on mobile devices. In addition, there is a collaborative community for users to use their projects.

Thus, for the design and implementation of this AR-based manufacturing system, the Aumentaty program was used. This software has a content manager called Creator, which is used to easily build the projects (https://www.ibm.com/topics/software-development) which can be shared in the cloud and then downloaded to any device for use. With its Scope add-on, the user can visualize the AR created.

The AR resources are shown in the form of 3D images and video, demonstrating how to do the assembly step by step, and whose objective is to give the construction instructions to people, so that the safety of doing the job most correctly is implicitly adhered to. The videos should be short since they are only part of the help offered with the AR: the duration for each video is suggested as a maximum of 2 minutes. The implementation of the time to review the AR is a resource that becomes a benefit to correctly assemble many pieces.

The markers are the words that compose the name or number of the specific garment to concatenate the name with its respective image. The people who will use the system will be the employees working in the garment manufacturing companies; the system is easy to use; it contains virtual elements that integrate with the physical reality, providing the user an easy and safe interaction, in addition, it is not necessary to have a previous knowledge of this technology; that happens because the movements produced are natural, easy to learn and could be as simple as moving a body part or an object (Cordeiro, Rui and Nuno, 2015). So it will not be necessary to have a training course as such, but rather, a series of short instructions to handle the system, which is built to be intuitive.

In the maquiladoras, the construction elements will be received, including the elements of the AR-based system. The workers will assemble and make use of this system when necessary, i.e., when they do not remember the process, do not know it in detail, or need to be sure of some details. When the user needs to consult how to make a model, then it will be enough to place the garment on a table, open the application on the mobile device, and focus the camera on it. As a result, the system will recognize the section of the garment and will superimpose the associated AR, so that the worker's doubts about the assembly process can be resolved instantly. Figure 6 shows how a sleeve is assembled. This type of AR was designed to show the basic shapes of a product, such as the combination of colors, measurements, sizes, placement of overlapping ornaments, size of stitching, thread color, etc.

Figure 6. The system with AR to analyze a sleeve

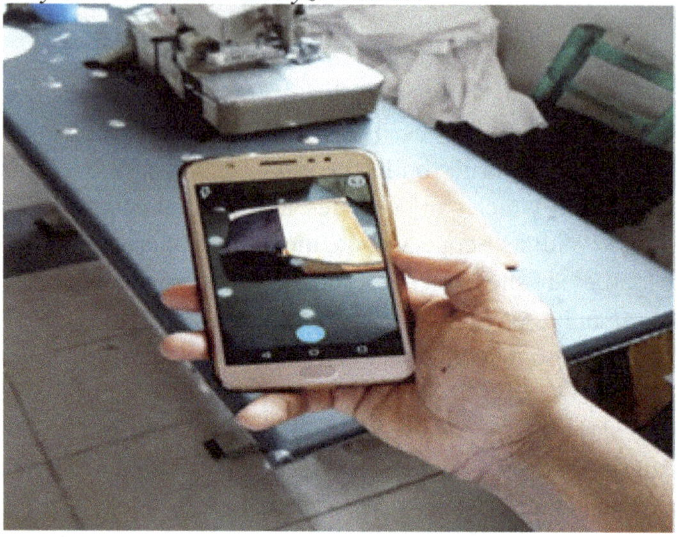

(Author's own elaboration)

Figure 7 shows the implementation of the system through the digital resource in video format; in this case, the user focuses the camera of the mobile device on the collar of the garment, which serves as a "trigger" of the AR by superimposing it as an aid, in the form of an explanatory video, which shows through images and audio the process of assembling a collar on a shirt. Of course, this form of AR will depend on the complexity of the process.

Figure 7. A system with AR to show the assembly process through a video

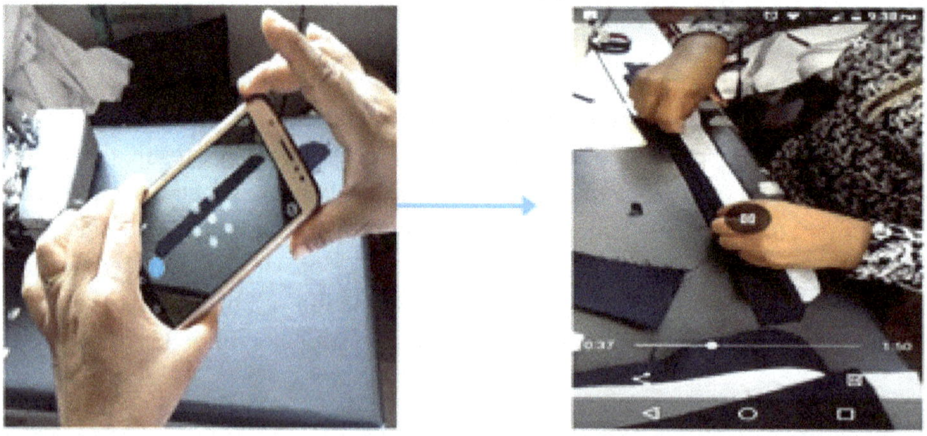

(Author's own elaboration)

Employing a code that should be the name or number assigned by the factory to the design and which will be the "trigger" of the AR, the prototype of the shirt will be displayed for the users to have it and consult it every time they need it. This procedure can be seen in Figure 8. In this procedure, the list has a series of names of the garments that are made in that workshop, so when the camera of the mobile device focuses on the name "Bona type shirt", then the AR superimposes the image of the shirt, where the complete design is shown.

Figure 8. The system with AR is associated with a marker to display a shirt

(Author's own elaboration)

The AR system was applied to workers of a microenterprise dedicated to sewing, that is, this company is dedicated to the manufacture of uniforms for various commercial companies. This AR system was used for one week, and at the end, an interview was conducted (see Annex). This interview was conducted by means of a previously designed questionnaire, which consisted of 15 questions aimed at obtaining the users' perspective, i.e., both for the workers and for the company's managers. The main results of these interviews are presented below:

1. Regarding the efficiency of the system, the users agreed that it is efficient because it helped them to perform their work better, since when the user has a doubt about some detail in the assembly, he then proceeds to place his cell phone on the garment and focus the camera so that the AR system shows the 3D image that includes the specifications and, in this way, he does not waste too much time asking the people in charge of supervising the garments, who in the worst cases are far away because the employee works in a branch of the workshop or even from home.
2. Regarding access to the AR material, the interviewees responded that they have stored the application on their phone and that every time they are given new designs, they are also trained and taught how to download them. This seemed very good to them because they said that if they had to store them in the memory of their devices (photos and videos), they would get confused with the models, and their search would be broader. This dynamic, in the short term, would become complicated and tedious for them. Instead, with AR, they can associate it with the garment in real-time.
3. Regarding the operation of the AR, as a support tool to increase the quality of the effective way of sewing, the results showed that with this new technology, it is possible to modify the seams in a shorter time in case there is any error. In the best of cases, this technological tool can be used to consult and verify the form of the assembly, even before starting it. In addition, if a piece has been damaged, this technology offers the possibility of not damaging the garments, which results in a correct assembly and, consequently, allows the production of higher-quality garments.
4. Regarding the AR materials, the videos showed the workers the process of constructing a garment piece, including folds, seam adjustments, stitch size, thread color, etc., providing them with an excellent understanding of what the garment designer requires; interviewees felt that they consulted it whenever they needed it and found the images attractive and easy to watch.
5. Regarding the use of videos, interviewees noted the advantage of viewing them with AR, since they felt that if they had to search for them on their devices without using AR, they would have to locate them in the gallery where they have

their personal videos and photos. This dynamic would imply a more extensive search and, consequently, loss of time, which would generate, in turn, a search conflict.

6. Regarding the ease of use of the system, the users agreed that it is not very difficult to use; rather, it seemed to them to be fast and attractive.
7. About the compatibility of the system, as most of the workers have a smartphone and all have with Android system, none of them had problems downloading or using the application. In this context, it should be clarified that the system also works with iOS.
8. Regarding motivation, users were encouraged, happy, and excited with the application. In this sense, they felt that with it, they consider themselves self-sufficient concerning the assembly of the garments because although they were trained in cutting and sewing, it is always necessary that their doubts are clarified, and this system does it instantly and correctly.
9. Regarding cost, the interviewees said they had no problems, since the tool is free and freely available, which is convenient for both the owners of the maquiladora companies and their employees.
10. The time investment in using the application is not expensive, since it only takes a maximum of 3 minutes if it is a detailed process, and one minute if it is a process shown with the AR through image format.
11. In terms of learning, AR is a way to teach and train users, since the interviewees agreed that they learned processes that they did not know in detail

Therefore, the hypothesis put forward in this work is confirmed: *AR is a technological tool capable of improving the process associated with manufacturing in a garment factory.*

And to answer the question: *What factors affect the improvement of a manufacturing system in a factory using augmented reality?* The following arguments were obtained:

- Easy and fast consultations. For the worker, when using the AR, he/she consult his/her manufacturing doubts in time and form.
- Time. When the employee uses the AR, he/she does not have to ask the supervisor about the question he/she has, which avoids wasting time in the transfer or waiting for the supervisor to attend to him/her.
- Optimization of material resources. When errors are minimal or nonexistent, there is no waste of material.
- Motivation. When a worker feels confident to do things or to have an element of help at hand, he/she does the manufacturing process with more encouragement and, consequently, have higher productivity.

- Digital resources. Images demonstrate how the garment or its parts are displayed, giving the user a comparison of what they are doing versus what they should be doing.
- The use of videos. These media show all the details of a procedure, obtaining abstract details, which give visual, auditory, and kinesthetic teaching to the users.

Evaluation of the System Based on AR

The system, which was developed using new AR technology, was validated through a series of questions asked of several people in several manufacturing centers. People were randomly selected until they made up the statistically significant sample size (384 people): the total sample size was calculated with 95% confidence, 5% error, and 50% homogeneity (Ros and Guillaume, 2020)., calculated using simple random sampling and without the knowledge of the total population (Wu and Thompson, 2020; Berndt, 2020). In this test, a series of questions was asked to evaluate the presentation and the search for a product through AR. This evaluation contains questions with dichotomous answers (the questions were implemented with JavaScript events performed on the SDK platform), which served as input so that, through the SDK platform, the score obtained for each section or section of questions is displayed.

After randomly selected individuals had the opportunity and experience of learning about the AR-based assembly system, they were given a survey. The purpose of this survey was to know three evaluation parameters, consistent with the objective of this research, and whose concept implies that people evaluate and know an assembly procedure through an AR-based system: presentation, novelty, and ease of use. Likewise, the level of knowledge retention was contrasted concerning the mentioned variables, which was carried out by comparing the two ways in which people appreciated an assembly procedure, that is, with and without AR. Table 1 contains the responses of the selected and surveyed people regarding 1) the score given to the presentation of an assembly procedure through the AR-based system; 2) what was their assessment of the novelty of an assembly procedure through an AR-based system and; 3) how accessible or complicated was the handling of the AR-based system. Finally, all the selected people were asked if the new way of assembly motivated them to use an AR system to view the entire assembly procedure of a garment and the result was that 91% (i.e., 9 out of 10 people) responded that the new way of appreciating and assembling a garment did motivate them to use an AR system.

Table 1. Results for the AR-based system

Concept	with AR	without AR
Presentation	91.3%	33.0%
Novelty	96.4%	37.5%
Easy to use	94.8%	73.6%
Motivation	91.0%	60.0%

Source: Author's own elaboration.

After the randomly selected workers had the opportunity and experience to learn about the AR system, another survey was administered to them. The purpose of this brief survey was to learn three evaluation parameters, consistent with the objective of this research, and where the purpose involved evaluating and learning an innovative way of learning the assembly of a garment using an AR system. The first result for this survey was that Cronbach's alpha (Taber, 2018) had a value of 83.1%, which implies that the questionnaire meets the objective of the study for which it was designed and, therefore, is valid and consistent. Consequently, the results for the three criteria evaluated utilizing this research instrument were as follows:

Ease of Use

With the technological resources that were implemented, it was observed that intuitive and easy-to-handle elements can be generated; this result was demonstrated with 93.8%, since when compared to the percentage obtained with the resources without AR, it seems that it should be equally or less easy than using paper (60% ease), however, the workers evaluated it according to what they learned. This result asserted that AR facilitates the imparting of knowledge to workers and its acquisition by them, which was reflected in better job performance. Therefore, the implementation of technological resources in the work area helped to improve the process of assembling a garment, increased motivation, and facilitated the work (individual and collaborative) of each worker.

Motivation

In this context, it was found that AR motivated 94.7% of the workers with elements that served to use their primary senses, which was verified in the interviews, where the workers answered that they liked the images, the audio, and the interaction with AR through the devices and, consequently, they were excited by the use of these systems superimposed in real contexts. In their work area, workers were cheerful, curious, and eager to use the technological tools, data that were pooled together with the interviews and, from which, it was determined that workers liked the AR elements in 95.4%, the structure of the system in 93.0%, they would like to learn other topics with AR in 95.3% and that they felt excited in 95%. These results allow us to argue that AR generates contexts where dopamine overflows to include the desire to continue learning, to appropriate this knowledge, and to store it in their long-term memory.

Performance

The workers obtained a performance of 86.4% because there were many words and actions to learn in a short time. Therefore, this result allows arguing that the learning that is obtained and that remains engraved in the mind in the long term, is easier to obtain when it is understood in a 3D model than when using flat resources (with 2D figures). This immediate evaluation of the worker allows decisions to be made at the appropriate time for improvement. in addition, it optimizes the participation of workers, and the benefit is reflected in subsequent sessions -to improve the assembly of a garment in the shortest possible time-. The evaluation shown by AR-based systems is carried out at the time of training or the end of training, allowing the worker to self-evaluate and, when necessary, continue training. This result emphasizes that using an AR-based system results in a 26% improvement in workers' performance. Therefore, there is statistical evidence that the evaluation through an AR-based system helps workers to generate challenges with themselves, based on the synergy of real and digital contexts, which allows them to increase their performance.

Meaningful Learning

The neural maps made during the activity allowed us to detect a significant learning of 89.3%. The second contrast on significant learning refers to the fact that the teaching of cognitively complex knowledge, so often disconnected from the emotional system, is a mistake, since nothing is learned if it is not loved or does not mean something. Thus, it was found that the significant learning of 89.3% achieved in the tests, occurred when the workers liked and understood what they were doing; the responses of the significant learning category showed that the analysis of the content was easy for them (87.0%); the system showed the learning sequences with an intuitive logic (88.0%); the learning was considered to improve in their work level (90.0%) and that it was easy to learn with a technology (92.0%). Consequently, it can be deduced that AR activates neural networks to generate thinking based on logical sequences and, therefore, more complex knowledge is understood.

FUTURE RESEARCH

This type of system can be implemented in any manufacturing company, adapting it in such a way that it becomes an indispensable aid in a short period. AR-based technology is developing new ways of seeing the world, including its interaction, which is beneficial for manufacturing companies. In the not-too-distant future, it will be possible to make use of this interaction to bring AR closer and closer to physical reality.

AR is a tool that has been applied to various topics using information systems to detail queries with results classified according to various input parameters. Systems that use AR could be viewed from different perspectives, generating systems that are an alternative in the locations of some garments in department stores or materials in a factory, but also in crowded spaces such as stadiums, archaeological sites, museums, or any type of tourist site. The idea of using AR is to generate, through user motivation, high expectations of use through new alternatives for manufacturing, tailoring, design, assembly, and marketing, to name just a few.

CONCLUSION

AR is a technology that applies to various areas of industry, such as maintenance, quality control, sales, and design; in this work, it was applied to the manufacturing process and, after analyzing the response to its implementation, it can be concluded that this technology provides the user with several benefits, among which the following can be mentioned: the feeling of being motivated, encouraged and confident of doing the assembly of a product correctly; those aspects of well-being are derived from the quick and instantaneous consultations, which serve to avoid wasting time in consulting specialized personnel and having the system at hand to consult any detail of an AR-based procedure.

Material losses should be low or null, since if the AR is used for any doubt, there will be no way to make a mistake and, consequently, no wasted elements will be damaged as a result of the assembly correction. Also, the consultation time is relatively low, which means that the investment time is less compared to its profit, and, consequently, the quality obtained is high. For the company, the decrease in production time benefits them, since the higher the production, the higher the monetary profit is expected. At the same time, the investment resources are few because if it is taken into account that only the videos and 3D images must be reproduced, then this dynamic is justified, mainly because currently most of the workers use a smartcell phone, that is, the only cost is the internet, which in any case is already a service that is essential in every company and person.

The AR makes the answer to a query about a procedure almost instantaneous, otherwise, one would have to search in the gallery for the photo or video that explains that procedure, which would represent a higher investment of time and an even more tedious process. In this way, the AR system presented in this research work allows showing the sequence of the construction of the assembly pieces in a visual, animated, and auditory way, making it understandable to anyone, even with little experience in cutting and sewing garments.

Mobile devices are frequently used, so implementation costs do not require the investment of human resources or infrastructure; in addition, mobile devices are used by people daily, so these modern technologies must be adapted to manufacturing processes to obtain a useful benefit from their use. Quality control can also be supported by this tool, precisely to control processes and make them better. This type of system is a way for companies to be at the forefront of the technological world; its support through these technologies will be reflected in increased production, but, above all, will be a reflection of a total attachment to digital trends worldwide. State-of-the-art technology will, in turn, enable the creation of new technologies to support these types of manufacturing-related processes, and it is expected that this support will be on a more frequent basis to achieve and provide full automation. The cost-benefit of this implementation will be reflected in delivery times and, consequently, in more dynamic payment times as well as in an increase in production.

REFERENCES

Adebowale, O. J., & Agumba, J. N. (2024). Applications of augmented reality for construction productivity improvement: A systematic review. *Smart and Sustainable Built Environment*, *13*(3), 479–495. DOI: 10.1108/SASBE-06-2022-0128

Adel, A. (2022). Future of industry 5.0 in society: Human-centric solutions, challenges and prospective research areas. *Journal of Cloud Computing (Heidelberg, Germany)*, *11*(1), 40. DOI: 10.1186/s13677-022-00314-5 PMID: 36101900

Agnihotri, R., Chaker, N. N., Dugan, R., Galvan, J. M., & Nowlin, E. (2023). Sales technology research: A review and future research agenda. *Journal of Personal Selling & Sales Management*, *43*(4), 307–335. DOI: 10.1080/08853134.2023.2260108

Aguinis, H. (2024). *Research Methodology. Best Practices for Rigorous, Credible, and Impactful Research*. SAGE Publications.

Ahmad, N., & Zulkifli, A. M. (2022). Internet of Things (IoT) and the road to happiness. *Digital Transformation and Society*, *1*(1), 66–94. DOI: 10.1108/DTS-05-2022-0009

Al-Ansi, A. M., Jaboob, M., Garad, A., & Al-Ansi, A. (2023). Analyzing augmented reality (AR) and virtual reality (VR) recent development in education. *Social Sciences & Humanities Open*, *8*(1), 100532. DOI: 10.1016/j.ssaho.2023.100532

Alt, R. (2020). Evolution and perspectives of electronic markets. *Electronic Markets*, *30*(1), 1–13. DOI: 10.1007/s12525-020-00413-8

Alves, C., & Machado, J. (2024). Immersive Shopping Experiences: The Role of Augmented Reality in E-Commerce. In: Rocha, Á., Ferrás, C., Hochstetter Diez, J., Diéguez Rebolledo, M. (Eds.) *Information Technology and Systems*. ICITS 2024. Lecture Notes in Networks and Systems, vol. 933. Springer, Cham. https://doi.org/ DOI: 10.1007/978-3-031-54256-5_19

Aquino, S., Rapaccini, M., Adrodegari, F., & Pezzotta, G. (2023). Augmented reality for industrial services provision: The factors influencing a successful adoption in manufacturing companies. *Journal of Manufacturing Technology Management*, *34*(4), 601–620. DOI: 10.1108/JMTM-02-2022-0077

Arena, F., Collotta, M., Pau, G., & Termine, F. (2022). An Overview of Augmented Reality. *Computers*, *11*(2), 28. DOI: 10.3390/computers11020028

Bellalouna, F. (2021). The Augmented Reality Technology as Enabler for the Digitization of Industrial Business Processes: Case Studies. *Procedia CIRP*, *98*, 400–405. DOI: 10.1016/j.procir.2021.01.124

Berndt, A. E. (2020). Sampling Methods. *Journal of Human Lactation*, *36*(2), 224–226. DOI: 10.1177/0890334420906850 PMID: 32155099

Cannavacciuolo, L., Ferraro, G., Ponsiglione, C., Primario, S., & Quinto, I. (2023). Technological innovation-enabling industry 4.0 paradigm: A systematic literature review. *Technovation*, *124*, 102733. DOI: 10.1016/j.technovation.2023.102733

Carthy, S., Cormican, K., & Sampaio, S. (2021). Knowing me knowing you: Understanding user involvement in the design process. *Procedia Computer Science*, *181*, 135–140. DOI: 10.1016/j.procs.2021.01.113

Chen, W., Li, Y., Tian, Z., & Zhang, F. (2023). 2D and 3D object detection algorithms from images: A Survey. *Array (New York, N.Y.)*, *19*, 100305. DOI: 10.1016/j.array.2023.100305

Cordeiro, D., Rui, J., & Nuno, C. (2015). ARZombie: A mobile augmented reality game with multimodal interaction. *2015 7th International Conference on Intelligent Technologies for Interactive Entertainment (INTETAIN), 1(4),* 22-31. http://dx.doi.org/DOI: 10.4108/icst.intetain.2015.259743

de Souza Cardoso, L. F., Martins Queiroz Mariano, F. C., & Zorzal, E. R. (2020). A survey of industrial augmented reality. *Computers & Industrial Engineering*, *139*, 106159. DOI: 10.1016/j.cie.2019.106159

Devagiri, J. S., Paheding, S., Niyaz, Q., Yang, X., & Smith, S. (2022). Augmented Reality and Artificial Intelligence in industry: Trends, tools, and future challenges. *Expert Systems with Applications*, *207*, 118002. DOI: 10.1016/j.eswa.2022.118002

Egger, J., & Masood, T. (2020). Augmented reality in support of intelligent manufacturing - A systematic literature review. *Computers & Industrial Engineering*, *140*, 106195. DOI: 10.1016/j.cie.2019.106195

Eswaran, M., & Raju Bahubalendruni, M. V. A. (2023). Augmented reality aided object mapping for worker assistance/training in an industrial assembly context: Exploration of affordance with existing guidance techniques. *Computers & Industrial Engineering*, *185*, 109663. DOI: 10.1016/j.cie.2023.109663

Feist, S., Jacques de Sousa, L., Sanhudo, L., & Poças Martins, J. (2024). Automatic Reconstruction of 3D Models from 2D Drawings: A State-of-the-Art Review. *Eng*, *5*(2), 784–800. DOI: 10.3390/eng5020042

Ghasemi, Y., Jeong, H., Choi, S. H., Park, K.-B., & Lee, J. Y. (2022). Deep learning-based object detection in augmented reality: A systematic review. *Computers in Industry*, *139*, 103661. DOI: 10.1016/j.compind.2022.103661

Gren, A., Jamarillo, B., Kiev, V., Shabrov, N., & Vasiliev, D. (2018). Development of digital simulation on the basis of technologies of virtual and augmented reality. *SHS Web of Conferences. 44*, 00036. https://doi.org/DOI: 10.1051/shsconf/20184400036

Gupta, C. P., & Goyal, K. K. (2020). *Computer Concepts and Management Information Systems*. USA. *Mercury Learning and Information.*, *ISBN-10*, 1683925866.

Hassan, R., Qamar, F., Hasan, M. K., Aman, A. H. M., & Ahmed, A. S. (2020). Internet of Things and Its Applications: A Comprehensive Survey. *Symmetry*, *12*(10), 1674. DOI: 10.3390/sym12101674

Ho, P. T., Albajez, J. A., Santolaria, J., & Yagüe-Fabra, J. A. (2022). Study of Augmented Reality Based Manufacturing for Further Integration of Quality Control 4.0: A Systematic Literature Review. *Applied Sciences (Basel, Switzerland)*, *12*(4), 1961. DOI: 10.3390/app12041961

Hoffmann, S., & Mai, R. (2022). Consumer behavior in augmented shopping reality. A review, synthesis, and research agenda. *Frontiers in Virtual Reality*, *3*, 961236. DOI: 10.3389/frvir.2022.961236

Jandyal, A., Chaturvedi, I., Wazir, I., Raina, A., & Ul Haq, M. I. (2022). 3D printing - A review of processes, materials and applications in industry 4.0. *Sustainable Operations and Computers*, *3*, 33–42. DOI: 10.1016/j.susoc.2021.09.004

Khalek, I., Chalhoub, J., & Ayer, S. (2019). Augmented Reality for Identifying Maintainability Concerns during Design. *Advances in Civil Engineering*, *2019*(1), 1–12. DOI: 10.1155/2019/8547928

Krajčovič, M., Matys, M., Gabajová, G., & Komačka, D. (2024). Utilization of Immersive Virtual Reality as an Interactive Method of Assignment Presentation. *Electronics (Basel)*, *13*(8), 1430. DOI: 10.3390/electronics13081430

Kubik, C., Hofmann, J., Molitor, D. A., Becker, M., & Groche, P. (2025). Reliable machine learning models for manufacturing processes. *International Journal of Advanced Manufacturing Technology*, *138*(11-12), 5913–5927. Advance online publication. DOI: 10.1007/s00170-025-15785-4

Laudon, K. C., & Laudon, J. P. (2020). *Management Information Systems. Managing the Digital Firm* (16th ed.). Pearson Education Limited.

Li, Y., Tian, X., Feng, Y., Yuan, G., Guo, X., Wang, D., He, B., Hu, L., & Jiang, G. (2025). Assembling in 3D printing: Functions and applications. *International Journal of Advanced Manufacturing Technology*, *138*(11-12), 5123–5135. Advance online publication. DOI: 10.1007/s00170-025-15860-w

Liang, Y.-Ch., Shiu, S.-Ch., & Liu, Ch.-W. (2025). Enhancing yield and process efficiency through dual internet of things and augmented reality for AI-driven human-machine interaction in centering mass production. *Journal of Manufacturing Systems*, *82*, 26–41. DOI: 10.1016/j.jmsy.2025.06.003

Lin, K.-Y., & Huang, T. K. (2024). Shopping in the digital world: How augmented reality mobile applications trigger customer engagement. *Technology in Society*, *77*, 102540. DOI: 10.1016/j.techsoc.2024.102540

Massa, E., & Ladhari, R. (2023). Augmented reality in marketing: Conceptualization and systematic review. *International Journal of Consumer Studies*, *47*(6), 2335–2366. DOI: 10.1111/ijcs.12930

Mekonnen, A. (2024). Augmented Reality (AR) in Retailing: Customers' Experience in Luxury Fashion. In Ozuem, W., Ranfagni, S., & Willis, M. (Eds.), *Digital Transformation for Fashion and Luxury Brands*. Palgrave Macmillan., DOI: 10.1007/978-3-031-35589-9_5

Molina Vargas, D. G., Vijayan, K. K., & Mork, O. J. (2020). Augmented Reality for Future Research Opportunities and Challenges in the Shipbuilding Industry: A Literature Review. *Procedia Manufacturing*, *45*, 497–503. DOI: 10.1016/j.promfg.2020.04.063

Morales Méndez, G., & del Cerro Velázquez, F. (2024). Augmented Reality in Industry 4.0 Assistance and Training Areas: A Systematic Literature Review and Bibliometric Analysis. *Electronics (Basel)*, *13*(6), 1147. DOI: 10.3390/electronics13061147

Morales Méndez, G., & del Cerro Velázquez, F. (2024). Impact of Augmented Reality on Assistance and Training in Industry 4.0: Qualitative Evaluation and Meta-Analysis. *Applied Sciences (Basel, Switzerland)*, *14*(11), 4564. DOI: 10.3390/app14114564

Parekh, P., Patel, S., Patel, N., & Shah, M. (2020). Systematic review and meta-analysis of augmented reality in medicine, retail, and games. *Visual Computing for Industry, Biomedicine, and Art*, *3*(1), 21. DOI: 10.1186/s42492-020-00057-7 PMID: 32954214

Paszkiewicz, A., Bolanowski, M., Budzik, G., Przeszłowski, Ł., & Oleksy, M. (2020). Process of Creating an Integrated Design and Manufacturing Environment as Part of the Structure of Industry 4.0. *Processes (Basel, Switzerland)*, *8*(9), 1019. DOI: 10.3390/pr8091019

Pereira, A. C., Alves, A. C., & Arezes, P. (2023). Augmented Reality in a Lean Workplace at Smart Factories: A Case Study. *Applied Sciences (Basel, Switzerland)*, *13*(16), 9120. DOI: 10.3390/app13169120

Plakas, G., Ponis, S. T., Agalianos, K., Aretoulaki, E., & Gayialis, S. P. (2020). Augmented Reality in Manufacturing and Logistics: Lessons Learnt from a Real-Life Industrial Application. *Procedia Manufacturing*, *51*, 1629–1635. DOI: 10.1016/j.promfg.2020.10.227

Pressman, R., & Maxim, B. (2020). *Software engineering. a practitioner's approach* (9th ed.). Edit. McGraw-Hill.

Qiu, H., Yu, Z., Zhao, T., Zhang, Q., Xu, M., Li, P., Li, T., Bao, W., Chai, Y., Chen, S., Chen, Y., Cheng, H.-M., Dai, D., Di, Z., Dong, Z., Duan, X., Feng, Y., Fu, Y., Guo, J., & Wang, X. (2024). Two-dimensional materials for future information technology: Status and prospects. *Science China. Information Sciences*, *67*(6), 160400. DOI: 10.1007/s11432-024-4033-8

Ramos, R., Rita, P., & Vong, C. (2024). Mapping research in marketing: Trends, influential papers and agenda for future research. *Spanish Journal of Marketing - ESIC*, *28*(2), 187–206. DOI: 10.1108/SJME-10-2022-0221

Reljić, V., Milenković, I., Dudić, S., Šulc, J., & Bajči, B. (2021). Augmented Reality Applications in Industry 4.0 *Environment. Applied Sciences (Basel, Switzerland)*, *11*(12), 5592. DOI: 10.3390/app11125592

Riar, M., Korbel, J. J., Xi, N., Meywirth, S., Zarnekow, R., & Hamari, J. (2023). Augmented Reality in Interactive Marketing: The State-Of-The-Art and Emerging Trends. In Wang, C. L. (Ed.), *The Palgrave Handbook of Interactive Marketing*. Palgrave Macmillan., DOI: 10.1007/978-3-031-14961-0_14

Roberto, R., Newnes, L., & Shokrani, A. (2024). Challenges to Deploy Augmented Reality in Manufacturing. *Procedia CIRP*, *130*, 381–386. DOI: 10.1016/j.procir.2024.10.104

Ros, F., & Guillaume, S. (2020). *Sampling Techniques for Supervised or Unsupervised Tasks*. Springer Nature., DOI: 10.1007/978-3-030-29349-9

Roschli, A., Gaul, K. T., Boulger, A. M., Post, B. K., Chesser, P. C., Love, L. J., Blue, F., & Borish, M. (2019). Designing for Big Area Additive Manufacturing. *Additive Manufacturing*, *25*, 275–285. DOI: 10.1016/j.addma.2018.11.006

Sakr, A., & Abdullah, T. (2024). Virtual, augmented reality and learning analytics impact on learners, and educators: A systematic review. *Education and Information Technologies*, *29*(15), 19913–19962. DOI: 10.1007/s10639-024-12602-5

Santi, G. M., Ceruti, A., Liverani, A., & Osti, F. (2021). Augmented Reality in Industry 4.0 and Future Innovation Programs. *Technologies*, *9*(2), 33. DOI: 10.3390/technologies9020033

Schifferstein, H. N. J., Lemke, M., & de Boer, A. (2022). An exploratory study using graphic design to communicate consumer benefits on food packaging. *Food Quality and Preference*, *97*, 104458. DOI: 10.1016/j.foodqual.2021.104458

Schumann, M., Fuchs, C., Kollatsch, C., & Klimant, P. (2021). Evaluation of augmented reality supported approaches for product design and production processes. *Procedia CIRP*, *97*, 160–165. DOI: 10.1016/j.procir.2020.05.219

Shyr, W.-J., Tsai, C.-J., Lin, C.-M., & Liau, H.-M. (2022). Development and Assessment of Augmented Reality Technology for Using in an Equipment Maintenance and Diagnostic System. *Sustainability (Basel)*, *14*(19), 12154. DOI: 10.3390/su141912154

Siriborvornratanakul, T. (2018). Enhancing User Experiences of Mobile-Based Augmented Reality via Spatial Augmented Reality: Designs and Architectures of Projector-Camera Devices. *Advances in Multimedia*, *17*, 1–17. Advance online publication. DOI: 10.1155/2018/8194726

Song, Y., Koeck, R., & Luo, S. (2021). Review and analysis of augmented reality (AR) literature for digital fabrication in architecture. *Automation in Construction*, *128*, 103762. DOI: 10.1016/j.autcon.2021.103762

Sun, Y., Shahzad, M., Ali, A., & Razzaq, A. (2025). Synergistic effect of green knowledge sharing and green creative climate for circular economy practices: role of artificial intelligence information quality. *Journal of Manufacturing Technology Management*, Vol. ahead-of-print No. ahead-of-print. https://doi.org/DOI: 10.1108/JMTM-09-2024-0503

Taber, K. S. (2018). The Use of Cronbach's Alpha When Developing and Reporting Research Instruments in Science Education. *Research in Science Education*, *48*(6), 1273–1296. DOI: 10.1007/s11165-016-9602-2

Tan, Y.-C., Chandukala, S. R., & Reddy, S. K. (2022). Augmented Reality in Retail and Its Impact on Sales. *Journal of Marketing*, *86*(1), 48–66. DOI: 10.1177/0022242921995449

Thang, T. Q., & Kim, J. (2025). Applications of 3D Printing in Paper-Based Devices for Biochemical and Environmental Analyses. *Chemosensors (Basel, Switzerland)*, *13*(3), 89. DOI: 10.3390/chemosensors13030089

Turovets, Y., Vishnevskiy, K., Tokareva, M., & Kukushkin, K. (2019). Technology foresight for digital manufacturing: Russian case. *IOP Conference Series. Materials Science and Engineering*, *497*, 012062. https://iopscience.iop.org/article/10.1088/1757-899X/497/1/012062. DOI: 10.1088/1757-899X/497/1/012062

Vilar, E., Filgueiras, E., & Rebelo, F. (2022). *Virtual and Augmented Reality for Architecture and Design*. CRC Press. DOI: 10.1201/9781003051381

Voinea, G.-D., Gîrbacia, F., Duguleană, M., Boboc, R. G., & Gheorghe, C. (2023). Mapping the Emergent Trends in Industrial Augmented Reality. *Electronics (Basel)*, *12*(7), 1719. DOI: 10.3390/electronics12071719

Wallace, P. (2021). *Introduction to Information Systems* (4th ed.). Pearson.

Wu, Ch., & Thompson, M. E. (2020). *Sampling Theory and Practice*. Springer Nature Switzerland., DOI: 10.1007/978-3-030-44246-0

Xiong, J., Hsiang, E. L., He, Z., Zhan, T., & Wu, S.-T. (2021). Augmented reality and virtual reality displays: Emerging technologies and future perspectives. *Light, Science & Applications*, *10*(1), 216. DOI: 10.1038/s41377-021-00658-8 PMID: 34697292

Xu, Z., Feng, Z., Jelodar, M. B., & Guo, B. H. W. (2024). Augmented reality applications in construction productivity: A systematic literature review. *Advanced Engineering Informatics*, *62*(Part C), 102798. DOI: 10.1016/j.aei.2024.102798

APPENDIX: QUESTIONNAIRE

The questionnaire was administered after the workers had been trained in the use of AR. Select the appropriate option:

1. What is the procedure you have to do?
   a) Sleeve b) Hem c) Finishing close e) Overlock f) Other__________
2. Do you know what RA is?
   a) Yes____ b) No_____
3. Do you know how to use AR on your cell phone?
   a) Little____ b) A lot____ c) No_____
4. Have you consulted any procedures with RA?
   a) Yes____ b) No____
5. What did you do before knowing about AR when you needed to ask a question?
   a) You went to ask the supervisor.
   b) You did it with what your memory allowed you to remember.
   c) You looked for a picture or a note you had taken about the proceedings.
6. How long does it take you to solve the doubt in the traditional way you used to do it?
   a) 3 Minutes b) 5 Minutes c) 10 Minutes d) Over 10 Minutes
7. How long does it take you to open the application, focus on the garment, and get the result?
   a) 3 Minutes b) 5 Minutes c) 10 Minutes
8. What is the advantage of the AR-based method?
   a) Speed c) Quality b) No travel d) Instant solution
9. If you are provided with the 3D images and videos to store on your phone, how long would it take to complete the consultation?
   a) 3 Minutes b) 5 Minutes c) 10 Minutes
10. Is the information contained in AR that displays videos useful?
    a) Yes b) No c) Middle
11. ¿You found it easy to use the system with AR?
    a) Si b) No c) Middle
12. ¿What type of operating system do you have on your mobile device??
    a) Android b) iOs c) Other
13. ¿You liked using the app with AR?
    a) Yes b)No c)Middle
14. ¿How much money does it cost to use the AR app?
    a) Nothing b) $5 or less c) A lot d) I don't know
15. In summary, what are the benefits that this application provides you? Choose 1 or several options, writing an X in front of them or underlining them.

Shorter production time
Less material waste
Shorter consultation time
Learning about processes

Chapter 13
Advancing Marketing in Society 5.0:
The Strategic Role of Predictive Analytics in Consumer Behaviour Analysis

Parihar Suresh Dahake
https://orcid.org/0000-0002-2286-4120
Ramdeobaba University, Nagpur, India

Nihar Suresh Dahake
https://orcid.org/0009-0007-8971-9247
G.H. Raisoni College of Engineering, Nagpur, India

Saket Narendra Bansod
https://orcid.org/0000-0002-2094-2084
Dr. Ambedkar Institute of Management Studies and Research, Nagpur, India

Prachita S. Patil
Dr. Ambedkar Institute of Management Studies and Research, Nagpur, India

ABSTRACT

Businesses are progressively utilising predictive analytics to enhance their comprehension and anticipation of client behaviour within the framework of Society 5.0, a human-centric society that integrates the internet with the physical realm. This paper examines the vital role of predictive analytics in the progression of marketing innovations, emphasising marketers' ability to anticipate trends, personalise experiences, and develop data-driven marketing strategies. This study will examine how predictive analytics technologies are transforming marketing strategies in Society 5.0

DOI: 10.4018/979-8-3373-1082-4.ch013

by providing enhanced customer insights and facilitating more accurate targeting. This study is distinctive since it examines the convergence of AI-driven analytics and marketing strategies within the Society 5.0 framework, which seeks to harmonise economic advancement with societal welfare.

INTRODUCTION

In the age of big data, website visitors generate a digital footprint. Organisations utilise digital footprints to analyse and predict consumer behaviour (Guha & Kumar, 2018). Commercialisation is a critical phase in technological innovation, as it involves significant risks and expenses (Chiesa & Frattini, 2011). Over the past decade, technology and data collection methods have produced vast quantities of data on several subjects at an accelerated rate. As corporate and startup pressures escalate in the market, standardisation is essential for enhanced measurement traceability and transferability (Pathare et al., 2013).

Predicting consumer behaviour to enable proactive marketing actions is crucial for strategizing how to impact the decision-making process in purchasing behaviour, as determined by the decision-making entity inside our selected society. The amalgamation of innovation and knowledge has emerged as a fundamental principle in knowledge- and labour-intensive enterprises (Scuotto et al., 2017). The objective is to conduct research on the development and importance of predictive analytics as a mechanism for achieving marketing success in our selected Society 5.0. The market's intense rivalry, coupled with the advent of developing technologies like artificial intelligence, presents numerous commercial opportunities. Predictive models enhance understanding of consumer behaviour(P. S. Dahake & Somani, 2024) and it is significantly Targeting of promotional offers specific to segments(Lazarus et al., 2004) A major commercial opportunity arises from the ability to utilise predictive models that leverage data to implement changes that enhance operational efficiency and eventually generate value for the organisation. Predictive models serve as an essential instrument for businesses to attain their objectives, optimise operations, and generate new prospects, while also guiding managers in making improved strategic and tactical decisions, minimising waste, and augmenting earnings. As cultures and technologies advance, marketing concepts evolve through different eras, culminating in Marketing in Society 5.0. Marketing in Society 5.0 is the framework wherein products and technologies from MarTech-driven industrial revolutions are converted into concurrent human-centric marketing requirements via the application of big data computing and artificial intelligence.

Predictive analytics reveals connections in data for precise predictions(Siegel, 2015) and Marketing 5.0 pertains to solutions engineered to integrate supplied data into recommendations, advertisements, and offers that dynamically adjust using MarTech methodologies, enhancing the relevancy and expediency of recognising consumer journeys while simultaneously fostering customer experiences and marketing satisfaction. The dynamic signifies a significant shift towards a more human-centric perspective in societal thinking, with criteria increasingly necessitating that marketing serves as a leisure facilitator rather than only targeting individuals (Dar et al., 2021). Innovative experience design will progressively emerge as a crucial element of the core competencies of tourism and hospitality businesses (Williams, 2006). Currently, innovations that achieve a fundamental equilibrium of consumption for society in the realms of personal care, private label, consumer goods, and household management are influenced by the requirements and strategies of contemporary marketers. Furthermore, the study elucidates how firms have historically sought to examine consumer behaviour in light of the requisite transformation inside the Indian marketplace. Predictive analytics is essential for examining consumer behaviour in the e-commerce sector. The study elucidates the sociological stages and their importance. Predictive analytics facilitates an in-depth examination of the research (Magazzino et al., 2021).

A contextual framework is provided to attain a comprehensive understanding of marketing advances in Society 5.0. Marketers typically create innovative concepts, products, and services that address both current and future consumer needs and desires. Marketers assert that both the approach and the result of this innovative endeavour are distinctive and provide a competitive advantage in the market. Digitising all processes and data is the most effective method for technologies to actualise this advancement. It is not a disputed arena, as it integrates technology innovations with market dynamics. The assessment of marketing viewpoints about new products and services from Industry 4.0 enhances the understanding of the culture and context of Society 5.0 when reconsidering others to revert to established truths. The collaboration of technology and marketing can yield significant outcomes. The objective of the current study is not to critique marketing objectives, but to illustrate how historical practices guide businesses towards one conclusion rather than another. The transition from society 1.0 to society 5.0 primarily concerns the organisation of numerous Societys and the distinct ways each Society takes towards life and reality; yet, there exists considerable contention over societal adaptations and the indicators of being hastily driven by commercial interests. For several decades, societies have existed in the 4.0 era, characterised by industrialisation that has led to the widespread advancement of various technologies. The understanding of analytics is limited as the literature does not depict marketing decisions made with data analytics. Furthermore, only macro-analytic models are discussed in the context of predictive models for dynamic mixing of media channels in situations where random shocks are expected. In real-time analytics, the literature is biased toward non-marketing applications. The conceptual marketing analytics framework e-Maring is indicative of the significant omission of agile predictive models and real-time marketing metrics. Sensitivity analysis is vitally related to marketing decisions and yields better marketing resource allocation. Expansively collating marketing operational procedures, tracking and analysing archives of existing marketing actions, and conducting simulations can ensure credibility and efficacy of marketing decisions.

Only few studies focus on game theory able to capture the interactive nature of predictive marketing decisions as analytics may be misused to gain rival advantage by retaining information insecurity in environmental changes. Using wide-magnitude measures of relative and appropriate complexity of dynamic modelling exhibits the sophistication and breadth of marketing data analytics. Marketing analytics are in diverse stages of progress with different degrees of sophistication and talents. Statistical hypothesis tests and kick-start optimisations are run-of-the-mill, wisdom-based standard, and easily sourced. The recall of decision trees and Michalewicz function optimising evolutionary algorithm run-time were grounds of concern yet rare. Given the dearth of a systematic scalable inventory of marketing analytics, the efforts reported encourage analytic capability tracking and strive for at minimum sophistication and diversity. The motivational comparison of benign imitations or nurturing in-house home-grown alternatives may also lead to innovative data mining.

With accompanying ethical concerns, it is the hope that the transparent reclamation and exposition of marketing analytics will encourage fair play and proactively assuage fears of any self-serving predictive abuse. Customer profiling, spying alignment with micro-targeting, and privacy and trust issues in context-based processing of customer information capture the dark side of the predictive big data. Clearly understood value confers strategy legitimacy, imbued with consideration to helpful consumer welfare by preventing consumer abuse. With better shoring up of committed and capable long-termership, conjunction with sensitive cognitive off-sets, productive cross-benefits, and alleviation of agency risk and monitoring costs, marketing is to avail excess returns. By obtaining good yields of sullied big data and analytics, this consumer welfare-centric market institutional arrangement necessarily promotes growth and societal progress (Bashir Dar et al., 2021). With the transition from industrial society to an information-oriented society, the notion of Society 5.0 emerged. Society 5.0 entails using advanced technology for humanity's growth and sustainability. Consumers demand convenience, prosperity, health, and security, and companies are expected to provide goods and services aligned with these demands. The sum of capital and physical value, including equipment, labor, funding, and data, is emphasized in business growth within Society 5.0. Businesses need to analyze data strategically and establish proper marketing-mix allocation using predictive analytics to stay competitive.

In India, consumer behaviour analysis is complex due to a growing population, diverse economic status, regional culture, languages, customs, traditions, and values (Bashir Dar et al., 2021). Additionally, India has a unique presence of multinational companies in virtually all sectors, each competitive due to different operational fields, management strategies, prices, packages, locations, and advertising channels. Emphasizing consumer behaviour can generate significant revenue, so every Indian company uses multiple channels to get closer to customers and earn customer loyalty. The marketing department is responsible for the proper allocation of marketing-mix, targeting, privacy, and identification of predictive analytics forecasting variables. Indian marketing departments tend to be larger than expected, but the number of persons focused on actionable and strategic intelligence generation beforehand is small. Many firms lack a fully equipped analytics department, hiring outsiders or deferring questioning exclusively to vendors or the IT department. On the other hand, decision-makers working with outsourced agencies have a greater, though still informational, gap. This gap is more significant in India due to closer ties between many decision-makers and agencies, resulting in higher expectations for detailed market analyses instead of purely predictive ones. Companies also struggle with abnormal sales spikes during events and significant consumer population movement during festive periods for effective advertising. Therefore, there is an urgent need for predictive analytics, particularly for consumer behaviour analysis, in India.

BACKGROUND AND SIGNIFICANCE OF MARKETING INNOVATIONS IN SOCIETY 5.0

Society 5.0 is a prospective notion that embodies an ideal human-centric society anticipated for the fifth phase of future Society. It is anticipated to harmonise economic advancement with the settlement of social challenges through a system that intricately links digital and physical environments. AI can enhance marketing research, strategy, and actions by automating repetitive functions, processing data, and analyzing interactions and emotions(Huang & Rust, 2021) and The impetus for Society 5.0 is considered to be more ambitious than that of earlier technical innovations. The presentation seeks to elucidate seven traits that differentiate Society 5.0 from prior industrial and individual revolutions. These characteristics are categorised into two groups: the predominant aspects from the literature and insights from alternative clusters. Innovations in pressing areas, like as marketing, may only benefit society if consumers embrace and utilise them. Olshavsky and Spreng (1996). Product managers and researchers investigate optimal organisational approaches, tactics, and frameworks to advance product innovation. Szmigin and Foxall (1998) characterise innovation resistance as a rational reaction rather than a negative trait of target markets. It is anticipated that marketing professionals would be more driven to excel and develop by prioritising societal interests in their strategies. Artificial intelligence in marketing reveals ten key research themes, including understanding consumer sentiments, industrial opportunities, analysing customer satisfaction, improving market performance, and using AI for brand(Mustak et al., 2021) Also Behavioral reasoning theory (BRT) plays a crucial role in predicting consumer behavior(Sahu et al., 2020) Ultimately, there has been an emphasis on difference regarding society-oriented corporations. Consumer Particular 5.0 is significant in this context. As consumers are vital to society, predicting their behaviours will enable firms to more effectively invent, produce, and distribute items that meet their needs. Society 5.0 is a visionary notion for the fifth stage of societal evolution, anticipated to harmonise economic advancement with the resolution of social issues through a system that intricately integrates the internet and physical environments to serve the greatest number of individuals possible. In the contemporary market, buyers possess extensive knowledge. Conventional marketing research is insufficient, as numerous studies indicate that consumers often exhibit unpredictable behaviour. The primary grievances are its incapacity to predict future demand for new products and its emphasis on the marketer's current geographical market. Although historical sales data is crucial as a foundation, it is insufficient to depend solely on past information when tackling issues related to consumer behaviour in the digital age, including evolving habits, preferences, and behaviours of contemporary consumers and AI integration in organizational strategy can create business value through decision support, customer engagement, automation, and new products and services(Borges et al., 2021). Consumer conduct in the digital era reveals new phenomena that will surely influence consumer behaviour in the twenty-first century. The digital era is fundamentally driven by information, networks, and technology that provide unprecedented access to knowledge, cultures, and economies. Customer engagement in social media is driven by satisfaction, positive emotions, and trust, with Twitter showing the strongest impact on engagement and firm performance(De Oliveira Santini et al., 2020).Currently, the consumer, organisation, and network function as facilitators in the innovation process and are in a state of constant evolution (Hinsch et al., 2020). Establishing significant connections and involving consumers using augmented reality can stimulate customer enthusiasm and enhance marketing outcomes. They are interdependent entities that exert influence and offer feedback, potentially resulting in an altered society. A novel surge of innovation and customer service demands is permeating this dynamic, diversified, and always evolving landscape. Source expertise, trustworthiness, and homophily significantly influence perceived usefulness and credibility of electronic word of mouth communications, influencing consumer behavior and information adoption(Ismagilova et al., 2020). In the Societal Marketing the changes in societal norms and consumers' expectations, framing this paper, regarding the cases market of Nagpur, India, with the evolution of marketing along the ages from society 1.0 to society 5.0, the 5th stage, throughout the manuscript text predictive analytics has been discussed which can be crucial for consumer behavior analysis, so the paper should be important for those on the platform who formulate marketing strategies for companies or government organizations in any part of the globe, including China and Japan, and any city in India including Nagpur. In order to better understand, Predictive maintenance in Industry 4.0 enhances machine downtime, costs, control, and quality, but requires a multidisciplinary approach to effectively address Industry 4.0 challenges(Zonta et al., 2020). it is highly relevant to review the evolution of marketing from 1.0 to 5.0 along with the relevant social changes. Society 1.0 was characterized by the hunting method hence people depended on the things for hunting purposes. However, as the climate got better for farming, there was a shift to the agriculture method hence dependence now shifted to the earth for cultivation purposes. Society 2.0 was characterized by the overcoming of natural power, which was steam engine. Now since the machines were in place, products, which were in the form of hand-made, in a personal basis, started to be mass produced. Also during this time, with the advent of trains and transportation vehicles, there was a national market. The methodology that was put in place to gain market share within the national market was Society 2.0. Society 3.0 was characterized by the introduction of computer chips and the internet. Developments in computers and communication technologies helped companies move towards business agility. Given the rapid changes that have taken place, consumer expectations have also changed, moving towards customizable products. A methodology based on goods/ service alliance had been put into place. Society 4.0 is characterized by being the age of Artificial Intelligence, Big Data and IoT Technologies. Machines, equipped with an economic and data-based super structure, have come to binary events, thereby reducing the need for human interaction. Marketing is now in the form of a pull strategy, based on the imaginative needs of consumers. Marketing innovations have progressed alongside the emergence of a new society. Emerging lifestyles, elevated earnings, and enhanced consumer awareness are augmenting the demand for a consistent supply of novel food products year-round (Guiné et al., 2020). Predictive analytics are crucial for company success in the contemporary landscape, holding particular significance for marketers due to the applicability of similar cultural and social themes. Consumer views of two of the three evaluated innovation criteria, specifically relative advantage and compatibility, were strongly connected with their acceptance or adoption of marketing messaging (Roach, 2009). The predictive analytics' accuracy in business, highlighting accuracy overestimation(N.

S. Dahake et al., 2024) This study will delineate the importance of evolving technology breakthroughs and personalized marketing by contextualising the notion of personalization and offering a comprehensive analysis of its implementation in an effective marketing strategy. Indian culture, characterised by quickly evolving behaviours, is shaped by influences including social media, marketing techniques, and enhanced digital advertising competition. Consumer patterns derived from individuals' enduring, rational behaviour is exceptionally scalable and transferable. Moreover, comparability elucidates methods to augment the range and intricacy of hypotheses regarding human behaviour. Cultural determinants, such as those in India, can significantly influence environmental consumer behaviour, especially when consumers identify with these culturally sensitive, socially built domains. According to stereotype, South Asians (Indians) are resilient, empathetic, sociable, and capable of enduring adverse conditions in alignment with their traditions and values. Sentiment analysis seeks to affect customer behaviour by gathering data on their emotional responses and conducting controlled experiments. Marketers' efforts to manipulate consumers through the aggregation of extensive real-time databases of their online behaviour and communications, with demographic data, must be incorporated into any examination of emotional economics (Andrejevic, 2011). Comprehending worldwide communities is a feature of India's marketing strategy. Market orientation in an expanding market must prioritise traditional significance. This section will focus solely on the various factors that affect Indian consumers. Additional counsel may be offered to executives.

UNDERSTANDING PREDICTIVE ANALYTICS

Predictive Analytics can be described as a combination of methods that are used to scrutinize past events, organization performance, and data against various variables for the purpose of forecasting future predictions. Again, these could be carried out using various tools and techniques ranging from simple statistical tools like spreadsheets to complex data mining tools. The creators of PA or data scientists use discriminative supervised algorithms like regression, decision tree, support vector machines, and generative supervised algorithms like machine learning. Heuristic based algorithms like neural networks are also applied for predictions (Bashir Dar et al., 2021).

Analytical techniques specifically design for such numerical data mining are widely categorized into three categories like statistical, machine learning, and soft computing techniques. There are diverse algorithms under each category that are primarily targeted for predictions. For example, statistical techniques include regression modeling, time series modeling, and ARIMA. Soft computing is another broad technique used for predictions which includes genetic algorithms, fuzzy logic, and neural networks. It is important to note here, at least two commands from each technique have been chosen based on their popularity. A model comprises a set of equations combined with the maximum likelihood estimates of coefficients is termed as statistical modeling. Statistical modeling is carried out primarily using a programming language called SA somewhere or forecast software developed for statistics analysis.

MARKETING AND ANALYSING CONSUMER BEHAVIOUR IN THE DIGITAL WORLD.

With growing maturation of digital technology more accurate behavior prediction is now feasible too. Digital technology spurs an economic feasibility of small volume transaction data collection with high frequency level followed by remote ubiquitous computation processing resulting a fairly very accurate behavior prediction. Consumer behavior can also be forecast very accurately with data from various different industry sectors. With the coming paradigm of Society 5.0, involvement of a vast digital environment is very much likely. Or it simply can be said that Society 5.0 will be conceived by the integration of the physical world with the cyber world via digital technology. And in this ever-changing world the effectiveness of classical theories and models regarding consumer behavior analysis along marketing policy, strategies, and tactic are now put into a critical condition. Mostly it's consumer motivation, perception, and environmental influences that determine what and how consumers will act. Understanding it is very useful for efficient marketing strategy.Explores machine learning's effectiveness in predicting online consumer behavior for personalized marketing(P. S. Dahake, Mohare, et al., 2023), Due to current complexity mostly classical theories play a vital role too. Among others fundamental are Nicosia model, Howard-Sheth buyer decision aggregation model, Fishbein model, Engel-Kollat-Blackwell model, Shih-Howard discovered preference model, and Consumer Decision Making Process, which of it are put into deep examination. Satisfaction is substantially influenced by all aspects of service innovation, such as product, process, organisational, and marketing innovations, which are positively correlated with perceived institutional image and loyalty(Cheng et al., 2019). And In the Indian context, consumers operate in a different scenario altogether. Indian consumers follow a logic of their own. Logic has never been too strong due to a dearth of information. No one knew much about the products we spend our hard-earned money on; but with the advent of the internet, the tables have turned but the use of predictive analytics allows it to provide light on cost projection, which in turn aids decision-making(Zekić-Sušac et al., 2021) and Predictive analytics technique uses past transactions to understand the dynamic between prosumer bids and market reactions(Chen et al., 2019). Also forecast the quality of video ads using a multimodal architecture that integrates user physiological data with publicly accessible online sentiment ratings(Gauba et al., 2017), The broad term 'industry' in Society 5.0 is now changed to 'society,' which may include a much larger group of members interacting with each other electronically. These days, most people spend a lot of time on social media. More and more, people are making purchases via social media(Chaudhary et al., 2021). The digital revolution has necessitated new systems and processes for engaging participants, "Predictive analytics revolutionizes marketing by enhancing customer insights, personalization, and engagement strategies(P. S. Dahake, Chandak, et al., 2023) whereas traditional marketing practices are becoming less effective and Better operations can't be achieved without cloud computing, high-performance computing, efficient data network management, and strong data analytics(Diamantoulakis et al., 2015).Consumers, currently, despite their personal and individual behaviours, are connected by their common approach toward technological immersion, marking the advent of a new environment that interconnects online and offline and thorough evaluation of their benefits, drawbacks, and functions is carried out when developing a forecasting model(Mamun et al., 2020).

REVIEW OF LITERATURE

Figure 1. All keywords analysis by VosViewer

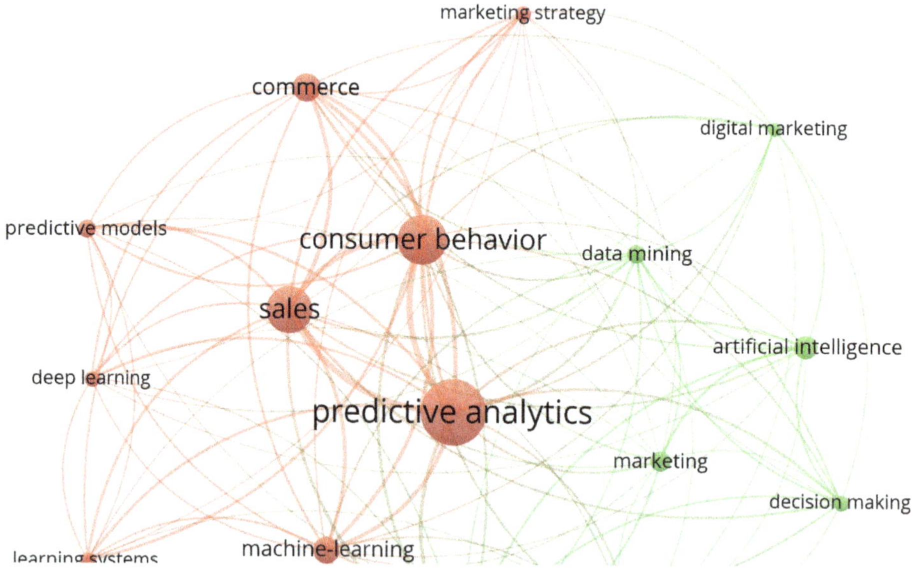

The network map's keyword analysis shows the relationships between main research issues related to marketing innovations, predictive analytics, and consumer behaviour. The map depicts various clusters, with each colour reflecting a thematic grouping of linked terms. "Marketing," "innovation," and "consumer behaviour" are prominent, highlighting fundamental difficulties with strong connections to other areas. These concepts underpin research conflicts, indicating their significance across investigations. Red Predictive Analytics Cluster. This cluster includes predictive analytics, artificial intelligence, big data, and sales, with an emphasis on data-driven marketing and consumer contact, forecasting, and decision-making. Green Cluster Innovation and Technology Adoption. Product innovation, design, and technology acceptance all point to the need for marketing research into technological breakthroughs and innovative management. These quotes demonstrate the importance of innovation in understanding consumers and developing marketing solutions. Commerce and Sustainability (The Yellow Cluster) Commerce, green marketing, and sustainability indicate a growing interest in environmentally responsible marketing, which reflects global sustainability trends. Digital Marketing and Emerging Technologies (purple cluster) Keywords such as machine learning, digital marketing, and social media highlight marketing's digital revolution and the increasing usage of AI and digital technology in marketing strategies. (Wang et al., 2019) examine the methodologies utilised or developed to manage each application, and on this node (Korfiatis et al., 2012) show that stylistic aspects have a significant impact on the helpfulness ratio of reviews, with very helpful reviews receiving higher rankings than less useful ones. To emphasise this point, (Chong et al., 2012) suggested that, the standard Technology Acceptance Model (TAM) and Diffusion of Innovation (DOI) model by including supplemental aspects such as cost, social impact, trust, and service diversity Furthermore, (W. B. Kim et al., 2025) investigate how social, cognitive, and hedonic consumer innovativeness influence global product purchases, excluding functional innovativeness; however, (Li et al., 2025) demonstrate that marketers must meticulously evaluate product attributes when using STH for new product recommendations to mitigate negative consequences. In a similar line (Sorescu & Spanjol, 2008), breakthrough discoveries can boost a company's value and economic rents while simultaneously increasing its risk. Incremental innovations boost regular earnings while leaving the firm's risk profile and economic rents unchanged. Furthermore, (Park & Jun, 2003) developed the concept in which cultural variations influence these influences, while Internet usage and perceived hazards influence Internet purchase behaviour. (Pham and Ho, 2015) Personal attributes have a smaller influence on adoption intention than product-related elements. (Lamy et al., 2012) Risk and perceived utility (PU) are the primary barriers and motivators. Examine a unique data collection that includes a comprehensive range of seven marketing-mix factors and private-label market share, based on twenty years of data from 106 consumer packaged goods categories in the United States. Notably (Fearne & Hughes, 1999), cost measurement and control have been improved to achieve higher efficiency, as has innovation in both product offers and service levels, as well as means of engaging with key consumers. Kang and Park-Poaps (2010) found a strong link between fashion innovation and hedonic shopping incentives. Fashion innovativeness showed a favourable link with adventure and concept shopping motivations, but a negative correlation with value shopping motivation. (Goldsmith et al., 1999). A robust and accurate self-report scale assessed fashion innovation, finding consumers who were most likely to buy new designs when they first hit the market. Similarly, Dahl and Hoeffler (2004) found that incremental products were better appraised when visualised using self-related imagery. Knowing which product category an incremental extension comes from allows people to envision how they will use the new product. However, according to (University of Antwerp et al., 2016), using fine-grained behaviour data, increasing data size across study is beneficial. Larger organisations may have more useful transaction data for targeted marketing. Crowd competitions (Garcia Martinez & Walton, 2014) can also swiftly improve the technical skills of a single dataset by integrating a varied range of individuals skilled in various analytical methodologies. (H. Kim et al., 2010) Investigate the relationship between consumer innovation, market mavenism, shopping delight, and beliefs, attitudes, and patronage intentions related to pop-up retail in order to generate a powerful outcome. (Toubia and Florès, 2007). Customer ratings influence the algorithms used to screen ideas. The literature contains little or no structured study of the content of marketing innovations in preparation for Society 5.0. The proposed study will be one of the first to attempt to bridge this gap. Specifically, the objectives of this study can be determined and achieved as follows: The study is developed to clarify which marketing innovations are used by companies. The significance of acquiring a deeper understanding of innovations is emphasised for them to be accepted and implemented successfully. Wondering what types of marketing ideas are popular today? It's interesting how significant they are and how broadly they are used. This is important for identifying which marketing innovations should be promoted, induced, and distributed in the marketing world. In Society 5.0, not only technological but also marketing innovations are key components of any efficient business committed to sustainability. Various studies evaluate the benefits and drawbacks of marketing innovations and produce conflicting findings. It is suggested that an innovation-driven strategy is more likely to succeed than one led by a corporation that is hesitant to take chances. A corporation can function considerably more responsibly and with consideration for its clients by incorporating the most recent marketing techniques. As a result, the organisation can increase its customer base and benefits. One of the most intriguing research problems in the marketing industry today is about the development of the most effective marketing innovation that has been and can be used to analyse customer behaviour. Indeed, strategic development is intimately tied to the ability to forecast how people will respond in specific scenarios; it can result in excellent marketing research and client loyalty to the organisation.

STRATEGIC IMPORTANCE OF PREDICTIVE ANALYTICS IN MARKETING

The business world is changing rapidly. To catch up with this change and to stay ahead of your competition, marketing strategies in today's megacities such as Nagpur need to advance to Society 5.0. This advancement in marketing strategy can be made by implementing predictive analytics as a process of examining big data to identify patterns of a future outcome based on that analysis (Bashir Dar et al., 2021). As more companies embrace data-led decision-making, predictive analytics has become more readily available and powerful. In a world where consumers expect instant and hyper-personalised engagements, having the ability to anticipate consumer behaviour is a unique ability and potentially game-changing for the business. Fortunately, the complexity of predictive analytics stays hidden under the bonnet of the technology, with consumers experiencing only the perfectly tailored recommendations and interactions they encounter every day.

If a rigorous analysis of the data, when global warming or human activity is excluded as possible explanations, there are no indications left to support such skepticism. Therefore, the premonition of future events, based on prognostic signs of the weather, can be considered a correct application of predictive analytics. In its simplest form, predictive analytics employs algorithms and historical data to model potential outcomes either predicting that such outcomes will occur in the first place or suggesting different possible results and accounting for these results to inform a decision-making process. With data becoming all-pervading in the age of Society 5.0, predictive analytics is perhaps the most vital element needed in a business's armoury to succeed. By delivering the right message to the right audience at the right time, predictive analytics has subtly pervaded shopping, entertainment and lifestyles, adjusting them so that personalisation has now become the norm. This indirectly improves the consumer experience, which in turn means better customer returns and loyalty. For businesses, there is no better competitive advantage to have in an ever-increasingly crowded marketplace. The average consumer, exposed to nearly 15,000 digital advertisements per day, however, will only engage with around 6-7. Therefore, more it is essential to be one of these selected adverts. Through predictive marketing and targeting, campaigns can be made as focused as possible, yielding greater efficiency and reducing costs. Further strategies and techniques can also be encapsulated here, as companies start exploring patterns in consumer behaviour in Society 5.0, to facilitate more meaningful, engaging consumer or B2B interactions.

THE IMPORTANCE OF DATA IN PREDICTIVE ANALYTICS

Data is crucial for any entity as organizations rely on data-driven decision-making. Structured, statistically clean, and purposeful data is essential for any analytics endeavor. Right from the need assessment and exploratory research stages, data acquisition should be planned optimally with costs, time, and skill acquisition in mind (Bashir Dar et al., 2021). This holds true for every design input of any predictive analytical effort. As predictive analytics is mostly pattern following, the challenge of adequate adherence of data to the model assumptions is paramount. The quality of the analyzed data should be optimal to get accurate prediction outcomes. Along with the quality, some additional prerequisites needed are sufficient, current, relevant, clean, dynamic, compliant, and stored at the right place. In order to acquire it, awareness about the general availability of many big datasets is needed along with their pros and cons. Any entity can purchase data if it is not available free of cost. Several database solution providers, many of whom run businesses around data licensing, provide customer data libraries based on demographic variables and user-generated data sources by APIs. Compliance with the licensing agreement and data protection laws is non-negotiable with such purchases. Micro the ethical acquisition of data should also be ensured. Accuracy, integrity, decay and corruption, mining privacy, permissible access, further use, and privacy control comprise the basic ethical premises for any personal customer data usage. In the regional context, awareness and ownership with such personal data issues are either non-existent or mismanaged. The onus of awareness is more on organizations claiming to be protective about the customer privacy and never on organizations sharing such personal data without consent. Hence, the ethical questioning of this usage with multifaceted involvement is very pertinent which is presently not available in the Indian context.

TECHNOLOGICAL ADVANCEMENTS IN ANALYTICS

Technological advancements have fostered the continuous growth of big data sources, spurring research in data storage, sharing, mining, information generation, and value extraction. With the passive flow of data from the web being easier to process, an avalanche of traditional, heterogeneous, and unstructured data sources has accumulated, creating storage and computing issues for both practitioners and analysts. Advanced streaming techniques for social media data mining, step-wise data extraction, and identification of deliverable formats are some concerns present in the marketing and general domains. Although these incidental sources of data cover much of humanity's thoughts, behaviour, and socio-economic indicators, much research has been done to turn these data sources into behavioral and multi-relational analytics. While technical aspects of analytic extraction and processing help parse extracted pieces of social information or build cutting-edge tools, how these deliverables can be produced, interpreted, and utilized remains under-researched and unclear. Moreover, the linkages between different Analytic sets, or between the Analytic and Insight units, remain hazy (Bashir Dar et al., 2021). While predictive and complex analytics are in widespread use among early-adopters, practitioners face massive issues in internal readjustments and read-through of their outputs. Research efforts are necessary to study the operational stream, volume, formats, and visualization tools of Analytic units and translate these into usable structures and procedures for analytical functions and specialized departments. From a business-wide perspective, marketing experts have researched the dissemination of data analytics culture and models among various functions, organizational levels, and non-expert users, which are all subject to integration and horizontal alignment with complex operational features of the data-rich organizations. Traditional managerial and analytic know-hows and research efforts from other sciences have failed to provide comprehensive insight into the engine room of big data organizations, which live in and breathe continually changing information flows that affect human lives. Marketers should take the lead in this important area, and marketing intelligence models that account for speed, precision, and scope might offer useful references for approaching the analytics ecosystem. Moreover, new knowledge-flows full of subtlety and intricacy should be studied and modeled, which will expect partnerships and cross-disciplined efforts beyond traditional marketers.

Artificial Intelligence and Machine Learning

In the vision of Society 5.0, where virtual, digital, and physical spaces converge, various evolutions are taking place in technology in general and artificial intelligence in particular. Artificial Intelligence (AI) is considered a key driver of Society 5.0. AI will be popularly registered to improve the convenience of life and work and prevent different characteristics of anxiety regarding care, capital, cybersecurity, crime, trust, and natural disasters. AI is broadly used from a marketing perspective. However, the philosophy of AI-based marketing in Society 5.0 should be investigated comprehensively and systematically. Zero-Based Budgeting (ZBB) and AI-based Marketing are discussed as two significant topics independently and somewhat separately in the literature (Bayoude et al., 2018). ZBB is a comprehensive method of the retrospective resource allocation process. ZBB is used in cost-cutting measures and revenue allocation decisions in enterprises and commercial institutions. Empirical studies indicate significant associations between ZBB and customer satisfaction with the allocation of resources (ZBB) as well as customer loyalty. AI-based Marketing involves the formulation and implementation of marketing policies and programs that are based on the insights gained from AI data analyses of the behavior of present and prospective customers. Data-driven Competitive Intelligence (CI) involves the modeling of commercial competitors by gathering, processing, analyzing, and storing data. CI assists in preparing Strategic Options and the Advocacy of Strategic Options in marketing strategy analysis. Data-driven CI prepares marketing strategy analysis to select the best Strategic Options. SI drafting and discussion is vital to outline the Strategy Policies. At the point of Implementation Monitoring and Control, the strategies progress, evolving targets and means, are tracked to maintain the alignment with the 5S performance goals. The first dynamic progression is the Refinement of Strategy Analysis and Outlining Policies.

Big Data Technologies

The contemporary world is witnessing explosive growth of data originated from various sources such as the Internet of Things (IoT), social media, video recordings, smartphones, and business transactions. This huge volume, variety, and velocity of data termed as big data is being generated at an unprecedented speed. The high-growth sectors generating massive amounts of data are customers, health, and environmental data. As a complementary resource with physical assets, big data has been touted as a game-changer for industries. Advancements in computer technologies and algorithms are so remarkable that big data has been regarded as the new oil, an engine of growth, and a national resource. As a result of big data proliferation, a wealth of information is readily available to consumers and firms alike for the personalisation of product/service offerings, analysing consumer behaviour and decision-making, better understanding social phenomena, comparing information, and for many more other needs. The success of big data adoption thus depends on the capabilities of firms to avail substantial business benefits from it (Bashir Dar et al., 2021). The coming dawn of Marketing 4.0 has reshaped the marketing landscape, compelling researchers and practitioners to look at marketing efforts from digital and social media perspectives. Firms are now beginning to embrace the interaction between the physical and digital worlds to accommodate and communicate with consumers. Marketing efforts have shifted to driving customer engagement, seeking to understand customer behaviour along the product/service lifecycle. Within this context, marketing analytics has emerged as a source of competitive advantage, offering firms the ability to extract actionable insights from large and complex data sets (Catherine Read Brewis, 2016). The information facilitated by business analytics can help firms better understand marketing phenomena and consumer behaviour, leading to improved marketing decisions and business performance. Marketing analytics can help address the challenges faced by marketing in this rapidly evolving environment. The threefold purpose of the study is to advance understanding of big data technologies in marketing, identify current trends, gaps, and future research priorities in big data technologies in marketing, and provide useful knowledge for further advancement in the field. Further, the importance of big data technologies in marketing is reflected on the applicability of the identified articles, as well as the recommendations in the identified articles.

FRAMEWORK FOR PREDICTIVE ANALYTICS IN MARKETING

In the transition to Society 5.0, marketing plays an important role in contributing to the consumer welfare, economic development, and environmental sustainability of society as a whole. Marketing has the potential to empower the masses by creating aspirations that instigate growth in consumer demand and eventually in demand-side industries. In addition, marketing has the potential to change the lives of masses by addressing heretofore neglected issues such as low literacy, low income, hygiene poverty, and pollution (Bashir Dar et al., 2021). Marketing in Society 5.0 could be empowered to develop scalable, replicable, and low-cost solutions to existing issues of the masses and protect the ethicality of the marketing mix. However, the marketing managers would need to identify macro issues in India that are being faced by the masses and the role that marketing can play in addressing them. As the risks associated with consumer behaviour analysis is mostly reductionist, addressing the issue of consumer behaviour analysis may be the right starting point.

Predictive analytics in the simplest sense refers to the use of historical data and past experience to predict, with a high degree of fulfilment, the future. Its inevitable adoption in marketing is plausible as the price of computing and storage has reduced multi-fold in the past decade. With the proliferation of the internet; online shopping along a wide number of categories; social media; and the availability of large amounts of data on consumer web behaviour, it is possible to collect huge amounts of data about consumers. Predictive analytics can reveal what a consumer is likely to buy and when, sharpening the effectiveness of target marketing. Most of the understanding of consumer behaviour is either descriptive or optimum. This leaves a huge space for data-rich environments, which calls for predictive analytics.

RESEARCH METHODOLOGY

The goal of this study is to investigate the role of predictive analytics in analysing consumer behaviour and marketing approaches in Society 5.0, using a mixed-methodologies approach that combines quantitative and qualitative data collection methods. We will perform a cross-sectional study to collect data over a set time period in order to examine the present influence of predictive analytics on marketing. Respondents for both quantitative and qualitative data will be chosen via purposive sampling. The quantitative research will focus on companies and marketers who use predictive analytics solutions. The qualitative component will identify key stakeholders, such as data analysts and marketing experts involved in decision-making and analytics. To ensure that the results are both generalisable and diversified, quantitative data will be gathered from at least 100 digital marketing campaigns across a variety of industries. In-depth interviews with 10-15 marketing experts and data analysts will be used to collect qualitative data on real-world predictive analytics applications. Data from digital marketing initiatives will be gathered, with a special emphasis on crucial indicators, including customer engagement, conversion rates, and return on investment (ROI). The predictive analytics software's effectiveness will be evaluated. Semi-structured interviews will be undertaken to get expert viewpoints on the potential and challenges of using predictive analytics in marketing operations. The Orange Data Mining program will implement predictive modelling and data analytics. This application will assess the accuracy of prediction models in understanding customer behavior using data visualization, categorization, and regression analysis. Predictive models: Orange Predictive models will be developed using software and machine learning approaches such as random forests and decision trees. These models will assess the accuracy of projecting consumer behaviour using historical data from marketing efforts.

Research Questions:

1) In what ways does predictive analytics technology influence the examination of consumer behaviour in Society 5.0?
2) How can marketers leverage predictive analytics to develop innovative marketing strategies?
3) What are the principal advantages of predictive analytics in enhancing marketing strategies, such as improved consumer engagement and return on investment (ROI)?

Objectives of the Study:

1) To examine the effect of predictive analytics on consumer behaviour in the context of Society 5.0.
2) To assess predictive analytics' accuracy and efficacy in optimising resource allocation and increasing ROI for marketing efforts.
3) Examine the impact of predictive analytics on balancing technology innovation, customer needs, and societal well-being.

THE PREDICTIVE ANALYTICS IN ANALYSING CONSUMER BEHAVIOUR FOR MARKETERS

Figure 2. Orange workflow for using different predictive models

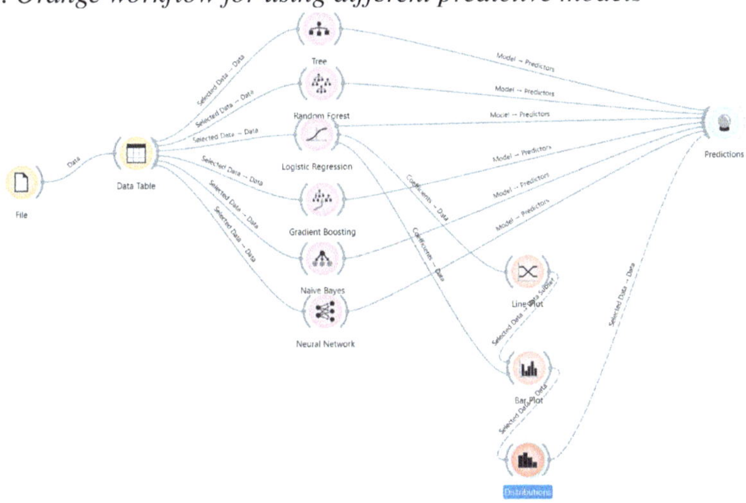

(Predictive Analytics by Orange)

Figure 3. Orange Tree Diagram to check accuracy of predictive models for Strategic Decisions

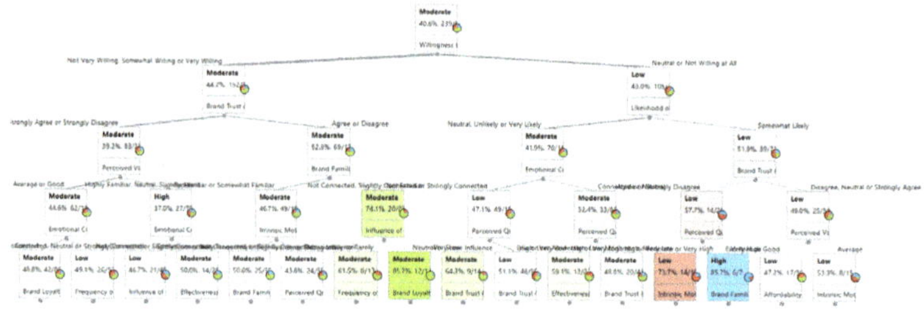

(Predictive Analytics by Orange)

Figure 4. Logical regression model for the predictive analytics

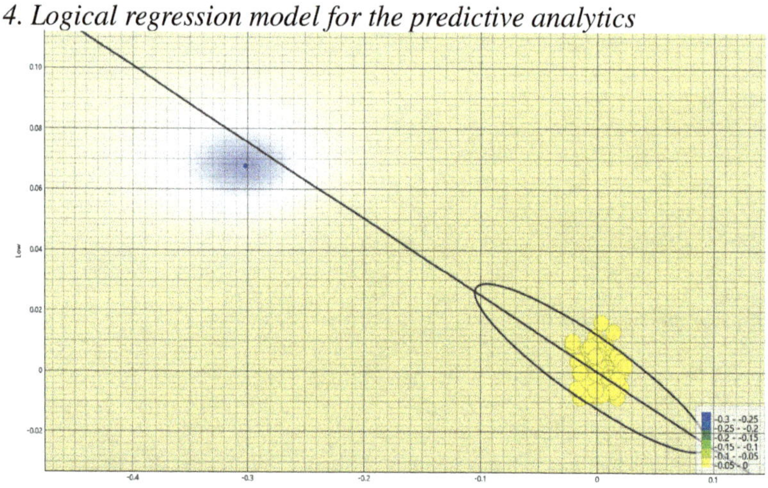

(Predictive Analytics by Orange)

Figure 5. Colliniarity model for the predictive analytics

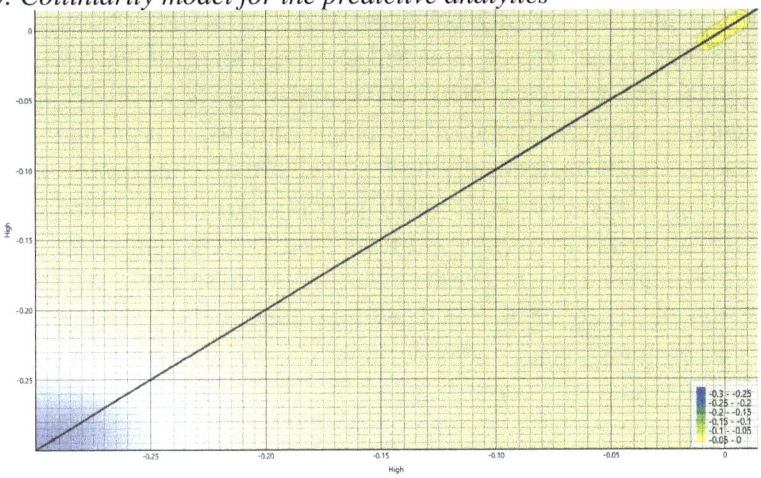

(Predictive Analytics by Orange)

The following table gives a summary of the results.

Table 1. Accuracy measures for the first target variable

Model	AUC	CA	F1	Prec	Recall	MCC
Logistic Regression	0.957	0.9305	0.9253	0.9553	0.9783	0.9553
Random Forest	0.9578	0.9383	0.938	0.9395	0.9355	0.9373
Gradient Boosting	0.9732	0.9857	0.9855	0.9858	0.9853	0.9798
Naive Bayes	0.7583	0.753	0.7533	0.9533	0.9539	0.9233
Neural Network	1	1	1	1	1	1

The performance metrics of the predictive models indicate a robust ability to analyse customer behaviour successfully within the framework of Society 5.0 marketing tactics. The Neural Network model got flawless scores on all performance metrics (AUC: 1, CA: 1, F1: 1, Precision: 1, Recall: 1, MCC: 1), establishing it as the most robust and dependable model for predictive analytics in consumer behaviour analysis. Gradient Boosting demonstrated remarkable performance, achieving an AUC of 0.9732 and an accuracy of 98.57%, highlighting its capability to capture complex consumer decision-making patterns. Random Forest and Logistic Regression exhibited commendable performance, achieving AUC ratings of 0.9578 and

0.957, respectively, so demonstrating their efficacy in predictive modelling. Logistic Regression, despite its simplicity, achieved good accuracy (93.05%) and precision (95.53%), rendering it an interpretable and scalable choice for marketing analytics. The suboptimal performance of Naïve Bayes (AUC: 0.7583) indicates that it may not be suitable for intricate customer behaviour patterns, as it presupposes feature independence, a condition frequently absent in actual marketing contexts. According to these findings, Neural Networks and Gradient Boosting are strongly endorsed for predictive analytics in Society 5.0 marketing tactics, guaranteeing enhanced accuracy, precision, and recall. These models will yield strategic insights into consumer preferences, enabling personalised marketing tactics, targeted campaigns, and enhanced customer engagement techniques.

SUMMARY OF KEY FINDINGS

As this chapter concludes, AI growth has been tremendous during the last few years, and its marketing applications across front and back-end activities are highly significant. Consumer behavior is evolving continuously in ways that are becoming increasingly difficult for organizations to predict and navigate in an actionable way. However, that challenge is being met with newer tools like predictive analytics. The focus of this text is on business analytics, and particularly on predictive analytics, how businesses are deploying these solutions, and which new applications are likely to emerge in view of these advancements. Predictive analytics-based marketing is today being applied across a wide spectrum of sectors, and integrating it with AI takes its undertakings even farther. Organizations can exploit predictive analytics and AI through a wider range of technologies than ever before, and customer data coupled with various analytical tools and capabilities, which are constantly evolving, offer novel opportunities for companies. AI is an integral part of modern business analytics and helps businesses not only to predict future outcomes but also to automate or semi-automate corrective action suggestions through various strategies.

PRACTICAL IMPLICATIONS FOR MARKETERS

This study has multiple implications for both strategic marketing management and industry. Indeed, it has increasingly been argued that big data analytics capabilities create a basis for innovation and competitive advantage for companies. Given rapidly increasing data volumes and the growing complexity and variety of data, associations apply data analytics to improve marketing operations. By clustering consumers based on past consumption behavior, predictive analytics identify group-specific patterns, test group differences regarding online and mobile device use, and estimate the influence of the results for strategic decision-making. Classification techniques allow the selection rules to discern two different clusters from the dataset, season-based attributes, and varying purchasing behavior within these clusters over time

BEST PRACTICES FOR LEVERAGING PREDICTIVE ANALYTICS IN MARKETING

The following best practices are distilled from successful case studies and are recommended for marketers wanting to leverage predictive analytics as a tool to enhance their marketing effectiveness: Prioritize data quality to create a deep understanding of consumers. Quality data leads to actionable insights that improve marketing effectiveness. Transparency fosters trust. A proprietary platform is only essential to control data. For substantial benefits, data confidentiality and exclusivity may be used. Materialize findings quickly and at low cost by means of test and learn deployments. Have a start and stop date on research projects to build in the understanding that this will not be the final answer and that there will be a continuing process of enhancement as the results commence and more findings are required. Expand competencies or initiate a partnership with universities or consultancies to enhance capacities for doing predictive analytics. Develop a company-wide culture of evidence-based decision-making in the long term. Open, clear, and focused structures encourage free exchanges of thoughts, exposing the company to fresh strategies and encouraging innovation. In conclusion, marketers and organizations must use this information to integrate predictive analytics into their marketing strategies. By doing so, they can enhance the capabilities of their marketing staff. This will make marketers more capable of executing marketing strategies and delivering customer value, which in turn will support the marketing concept objectives of enhancing consumer standings and, consequently, industry profitability. This study contributes to the still limited amount of academic research on the practice of predictive analytics in marketing. The study is unique because it aggregates best practices from influential business people, senior managers, and marketing personnel.

ETHICAL CONSIDERATIONS IN PREDICTIVE ANALYTICS

Predictive modeling and analysis have become a key tool for marketing and companies increasingly use predictive analytics to inform their decisions about consumer acquisition, retention, and investments. As an approach to economic analysis and benchmarking for predictive analytics, this paper addresses the strategic role of predictive analytics in consumer behavior analysis of Nagpur. It aims to advance marketing in society 5.0. The strategic importance of consumer behavior analysis and its implications for public policy, market structure, and competition are discussed. Moreover, the predictive model as applied by marketers and economists often serves to distinguish the market mechanism and their implications for market policy. Two types of predictive models are setting up, and an optimal choice of description for predictive models encased in demand analysis is addressed. Next, a focus on data, methodology, and forecast accuracy, with lots of room for improvement, and the market analysis and business scope is concluded. Predictive analysis and consumer behavior have recently emerged as important tools for marketers. They aim to anticipate marketing effectiveness or customers' future actions by analyzing a plethora of customer and market data. Customers' actions and interactions, tracked by companies, are then described, in turn, by predictive models, i.e., models forecasting, e.g., purchase propensity, churn probability, customer lifetime value, etc. These predictive models are applied by marketers to inform their decisions about acquisitions, retention actions, pricing, channelling, and so forth. But as the use of this approach becomes mainstream, questions arise both about the strategic implications for competition and the ethical issues arising from the massive availability of data on consumers' actions. These data, in fact, mostly come from companies themselves and thus often provide information on competitors' customers.

FUTURE TRENDS AND IMPLICATIONS

Introduction Given the pace of advancements in technology and data analytics, predictive analytics is anticipated to play a more significant strategic role in consumer behavior analysis and marketing practices. With the advancements in technology, data analytics often using predictive modeling is expected to rapidly progress and will take the marketing practices one step ahead of the traditional practices. The marketing practices in the Nagpur, India, landscape will soon become utterly adaptive with these techniques. These predictive modeling techniques will be evolving and expected to show inkling toward multiclass classification and will enable marketers to anticipate the campaigns and advertisements that the customers are receptive to before launching them. Also, the era of advertisements will soon confine to hyper-targeting and will be supported through push-notifications on smart devices and location-based advertisements. Taking the insights from Big Data, the Nagpur marketers need to pivot—even cannibalize—what they had acquired and adopted for years. Naïve marketing practitioners who ignore these insights will soon find themselves trailing far behind nimbler competitors. Echoing (Bashir Dar et al., 2021), since analytics has been widely accepted by strategic planners, marketers need to do them or be prepared to be left behind. Consumer behavior analysis given the high degree of uncertainty and evolving societal norms is forever demanding. Shaping profound corporate strategies with a few sophisticated techniques and methods each season make the business landscape utterly restrictive. If the predictions prove to be disjointed with outcomes, corporations will be inclined to plan strategies around inferred outcomes much different from the real ones. Marketers and corporations, consequently, may end up in a quagmire.

EMERGING TECHNOLOGIES IN PREDICTIVE ANALYTICS

The marketing era has evolved through five stages, and the 5th stage is said to be society-oriented. There are several articles mentioned about the development of marketing from stage 1.0 to era 4.0, including the criticism faced by this era. The rise of a hand-crafted digital analytics tail, tracing the web activity of individuals and matching it to their names and addresses, suggests further criticisms, and it carries the risk of a consumer backlash. In 2024 World Marketing Congress, the birth and development of the five stages of marketing were reviewed, focusing on the advancement of marketing to surpass the 4.0 era with the important question of how certain this society-oriented marketing, known as Marketing 5.0, adopted a more comprehensive view with this article. This subsection introduces and explores developing technologies, and the trends they are set to influence, that would shape the future of marketing through predictive analytics. Wave predictions are made for technology applications, commercial implementations, and the affecting environment every 5, 2, and 1 year ahead. Soothsaying is focused on 2 facets. First, technology emergences that are intrinsically novel and would lead to unique marketing adaptations. Second, technologies would bring about widespread changes and inducements in predictive analytics applications. The environment in which predictions are being made mandates a focus on the more dynamic aspects of tech development. Consequently, more conventional or relatively stable technology can evolve but is assumed knowledge and not a salient input to the future-facing discussion is left unmentioned. It encompasses innovations likely to affect or are expected to derive advantage from the deployment of predictive analytics-based forecasting methods. An indicative selection of emerging technologies that marketers will deal with in the next era includes blockchain-based smart contracts enabling advanced audit seeding, self-enforcing payments, and heightened consumer control. Smart clothing that amalgamates wearable technology with predictive data analysis methods to improve user health and fitness outcomes. Other techs to keep an eye on are tools facilitating nano-targeted personalized advertising, improved non-intrusive biometrics for market research, and robotized AI customer service for increased accessibility and exceptionally rapid content delivery

RESULT & DISCUSSION

Innovation in marketing analytics has led to collaboration between data scientists and marketing managers. Machine learning models are crucial for explaining the why of prediction or classification. Recent improvements in prediction interpretation methods have heightened interest in providing practical interpretive solutions to firms (Bashir Dar et al., 2021). However, it is important to clarify management's genuine interest and needs in understanding the predictive models to ensure development of practical solutions. This research develops strategies for gaining managerial insights from existing predictive marketing analytics. The first understanding acquisition strategy takes an exploration approach to discover what existing models predict, where trainable solutions are used to help them understand the importance of predictions (D. Vollrath & G. Villegas, 2021). The data-driven knowledge aggregation strategy helps cross-business model explanations, where existing solutions are developed to transfer knowledge from predicted to predictor variable space. The mixed understanding enhancement strategy assists managers explain more by formulating predictive information into comprehensible and structured messages, where comprehensible messages are produced using existing and adapted interpretation methods and structured messages consist of visualizations, concise explanations, and how-to guidance. All three strategies are novel and have substantial strategic implications for firms.

Explorations of a past client churn prediction model of a telecommunication company demonstrate how to make existing predictive marketing analytics understandable for senior managers with diverse statistical knowledge. The results indicate that whilst an understanding acquisition strategy does not guarantee managerial understanding improvements, it provides research avenues to develop generic management needs clarification solutions. Hence, three valuable strategies are introduced for improving interpretation of existing predictive marketing analytics. The exploration strategy aims to help managers discover what predictive marketing analytics model predictions. The knowledge aggregation strategy helps explain predictions by expressing the marketing share of predictions in previous case choices. The understanding enhancement strategy aims to enhance each understanding type. Practical guidance on how to implement understanding improvement strategies reduces existing predictive marketing analytics' black-box character and is expected to expedite further conceptualization and development of interpretive marketing analytics.

CONCLUSION

This study highlights an emerging concern toward using predictive analytics in understanding consumer behavior to retail vendors in stores, especially in Nagpur, Maharashtra Province in order to survive in extremely competitive consumer goods markets. It also offers a solution to retail vendors in stores to enhance their adaptation of using predictive analytics. Various shopping behaviors of customers (especially in Nagpur region) were analyzed using deep learning neural networks. Analyzing customer behavior in store shopping from Historical transaction data on customer purchase Ration between January and February 2025 in Nagpur region showed that the common constraint is to use traditional statistical methods in Orange Software. The results of this study show that classification models are better than statistical models in analyzing consumer behavior data. Predictive analytics plays an important role in understanding consumer behavior and advancing marketing strategies. In businesses, consumer behavior analysis helps to shape the way a store is initiated or maintained by better understanding shopping behavior in the store. Even arranging spaces well in store can be an important consideration. These business decisions are faced by many vendors, especially vendors of consumer goods or groceries in store who in some cases can become a compete against one another. Thus, using data-driven approaches is essential for businesses and marketers to remain competitive and relevant in consumer behavior analytics (Bashir Dar et al., 2021). Numerous research in consumer behavior analysis has focused on collecting and manipulating structured data using various statistical methods. However, there are various unstructured and/or semi-structured data sources, including social networks, blog posts, or reviews with less statistical manipulation. The resulting sentiment analysis can analyze market segment orientations to certain goods/services, assess the effectiveness of word-of-mouth campaigns, or even assist in managing the reputation of a company. The key findings point to the need of transparent channels for the consumer behaviour understanding use, stimulating the ways marketing and personalised consumer experience are handled. As big data expands, harvesting is an undying process. Marketeers, a core player with increasing data are encouraged to navigate through the challenges by further exploring data and capitalising on the opportunity of a more detailed customer profile identification, they are urged to do it by investing in understanding their own consumer, and use the harvested data responsibly.

FUTURE SCOPE OF THE STUDY

This study attempts to delineate predictive analytics as a research-concept and investigates its strategic role in the analysis of consumer behaviour in India. Available literature was systematically reviewed and synthesized, revealing that earlier works were predominantly exploratory studies describing predictive analytics definitions & methodologies. To bridge this gap, the air of research was categorized and presented as a conceptual framework showing predictive analytics and its various approaches (descriptive, predictive, and prescriptive) that can be operationalized by the marketing managers. The capability of predictive analytics is tested in consumer behaviour analysis domain encompassing purchasing decisions, attitudes evaluation, target market selection, unwanted forecasting, churn estimation, and provision breakthroughs (Bashir Dar et al., 2021).

Advent of the digital space and smart devices has transformed the marketing paradigms with real-time and ubiquitous access to consumers (consumer-in-the-loop). To leverage this, organizations have been investing in analytics viz. descriptive analytics using summary statistics; diagnostic analytics using manual modelling & simulation; predictive analytics using machine learning techniques; and prescriptive analytics using optimizations. Of them, the most mature yet underrepresented area is predictive analytics, which is tested to be the most economically impactful form of analytics.

Utilizing various data types and enabling decisions in consumer behaviours across industries, predictive analytics is researched as a tactic in marketing field. It is of significance for the developing economy like India, for which the present study can be viewed as inaugural. Further research avenue focuses on basic consumer behaviour responses, where the techniques and methods are more accounted for. Methodologically, exploratory case-study design is followed using multiple sources of data with model robustness ensured through triangulation.

REFERENCES

Andrejevic, M. (2011). THE WORK THAT AFFECTIVE ECONOMICS DOES. *Cultural Studies*, *25*(4–5), 604–620. DOI: 10.1080/09502386.2011.600551

) Bashir Dar, I., Bashir Khan, M., Zahid Khan, A., & G. Mujtaba, B. (2021). A qualitative analysis of the marketing analytics literature: where would ethical issues and legality rank?.

Bayoude, K., Ouassit, Y., Ardchir, S., & Azouazi, M. (2018). How machine learning potentials are transforming the practice of digital marketing: State of the art.

Borges, A. F. S., Laurindo, F. J. B., Spínola, M. M., Gonçalves, R. F., & Mattos, C. A. (2021). The strategic use of artificial intelligence in the digital era: Systematic literature review and future research directions. *International Journal of Information Management*, *57*, 102225. DOI: 10.1016/j.ijinfomgt.2020.102225

Catherine Read Brewis, C. (2016). To What Extent Are UK Companies Using Big Data to Inform Their Marketing Strategy?.

Chaudhary, K., Alam, M., Al-Rakhami, M. S., & Gumaei, A. (2021). Machine learning-based mathematical modelling for prediction of social media consumer behavior using big data analytics. *Journal of Big Data*, *8*(1), 73. DOI: 10.1186/s40537-021-00466-2

Chaudhary, K., Alam, M., Al-Rakhami, M. S., & Gumaei, A. (2021). Machine learning-based mathematical modelling for prediction of social media consumer behavior using big data analytics. *Journal of Big Data*, *8*(1), 73. DOI: 10.1186/s40537-021-00466-2

Chen, K., Lin, J., & Song, Y. (2019). Trading strategy optimization for a prosumer in continuous double auction-based peer-to-peer market: A prediction-integration model. *Applied Energy*, *242*, 1121–1133. DOI: 10.1016/j.apenergy.2019.03.094

Cheng, B. L., Cham, T. H., Dent, M. M., & Lee, T. H. (2019). Service innovation: Building a sustainable competitive advantage in higher education. *International Journal of Services. Economics and Management*, *10*(4), 289. DOI: 10.1504/IJSEM.2019.105014

Cheng, B. L., Cham, T. H., Dent, M. M., & Lee, T. H. (2019). Service innovation: Building a sustainable competitive advantage in higher education. *International Journal of Services. Economics and Management*, *10*(4), 289. DOI: 10.1504/IJSEM.2019.105014

Chiesa, V., & Frattini, F. (2011). Commercializing Technological Innovation: Learning from Failures in High-Tech Markets*: Commercializing Technological Innovation. *Journal of Product Innovation Management*, *28*(4), 437–454. DOI: 10.1111/j.1540-5885.2011.00818.x

Chong, A. Y.-L., Chan, F. T. S., & Ooi, K.-B. (2012). Predicting consumer decisions to adopt mobile commerce: Cross country empirical examination between China and Malaysia. *Decision Support Systems*, *53*(1), 34–43. DOI: 10.1016/j.dss.2011.12.001

Dahake, N. S., Suresh Dahake, P., & Gajghate, A. (2024). Beyond the Hype: A Critical Evaluation of Predictive Analytics Accuracy in High-Stakes Business Decisions. *2024 2nd DMIHER International Conference on Artificial Intelligence in Healthcare, Education and Industry (IDICAIEI)*, 1–6. DOI: 10.1109/IDICAIEI61867.2024.10842857

Dahake, P. S., Chandak, S., Mohare, R. V., Wadhwani, K., & Bhadade, P. (2023). The Crystal Ball of Marketing: How Predictive Analytics is Reshaping the Industry? *2023 Second International Conference On Smart Technologies For Smart Nation (SmartTechCon)*, 304–311. DOI: 10.1109/SmartTechCon57526.2023.10391334

Dahake, P. S., Mohare, R. V., & Somani, N. (2023). Exploring Machine Learning's Potential in Predicting Consumer Behavior for Captivating Marketing. *2023 Second International Conference On Smart Technologies For Smart Nation (SmartTechCon)*, 1347–1352. DOI: 10.1109/SmartTechCon57526.2023.10391558

Dahake, P. S., & Somani, N. (2024). Harnessing Predictive Analytics for Accurate Consumer Behaviour Forecasting: A Comprehensive Review. *2024 2nd DMIHER International Conference on Artificial Intelligence in Healthcare, Education and Industry (IDICAIEI)*, 1–6. DOI: 10.1109/IDICAIEI61867.2024.10842743

Dahl, D. W., & Hoeffler, S. (2004). Visualizing the Self: Exploring the Potential Benefits and Drawbacks for New Product Evaluation. *Journal of Product Innovation Management*, *21*(4), 259–267. DOI: 10.1111/j.0737-6782.2004.00077.x

De Oliveira Santini, F., Ladeira, W. J., Pinto, D. C., Herter, M. M., Sampaio, C. H., & Babin, B. J. (2020). Customer engagement in social media: A framework and meta-analysis. *Journal of the Academy of Marketing Science*, *48*(6), 1211–1228. DOI: 10.1007/s11747-020-00731-5

Diamantoulakis, P. D., Kapinas, V. M., & Karagiannidis, G. K. (2015). Big Data Analytics for Dynamic Energy Management in Smart Grids. *Big Data Research*, *2*(3), 94–101. DOI: 10.1016/j.bdr.2015.03.003

Fearne, A., & Hughes, D. (1999). Success factors in the fresh produce supply chain: Insights from the UK. *Supply Chain Management*, *4*(3), 120–131. DOI: 10.1108/13598549910279567

Garcia Martinez, M., & Walton, B. (2014). The wisdom of crowds: The potential of online communities as a tool for data analysis. *Technovation*, *34*(4), 203–214. DOI: 10.1016/j.technovation.2014.01.011

Gauba, H., Kumar, P., Roy, P. P., Singh, P., Dogra, D. P., & Raman, B. (2017). Prediction of advertisement preference by fusing EEG response and sentiment analysis. *Neural Networks*, *92*, 77–88. DOI: 10.1016/j.neunet.2017.01.013 PMID: 28254237

Gauba, H., Kumar, P., Roy, P. P., Singh, P., Dogra, D. P., & Raman, B. (2017). Prediction of advertisement preference by fusing EEG response and sentiment analysis. *Neural Networks*, *92*, 77–88. DOI: 10.1016/j.neunet.2017.01.013 PMID: 28254237

Goldsmith, R. E., Moore, M. A., & Beaudoin, P. (1999). Fashion innovativeness and self-concept: A replication. *Journal of Product and Brand Management*, *8*(1), 7–18. DOI: 10.1108/10610429910257904

Guha, S., & Kumar, S. (2018). Emergence of Big Data Research in Operations Management, Information Systems, and Healthcare: Past Contributions and Future Roadmap. *Production and Operations Management*, *27*(9), 1724–1735. DOI: 10.1111/poms.12833

Guiné, R. P. F., Florença, S. G., Barroca, M. J., & Anjos, O. (2020). The Link between the Consumer and the Innovations in Food Product Development. *Foods*, *9*(9), 1317. DOI: 10.3390/foods9091317 PMID: 32962007

Hinsch, C., Felix, R., & Rauschnabel, P. A. (2020). Nostalgia beats the wow-effect: Inspiration, awe and meaningful associations in augmented reality marketing. *Journal of Retailing and Consumer Services*, *53*, 101987. DOI: 10.1016/j.jretconser.2019.101987

Huang, M.-H., & Rust, R. T. (2021). A strategic framework for artificial intelligence in marketing. *Journal of the Academy of Marketing Science*, *49*(1), 30–50. DOI: 10.1007/s11747-020-00749-9

Ismagilova, E., Slade, E., Rana, N. P., & Dwivedi, Y. K. (2020). The effect of characteristics of source credibility on consumer behaviour: A meta-analysis. *Journal of Retailing and Consumer Services*, *53*, 101736. DOI: 10.1016/j.jretconser.2019.01.005

Kang, J., & Park-Poaps, H. (2010). Hedonic and utilitarian shopping motivations of fashion leadership. *Journal of Fashion Marketing and Management*, *14*(2), 312–328. DOI: 10.1108/13612021011046138

Kim, H., Fiore, A. M., Niehm, L. S., & Jeong, M. (2010). Psychographic characteristics affecting behavioral intentions towards pop-up retail. *International Journal of Retail & Distribution Management, 38*(2), 133–154. DOI: 10.1108/09590551011020138

Kim, W. B., Kim, C., & Kurata, K. (2025). Relationships among consumer innovativeness, learning, and global product purchases: Store manager perspectives in retailing. *Journal of Retailing and Consumer Services, 82*, 104084. DOI: 10.1016/j.jretconser.2024.104084

Korfiatis, N., García-Bariocanal, E., & Sánchez-Alonso, S. (2012). Evaluating content quality and helpfulness of online product reviews: The interplay of review helpfulness vs. review content. *Electronic Commerce Research and Applications, 11*(3), 205–217. DOI: 10.1016/j.elerap.2011.10.003

Lazarus, M. A., Blume, M. A. U., Brown, K. B., Caid, W. R., Dunning, T. E., Peranich, L. S., Russell, G. R., & Sitze, K. L. (2004). *Predictive modeling of customer financial behavior* (European Union Patent No. EP1050833A3). https://patents.google.com/patent/EP1050833A3/en

Li, Y., Jin, X., & Xue, H. (2025). Avoid following individual shopping time habit to promote new products: Evidence from a randomized field experiment. *Journal of Retailing and Consumer Services, 82*, 104058. DOI: 10.1016/j.jretconser.2024.104058

Mamun, A. A., Sohel, M., Mohammad, N., Haque Sunny, M., Dipta, D. R., & Hossain, E. (2020). A Comprehensive Review of the Load Forecasting Techniques Using Single and Hybrid Predictive Models. *IEEE Access : Practical Innovations, Open Solutions, 8*, 134911–134939. DOI: 10.1109/ACCESS.2020.3010702

Mustak, M., Salminen, J., Plé, L., & Wirtz, J. (2021). Artificial intelligence in marketing: Topic modeling, scientometric analysis, and research agenda. *Journal of Business Research, 124*, 389–404. DOI: 10.1016/j.jbusres.2020.10.044

Olshavsky, R. W., & Spreng, R. A. (1996). An Exploratory Study of the Innovation Evaluation Process. *Journal of Product Innovation Management, 13*(6), 512–529. DOI: 10.1111/1540-5885.1360512

Park, C., & Jun, J. (2003). A cross-cultural comparison of Internet buying behavior: Effects of Internet usage, perceived risks, and innovativeness. *International Marketing Review, 20*(5), 534–553. DOI: 10.1108/02651330310498771

Pham, T.-T. T., & Ho, J. C. (2015). The effects of product-related, personal-related factors and attractiveness of alternatives on consumer adoption of NFC-based mobile payments. *Technology in Society, 43*, 159–172. DOI: 10.1016/j.techsoc.2015.05.004

Roach, G. (2009). Consumer perceptions of mobile phone marketing: A direct marketing innovation. *Direct Marketing (Bradford)*, *3*(2), 124–138. DOI: 10.1108/17505930910964786

Sahu, A. K., Padhy, R. K., & Dhir, A. (2020). Envisioning the Future of Behavioral Decision-Making: A Systematic Literature Review of Behavioral Reasoning Theory. *Australasian Marketing Journal*, *28*(4), 145–159. DOI: 10.1016/j.ausmj.2020.05.001

Scuotto, V., Del Giudice, M., & Carayannis, E. G. (2017). The effect of social networking sites and absorptive capacity on SMES' innovation performance. *The Journal of Technology Transfer*, *42*(2), 409–424. DOI: 10.1007/s10961-016-9517-0

Siegel, E. (Ed.). (2015). *Predictive Analytics: The Power to Predict Who Will Click, Buy, Lie, or Die, Revised and Updated* (1st ed.). Wiley., DOI: 10.1002/9781119172536

Sorescu, A. B., & Spanjol, J. (2008). Innovation's Effect on Firm Value and Risk: Insights from Consumer Packaged Goods. *Journal of Marketing*, *72*(2), 114–132. DOI: 10.1509/jmkg.72.2.114

Szmigin, I., & Foxall, G. (1998). Three forms of innovation resistance: The case of retail payment methods. *Technovation*, *18*(6–7), 459–468. DOI: 10.1016/S0166-4972(98)00030-3

Toubia, O., & Florès, L. (2007). Adaptive Idea Screening Using Consumers. *Marketing Science*, *26*(3), 342–360. DOI: 10.1287/mksc.1070.0273

Vollrath, M. D., & Villegas, S. G. (2022). Avoiding digital marketing analytics myopia: Revisiting the customer decision journey as a strategic marketing framework. *Journal of Marketing Analytics*, *10*(2), 106–113.

Williams, A. (2006). Tourism and hospitality marketing: Fantasy, feeling and fun. *International Journal of Contemporary Hospitality Management*, *18*(6), 482–495. DOI: 10.1108/09596110610681520

Zekić-Sušac, M., Has, A., & Knežević, M. (2021). Predicting energy cost of public buildings by artificial neural networks, CART, and random forest. *Neurocomputing*, *439*, 223–233. DOI: 10.1016/j.neucom.2020.01.124

Zonta, T., Da Costa, C. A., Da Rosa Righi, R., De Lima, M. J., Da Trindade, E. S., & Li, G. P. (2020). Predictive maintenance in the Industry 4.0: A systematic literature review. *Computers & Industrial Engineering*, *150*, 106889. DOI: 10.1016/j.cie.2020.106889

Compilation of References

Abbas, A. (2024). AI for predictive maintenance in industrial systems. *International Journal of Advanced Engineering Technologies and Innovations*, *1*(1), 31–51. DOI: 10.31219/osf.io/vq8zg

Abdallah, A. (2020). The effect of corona pandemic on workers' psychology and performance in the manufacturing sector. *Systematic Reviews in Pharmacy*, *11*(12), 1593–1599.

Abollado, J. R., Shehab, E., & Bamforth, P. (2017). Challenges and benefits of digital workflow implementation in aerospace manufacturing engineering. *Procedia CIRP*, *60*, 80–85. DOI: 10.1016/j.procir.2017.02.044

Acemoglu, D., & Restrepo, P. (2020). Robots and jobs: Evidence from US labor markets. *Journal of Political Economy*, *128*(6), 2188–2244. DOI: 10.1086/705716

Achouch, M., Dimitrova, M., Ziane, K., Sattarpanah Karganroudi, S., Dhouib, R., Ibrahim, H., & Adda, M. (2022). On predictive maintenance in industry 4.0: Overview, models, and challenges. *Applied Sciences (Basel, Switzerland)*, *12*(16), 8081. DOI: 10.3390/app12168081

Adams, R., Bessant, J., & Phelps, R. (2006). Innovation management. *International Journal of Management Reviews*, ●●●, 1–45.

Adams, R., Martin, S., & Boom, K. (2016). University culture and sustainability: Designing and implementing an enabling framework. *Journal of Cleaner Production*, *171*, 434–445. DOI: 10.1016/j.jclepro.2017.10.032

Addepto. (n.d.). *Top 10 practical applications and use cases of AI in the aviation industry*. https://addepto.com/blog/top-10-practical-applications-and-use-cases-of-ai-in-the-aviation-industry/

Adebayo, R. A., Obiuto, N. C., Olajiga, O. K., & Festus-Ikhuoria, I. C. (2024). AI-enhanced manufacturing robotics: A review of applications and trends. *World Journal of Advanced Research and Reviews, 21*(3), 2060–2072. DOI: 10.30574/wjarr.2024.21.3.0924

Adebowale, O. J., & Agumba, J. N. (2024). Applications of augmented reality for construction productivity improvement: A systematic review. *Smart and Sustainable Built Environment, 13*(3), 479–495. DOI: 10.1108/SASBE-06-2022-0128

Adel, A. (2022). Future of industry 5.0 in society: Human-centric solutions, challenges and prospective research areas. *Journal of Cloud Computing (Heidelberg, Germany), 11*(1), 40. DOI: 10.1186/s13677-022-00314-5 PMID: 36101900

Aerospace Industries Association of the Philippines. (2013). *Aerospace manufacturing: An industry road map.*

Aerospace Industries Association of the Philippines. (2023). *Aerospace manufacturing: An industry road map.*

Aggarwal, K. K., & Agrawal, S. (2024). Artificial intelligence and its role in financial market. In *Global Financial Analytics and Business Forecasting*.

Aggarwal, K. K., Sharma, A., Kaur, R., & Lakhera, G. (2023). Algorithmic FinTech pioneering the financial landscape of tomorrow. In *Algorithmic Approaches to Financial Technology*. Forecasting, Trading, and Optimization., DOI: 10.4018/979-8-3693-1746-4.ch002

Agnihotri, R., Chaker, N. N., Dugan, R., Galvan, J. M., & Nowlin, E. (2023). Sales technology research: A review and future research agenda. *Journal of Personal Selling & Sales Management, 43*(4), 307–335. DOI: 10.1080/08853134.2023.2260108

Aguinis, H. (2024). *Research Methodology. Best Practices for Rigorous, Credible, and Impactful Research.* SAGE Publications.

Ahmad, N., & Zulkifli, A. M. (2022). Internet of Things (IoT) and the road to happiness. *Digital Transformation and Society, 1*(1), 66–94. DOI: 10.1108/DTS-05-2022-0009

Air Transport Action Group. (n.d.). *Economic growth.* https://aviationbenefits.org/economic-growth/

Al-Ansi, A. M., Jaboob, M., Garad, A., & Al-Ansi, A. (2023). Analyzing augmented reality (AR) and virtual reality (VR) recent development in education. *Social Sciences & Humanities Open, 8*(1), 100532. DOI: 10.1016/j.ssaho.2023.100532

Altair Engineering. (n.d.). *Guide to urban air mobility* [White paper].

Alt, R. (2020). Evolution and perspectives of electronic markets. *Electronic Markets*, *30*(1), 1–13. DOI: 10.1007/s12525-020-00413-8

Alves, C., & Machado, J. (2024). Immersive Shopping Experiences: The Role of Augmented Reality in E-Commerce. In: Rocha, Á., Ferrás, C., Hochstetter Diez, J., Diéguez Rebolledo, M. (Eds.) *Information Technology and Systems*. ICITS 2024. Lecture Notes in Networks and Systems, vol. 933. Springer, Cham. https://doi.org/ DOI: 10.1007/978-3-031-54256-5_19

Anderson, J., Smith, K., & Williams, M. (2021). Digital transformation in manufacturing: A systems perspective. *Manufacturing Engineering Review*, *45*(3), 123–138.

Anderson, P., Davis, L., & Thompson, R. (2023). Community infrastructure and workforce development in rural manufacturing. *Rural Development Quarterly*, *38*(2), 87–102.

Anderson, T., Johnson, M., & Lee, S. (2024). Cybersecurity challenges in digital manufacturing environments. *Industrial Security Journal*, *29*(4), 45–62.

Andrejevic, M. (2011). THE WORK THAT AFFECTIVE ECONOMICS DOES. *Cultural Studies*, *25*(4–5), 604–620. DOI: 10.1080/09502386.2011.600551

Antony, J., Jalušić, D., Bergweiler, S., Hajnal, Á., Žlabravec, V., Emődi, M., Strbad, D., Legler, T., & Marosi, A. C. (2024). Adapting to Changes: A Novel Framework for Continual Machine Learning in Industrial Applications. *Journal of Grid Computing*, *22*(4), 71. Advance online publication. DOI: 10.1007/s10723-024-09785-z

Aquino, S., Rapaccini, M., Adrodegari, F., & Pezzotta, G. (2023). Augmented reality for industrial services provision: The factors influencing a successful adoption in manufacturing companies. *Journal of Manufacturing Technology Management*, *34*(4), 601–620. DOI: 10.1108/JMTM-02-2022-0077

Arena, F., Collotta, M., Pau, G., & Termine, F. (2022). An Overview of Augmented Reality. *Computers*, *11*(2), 28. DOI: 10.3390/computers11020028

Arimoto, S., Kawamura, S., & Miyazaki, F. (1984). Bettering operation of robots by learning. *Journal of Robotic Systems*, *1*(2), 123–140. DOI: 10.1002/rob.4620010203

Arntz, M., Gregory, T., & Zierahn, U. (2016). The risk of automation for jobs in OECD countries: A comparative analysis. *OECD Social, Employment and Migration Working Papers*, No. 189.

Asif, S., Bueno, M., Ferreira, P., Anandan, P., Zhang, Z., Yao, Y., Ragunathan, G., Tinkler, L., Sotoodeh-Bahraini, M., Lohse, N., Webb, P., Hutabarat, W., & Tiwari, A. (2025). Rapid and automated configuration of robot manufacturing cells. *Robotics and Computer-integrated Manufacturing, 92*, 102862. DOI: 10.1016/j.rcim.2024.102862

Autor, D. H. (2015). Why are there still so many jobs? The history and future of workplace automation. *The Journal of Economic Perspectives, 29*(3), 3–30. DOI: 10.1257/jep.29.3.3

Bag, S., Gupta, S., & Luo, Z. (2020). Examining the role of logistics 4.0 enabled dynamic capabilities on firm performance. *International Journal of Logistics Management, 31*(3), 607–628. DOI: 10.1108/IJLM-11-2019-0311

Baily, M. N., Gersbach, H., Scherer, F. M., & Lichtenberg, F. R. (1995). Efficiency in manufacturing and the need for global competition. *Brookings Papers on Economic Activity: Microeconomics, 1995*, 307–358. DOI: 10.2307/2534776

Bales, R. R., Maull, R. S., & Radnor, Z. (2004). The development of supply chain management within the aerospace manufacturing sector. *Supply Chain Management, 9*(3), 250–255. DOI: 10.1108/13598540410544944

Balinbin, A. L. (2022, March 31). PHL aircraft maintenance market seen growing as travel industry recovers. *BusinessWorld*. https://www.bworldonline.com/economy/2022/03/31/439548/phl-aircraft-maintenance-market-seen-growing-as-travel-industry-recovers/

Bamber, P., Frederick, S., & Gereffi, G. (2016). *The Philippines in the aerospace global value chain*. Duke University, Center on Globalization, Governance & Competitiveness. https://hdl.handle.net/10161/12483

Bandura, A. (1997). *Self-Efficacy: The Exercise of Control*. W.H. Freeman and Company.

Baptista Torres da Rocha, A., Borges de Oliveira, K., Espuny, M., Salvador da Motta Reis, J., & José Oliveira, O. (2022). Business transformation through sustainability based on Industry 4.0

Barley, S. R. (1986). Technology as an occasion for structuring: Evidence from observations of CT scanners and the social order of radiology departments. *Administrative Science Quarterly, 31*(1), 78–108. DOI: 10.2307/2392767 PMID: 10281188

Bartik, T. J. (2018). *Who benefits from state and local economic development policies?* W.E. Upjohn Institute for Employment Research.

Bayoude, K., Ouassit, Y., Ardchir, S., & Azouazi, M. (2018). How machine learning potentials are transforming the practice of digital marketing: State of the art.

Beeson, P. E., & Husted, S. (1989). Patterns and determinants of productive efficiency in state manufacturing. *Journal of Regional Science*, *29*(1), 15–28. DOI: 10.1111/j.1467-9787.1989.tb01219.x

Bellalouna, F. (2021). The Augmented Reality Technology as Enabler for the Digitization of Industrial Business Processes: Case Studies. *Procedia CIRP*, *98*, 400–405. DOI: 10.1016/j.procir.2021.01.124

Bernabei, M., Eugeni, M., Gaudenzi, P., & Costantino, F. (2023). Assessment of smart transformation in the manufacturing process of aerospace components through a data-driven approach. *Global Journal of Flexible Systems Managment*, *24*(1), 67–86. DOI: 10.1007/s40171-022-00328-7

Berndt, A. E. (2020). Sampling Methods. *Journal of Human Lactation*, *36*(2), 224–226. DOI: 10.1177/0890334420906850 PMID: 32155099

Bertolini, M., Mezzogori, D., Neroni, M., & Zammori, F. (2021). Machine Learning for industrial applications: A comprehensive literature review. *Expert Systems with Applications*, *175*, 114820. DOI: 10.1016/j.eswa.2021.114820

Bescast. (n.d.). *Aerospace supply chain tiers explained.* https://www.bescast.com/blog/aerospace-supply-chain-tiers/

Bessen, J. E. (2019). Automation and jobs: When technology boosts employment. *Economic Policy*, *34*(100), 589–626. DOI: 10.1093/epolic/eiaa001

Bhatia, V., Kumar, A., Sidharth, S., Khare, S. K., Ghorpade, S. C., Kumar, P., & AlZohbi, G. (2024). Industry 4.0 in Aircraft Manufacturing: Innovative Use Cases and Patent Landscape. In *Industry 4.0 Driven Manufacturing Technologies* (pp. 103–137). Springer Nature Switzerland. DOI: 10.1007/978-3-031-68271-1_5

Bidollahkhani, M., & Kunkel, J. M. (2024). Revolutionizing system reliability: The role of AI in predictive maintenance strategies. arXiv preprint arXiv:2404.13454.

Bogue, R. (2019). Collaborative robots: A review of capabilities, applications and market prospects. *The Industrial Robot*, *46*(4), 404–408. DOI: 10.1108/IR-05-2019-0092

Borges, A. F. S., Laurindo, F. J. B., Spínola, M. M., Gonçalves, R. F., & Mattos, C. A. (2021). The strategic use of artificial intelligence in the digital era: Systematic literature review and future research directions. *International Journal of Information Management*, *57*, 102225. DOI: 10.1016/j.ijinfomgt.2020.102225

Bressanelli, G., Adrodegari, F., Perona, M., & Saccani, N. (2018). Exploring how usage-focused business models enable circular economy through digital technologies. *Sustainability (Basel), 10*(3), 639. DOI: 10.3390/su10030639

Briken, K., Moore, J., Scholarios, D., Rose, E., & Sherlock, A. (2023). Industry 5 and the Human in Human-Centric Manufacturing. *Sensors (Basel), 23*(14), 6416. DOI: 10.3390/s23146416 PMID: 37514710

Bronfenbrenner, U. (1979). *The ecology of human development: Experiments by nature and design*. Harvard University Press. DOI: 10.4159/9780674028845

Brooks, R. A. (1986). Achieving artificial intelligence through building robots. *Massachusetts Institute of Technology, Artificial Intelligence Lab*.

Brown, A., Wilson, K., & Davis, P. (2022). Manufacturing resilience during COVID-19: Lessons for digital transformation. *Industrial Management Review, 67*(4), 234–251.

Brynjolfsson, E., & McAfee, A. (2014). *The second machine age: Work, progress, and prosperity in a time of brilliant technologies*. W. W. Norton & Company.

Büchi, G., Cugno, M., & Castagnoli, R. (2020). Smart factory performance and Industry 4.0. *Technological Forecasting and Social Change, 150*, 119790. Advance online publication. DOI: 10.1016/j.techfore.2019.119790

Burdick, J. W. (1989). On the inverse kinematics of redundant manipulators: Characterization of the self-motion manifolds. In *Advanced Robotics: 1989: Proceedings of the 4th International Conference on Advanced Robotics* (pp. 339–346). Springer Berlin Heidelberg.

Bureau of Labor Statistics. (2023). *Employment projections: 2022-2032*. U.S. Department of Labor.

Butt, J. (2020). A conceptual framework to support digital transformation in manufacturing using an integrated business process management approach. *Designs, 4*(3), 17. DOI: 10.3390/designs4030017

Calì, M., & Presidente, G. (2025). Robots for economic development. *Labour Economics, 96*, 102731. DOI: 10.1016/j.labeco.2025.102731

Caloud, P., Carlsson, C., Latombe, J. C., Le Pape, C., & Yim, M. (1990). Indoor automation with many mobile robots. In *IEEE International Workshop on Intelligent Robots and Systems*. IEEE.

Cannavacciuolo, L., Ferraro, G., Ponsiglione, C., Primario, S., & Quinto, I. (2023). Technological innovation-enabling industry 4.0 paradigm: A systematic literature review. *Technovation*, *124*, 102733. DOI: 10.1016/j.technovation.2023.102733

Canton, H. (2021). Organisation for economic co-operation and development—OECD. In *The Europa Directory of International Organizations 2021* (pp. 677–687). Routledge.

Caputo, R. K. (2019). *Policy analysis for social workers*. SAGE Publications.

Card, D., & Krueger, A. B. (1994). *Minimum wages and employment: A case study of the fast-food industry in New Jersey and Pennsylvania*. Princeton University Press.

Cardoso, D., & Ferreira, L. (2020). Application of predictive maintenance concepts using artificial intelligence tools. *Applied Sciences (Basel, Switzerland)*, *11*(1), 18. DOI: 10.3390/app11010018

Carou, D., & Carou, D. (2021). Aerospace transformation through Industry 4.0 technologies. *Aerospace and Digitalization: A Transformation Through Key Industry 4.0 Technologies*, 17-46.

Carthy, S., Cormican, K., & Sampaio, S. (2021). Knowing me knowing you: Understanding user involvement in the design process. *Procedia Computer Science*, *181*, 135–140. DOI: 10.1016/j.procs.2021.01.113

Catherine Read Brewis, C. (2016). To What Extent Are UK Companies Using Big Data to Inform Their Marketing Strategy?.

Chafe, W. H. (1992). *The paradox of change: American women in the 20th century*. Oxford University Press. DOI: 10.1093/oso/9780195044195.001.0001

Chang, V., Walters, R. J., & Wills, G. (2013). The development that leads to the Cloud Computing Business Framework. *International Journal of Information Management*, *33*(3), 524–538. DOI: 10.1016/j.ijinfomgt.2013.01.005

Chatterjee, S., Chaudhuri, R., Vrontis, D., & Basile, G. (2022). Digital transformation and entrepreneurship process in SMEs of India: A moderating role of adoption of AI-CRM capability and strategic planning. *Journal of Strategy and Management*, *15*(3), 416–433. DOI: 10.1108/JSMA-02-2021-0049

Chaudhary, K., Alam, M., Al-Rakhami, M. S., & Gumaei, A. (2021). Machine learning-based mathematical modelling for prediction of social media consumer behavior using big data analytics. *Journal of Big Data*, *8*(1), 73. DOI: 10.1186/s40537-021-00466-2

Chen, F. Y. (1982). Gripping mechanisms for industrial robots: An overview. *Mechanism and Machine Theory*, *17*(5), 299–311. DOI: 10.1016/0094-114X(82)90011-8

Cheng, B. L., Cham, T. H., Dent, M. M., & Lee, T. H. (2019). Service innovation: Building a sustainable competitive advantage in higher education. *International Journal of Services. Economics and Management*, *10*(4), 289. DOI: 10.1504/IJSEM.2019.105014

Cheng, B., Wang, M., Yang, S., Kinshuk, , & Peng, J. (2011). Acceptance of competency-based workplace e-learning systems: Effects of individual and peer learning support. *Computers & Education*, *57*(1), 1317–1333. DOI: 10.1016/j.compedu.2011.01.018

Chen, K., Lin, J., & Song, Y. (2019). Trading strategy optimization for a prosumer in continuous double auction-based peer-to-peer market: A prediction-integration model. *Applied Energy*, *242*, 1121–1133. DOI: 10.1016/j.apenergy.2019.03.094

Chen, L., Martinez, R., & Johnson, K. (2021). COVID-19 acceleration of digital transformation in manufacturing. *Technology and Society Quarterly*, *42*(3), 178–195.

Chen, M., Rodriguez, A., & Kim, J. (2024). Artificial intelligence integration in manufacturing: Human factors and implementation strategies. *AI in Manufacturing Journal*, *8*(2), 67–84.

Chen, S., Ren, Z., Yu, X., & Huang, A. (2022). A Dynamic Model of Evolutionary Knowledge and Capabilities Based on Human-Machine Interaction in Smart Manufactures. *Computational Intelligence and Neuroscience*, *8584888*, 1–10. Advance online publication. DOI: 10.1155/2022/8584888 PMID: 35515504

Chen, W., Li, Y., Tian, Z., & Zhang, F. (2023). 2D and 3D object detection algorithms from images: A Survey. *Array (New York, N.Y.)*, *19*, 100305. DOI: 10.1016/j.array.2023.100305

Chiesa, V., & Frattini, F. (2011). Commercializing Technological Innovation: Learning from Failures in High-Tech Markets*: Commercializing Technological Innovation. *Journal of Product Innovation Management*, *28*(4), 437–454. DOI: 10.1111/j.1540-5885.2011.00818.x

Choi, Y. W., Lee, J., Lee, Y., Lee, S., Jeong, W., Lim, D. Y., & Lee, S. W. (2025). A vision-guided adaptive and optimized robotic fabric gripping system for garment manufacturing automation. *Robotics and Computer-integrated Manufacturing*, *92*, 102874. DOI: 10.1016/j.rcim.2024.102874

Chong, A. Y.-L., Chan, F. T. S., & Ooi, K.-B. (2012). Predicting consumer decisions to adopt mobile commerce: Cross country empirical examination between China and Malaysia. *Decision Support Systems, 53*(1), 34–43. DOI: 10.1016/j.dss.2011.12.001

Çınar, Z. M., Abdussalam Nuhu, A., Zeeshan, Q., Korhan, O., Asmael, M., & Safaei, B. (2020). Machine learning in predictive maintenance towards sustainable smart manufacturing in industry 4.0. *Sustainability (Basel), 12*(19), 8211. DOI: 10.3390/su12198211

Cordeiro, D., Rui, J., & Nuno, C. (2015). ARZombie: A mobile augmented reality game with multimodal interaction. *2015 7th International Conference on Intelligent Technologies for Interactive Entertainment (INTETAIN), 1(4),* 22-31. http://dx.doi.org/DOI: 10.4108/icst.intetain.2015.259743

Council on Social Work Education. (2022). *Educational policy and accreditation standards*. CSWE Press.

Cui, Y., Butler, R., Liu, Z., Jaremko, T., & Hinds, S. (2025). The Future of Predictive Maintenance for Variable Frequency Drives Based on AI/ML. *IEEE Transactions on Industry Applications*, 1–9. DOI: 10.1109/TIA.2025.3550104

Dahake, N. S., Suresh Dahake, P., & Gajghate, A. (2024). Beyond the Hype: A Critical Evaluation of Predictive Analytics Accuracy in High-Stakes Business Decisions. *2024 2nd DMIHER International Conference on Artificial Intelligence in Healthcare, Education and Industry (IDICAIEI)*, 1–6. DOI: 10.1109/IDICAIEI61867.2024.10842857

Dahake, P. S., & Somani, N. (2024). Harnessing Predictive Analytics for Accurate Consumer Behaviour Forecasting: A Comprehensive Review. *2024 2nd DMIHER International Conference on Artificial Intelligence in Healthcare, Education and Industry (IDICAIEI)*, 1–6. DOI: 10.1109/IDICAIEI61867.2024.10842743

Dahake, P. S., Chandak, S., Mohare, R. V., Wadhwani, K., & Bhadade, P. (2023). The Crystal Ball of Marketing: How Predictive Analytics is Reshaping the Industry? *2023 Second International Conference On Smart Technologies For Smart Nation (SmartTechCon)*, 304–311. DOI: 10.1109/SmartTechCon57526.2023.10391334

Dahake, P. S., Mohare, R. V., & Somani, N. (2023). Exploring Machine Learning's Potential in Predicting Consumer Behavior for Captivating Marketing. *2023 Second International Conference On Smart Technologies For Smart Nation (SmartTechCon)*, 1347–1352. DOI: 10.1109/SmartTechCon57526.2023.10391558

Dahl, D. W., & Hoeffler, S. (2004). Visualizing the Self: Exploring the Potential Benefits and Drawbacks for New Product Evaluation. *Journal of Product Innovation Management*, *21*(4), 259–267. DOI: 10.1111/j.0737-6782.2004.00077.x

Daniyan, I., Mpofu, K., Oyesola, M., Ramatsetse, B., & Adeodu, A. (2020). Artificial intelligence for predictive maintenance in the railcar learning factories. *Procedia Manufacturing*, *45*, 13–18. DOI: 10.1016/j.promfg.2020.04.032

Dao, V., Langella, I., & Carbo, J. (2011). From green to sustainability: Information Technology and an integrated sustainability framework. *The Journal of Strategic Information Systems*, *20*(1), 63–79. DOI: 10.1016/j.jsis.2011.01.002

Davis, R., & Thompson, J. (2024). Career counseling in the digital age: Approaches for rapid technological change. *The Career Development Quarterly*, *72*(3), 145–162.

Davis, S., Wilson, M., & Brown, K. (2024). Trauma-informed approaches to organizational change in manufacturing. *Organizational Psychology Review*, *41*(2), 89–106.

De Oliveira Santini, F., Ladeira, W. J., Pinto, D. C., Herter, M. M., Sampaio, C. H., & Babin, B. J. (2020). Customer engagement in social media: A framework and meta-analysis. *Journal of the Academy of Marketing Science*, *48*(6), 1211–1228. DOI: 10.1007/s11747-020-00731-5

De San Bernabe, A., Lakshminarayanan, S., Xing, K. C. H., Mutiargo, B., Suppiah, S., Xue, F., & Fu, S. (2024). Robots in manufacturing: Current technology trends. In *Digital Manufacturing* (pp. 39–83). Elsevier., DOI: 10.1016/B978-0-443-13812-6.00008-7

de Souza Cardoso, L. F., Martins Queiroz Mariano, F. C., & Zorzal, E. R. (2020). A survey of industrial augmented reality. *Computers & Industrial Engineering*, *139*, 106159. DOI: 10.1016/j.cie.2019.106159

Deaton, A. (2017). *The great escape: Health, wealth, and the origins of inequality*. Princeton University Press.

Deci, E. L., & Ryan, R. M. (2000). The "what" and "why" of goal pursuits: Human needs and the self-determination of behavior. *Psychological Inquiry*, *11*(4), 227–268. DOI: 10.1207/S15327965PLI1104_01

Decker, R., Haltiwanger, J., Jarmin, R., & Miranda, J. (2014). The role of entrepreneurship in US job creation and economic dynamism. *The Journal of Economic Perspectives*, *28*(3), 3–24. DOI: 10.1257/jep.28.3.3

Delmas, M. A., & Burbano, V. C. (2011). The drivers of greenwashing. *California Management Review*, *54*(1), 64–87. DOI: 10.1525/cmr.2011.54.1.64

Deming, D. J. (2017). The growing importance of social skills in the labor market. *The Quarterly Journal of Economics*, *132*(4), 1593–1640. DOI: 10.1093/qje/qjx022

Department of Trade and Industry. (n.d.). *Aerospace: Securing the future*. https://industry.gov.ph/industry/aerospace/

Devagiri, J. S., Paheding, S., Niyaz, Q., Yang, X., & Smith, S. (2022). Augmented Reality and Artificial Intelligence in industry: Trends, tools, and future challenges. *Expert Systems with Applications*, *207*, 118002. DOI: 10.1016/j.eswa.2022.118002

Dhanda, M., Rogers, B. A., Hall, S., Dekoninck, E., & Dhokia, V. (2025). Reviewing human-robot collaboration in manufacturing: Opportunities and challenges in the context of Industry 5.0. *Robotics and Computer-integrated Manufacturing*, *93*, 102937. DOI: 10.1016/j.rcim.2024.102937

Diamantoulakis, P. D., Kapinas, V. M., & Karagiannidis, G. K. (2015). Big Data Analytics for Dynamic Energy Management in Smart Grids. *Big Data Research*, *2*(3), 94–101. DOI: 10.1016/j.bdr.2015.03.003

Dogan, A., & Birant, D. (2021). Machine learning and data mining in manufacturing. *Expert Systems with Applications*, *166*, 114060. Advance online publication. DOI: 10.1016/j.eswa.2020.114060

Dogru, A. K., & Keskin, B. B. (2020). Artificial intelligence in operations management: Applications, challenges and opportunities. *Journal of Data, Information and Management. DOI, 10.*

Drouot, A., Zhao, R., Irving, L., Sanderson, D., & Ratchev, S. (2018). Measurement assisted assembly for high accuracy aerospace manufacturing. *IFAC-PapersOnLine*, *51*(11), 393–398. DOI: 10.1016/j.ifacol.2018.08.326

Dubey, R., Gunasekaran, A., Childe, S. J., Bryde, D. J., Giannakis, M., Foropon, C., Roubaud, D., & Hazen, B. T. (2020). Big data analytics and artificial intelligence pathway to operational performance under the effects of entrepreneurial orientation and environmental dynamism: A study of manufacturing organisations. *International Journal of Production Economics*, *226*, 107599. Advance online publication. DOI: 10.1016/j.ijpe.2019.107599

Edwards, M. (1984). Robots in industry: An overview. *Applied Ergonomics*, *15*(1), 45–53. DOI: 10.1016/S0003-6870(84)90121-2 PMID: 15676500

Egger, J., & Masood, T. (2020). Augmented reality in support of intelligent manufacturing - A systematic literature review. *Computers & Industrial Engineering*, *140*, 106195. DOI: 10.1016/j.cie.2019.106195

Eich, A., Klichowicz, A., & Bocklisch, F. (2023). How automation level influences moral decisions of humans collaborating with industrial robots in different scenarios. *Frontiers in Psychology*, *14*, 1107306. DOI: 10.3389/fpsyg.2023.1107306 PMID: 36968758

Elkington, J. (1997). The triple bottom line. *Environmental Management: Readings and Cases*, *2*, 49–66.

Endsley, M. R. (1995). Toward a theory of situation awareness in dynamic systems. *Human Factors*, *37*(1), 32–64. DOI: 10.1518/001872095779049543

Eriksson, K. M., Carlsson, L., & Olsson, A. K. (2022). To digitalize or not? Navigating and merging human- and technology perspectives in production planning and control. *International Journal of Advanced Manufacturing Technology*, *122*(11-12), 4365–4373. DOI: 10.1007/s00170-022-09874-x PMID: 35968033

Eswaran, M., & Raju Bahubalendruni, M. V. A. (2023). Augmented reality aided object mapping for worker assistance/training in an industrial assembly context: Exploration of affordance with existing guidance techniques. *Computers & Industrial Engineering*, *185*, 109663. DOI: 10.1016/j.cie.2023.109663

Fahrudin, A., Tiwari, S. P., & Rahmi, K. H. (2025). Psychological well-being of human content moderators and wellness strategies in AI-driven content moderation organizations. In *Content Moderation in the Age of AI* (pp. 221–250). IGI Global Scientific Publishing. DOI: 10.4018/979-8-3373-0335-2.ch009

Fang, H., Jiang, C., Hussain, T., Zhang, X., & Huo, Q. (2022). Input digitization of the manufacturing industry and carbon emission intensity based on testing the world and developing countries. *International Journal of Environmental Research and Public Health*, *19*(19), 12855. DOI: 10.3390/ijerph191912855 PMID: 36232155

Fatimah, Y. A., Govindan, K., Murniningsih, R., & Setiawan, A. (2020). Industry 4.0 based sustainable circular economy approach for smart waste management system to achieve sustainable development goals: A case study of Indonesia. *Journal of Cleaner Production*, *269*, 122263. DOI: 10.1016/j.jclepro.2020.122263

Fearne, A., & Hughes, D. (1999). Success factors in the fresh produce supply chain: Insights from the UK. *Supply Chain Management*, *4*(3), 120–131. DOI: 10.1108/13598549910279567

Feist, S., Jacques de Sousa, L., Sanhudo, L., & Poças Martins, J. (2024). Automatic Reconstruction of 3D Models from 2D Drawings: A State-of-the-Art Review. *Eng*, *5*(2), 784–800. DOI: 10.3390/eng5020042

First Philippine Industrial Park. (2023a, October). *Aerospace manufacturing industry* [Brochure]. https://fpip.com/sites/default/files/2023-10/FPIP%20-%20Aerospace%20Manufacturing%20Industry%20brochure.pdf

First Philippine Industrial Park. (2023b, October). *Electronics industry* [Brochure]. https://fpip.com/fpip-official/sites/default/files/2023-10/FPIP%20-%20Electronics%20Industry%20brochure.pdf

Fratocchi, L., Nassimbeni, G., Sartor, M., Ancarani, A., Di Mauro, C., Zanoni, A., ... & Vignoli, M. (2013). Manufacturing Back-shoring and the Global Fragmentation of Production: What it is Changing after the Financial Crisis?. 40th Acdemy of International Business UK & Ireland.

Frey, C. B., & Osborne, M. A. (2017). The future of employment: How susceptible are jobs to computerisation? *Technological Forecasting and Social Change*, *114*, 254–280. DOI: 10.1016/j.techfore.2016.08.019

Frosch, R. A., & Gallopoulos, N. E. (1989). Strategies for manufacturing. *Scientific American*, *261*(3), 144–153. DOI: 10.1038/scientificamerican0989-144

Gadekar, R., Sarkar, B., & Gadekar, A. (2022). Model development for assessing inhibitors impacting Industry 4.0 implementation in Indian manufacturing industries: An integrated ISM-Fuzzy MICMAC approach. *International Journal of System Assurance Engineering and Management*, •••, 1–26. Advance online publication. DOI: 10.1007/s13198-022-01691-5

Garcia Martinez, M., & Walton, B. (2014). The wisdom of crowds: The potential of online communities as a tool for data analysis. *Technovation*, *34*(4), 203–214. DOI: 10.1016/j.technovation.2014.01.011

Garcia, M., & Smith, J. (2023). Leveraging worker expertise in digital transformation: A strengths-based approach. *Human Resource Development Review*, *22*(4), 312–329.

Garcia, P., Johnson, L., & Davis, M. (2024). Digital assessment tools for workforce development: Innovations and applications. *Training and Development Journal*, *78*(5), 23–38.

Garre, P., Bharadwaj, V. N., Shashank, P. S., Harish, M., & Dheeraj, M. S. (2017). Applying lean in aerospace manufacturing. *Materials Today: Proceedings*, *4*(8), 8439–8446. DOI: 10.1016/j.matpr.2017.07.189

Gauba, H., Kumar, P., Roy, P. P., Singh, P., Dogra, D. P., & Raman, B. (2017). Prediction of advertisement preference by fusing EEG response and sentiment analysis. *Neural Networks*, *92*, 77–88. DOI: 10.1016/j.neunet.2017.01.013 PMID: 28254237

Gautier, M. (1986). Identification of robots dynamics. *IFAC Proceedings Volumes, 19*(14), 125–130. https://doi.org/DOI: 10.1016/S1474-6670(17)56341-3

Geissdoerfer, M., Savaget, P., & Evans, S. (2017). The Cambridge business model innovation process. *Procedia Manufacturing*, *8*, 262–269. DOI: 10.1016/j.promfg.2017.02.033

Germain, C. B., & Gitterman, A. (2017). *The life model of social work practice: Advances in theory and practice* (3rd ed.). Columbia University Press.

Ghasemi, Y., Jeong, H., Choi, S. H., Park, K.-B., & Lee, J. Y. (2022). Deep learning-based object detection in augmented reality: A systematic review. *Computers in Industry*, *139*, 103661. DOI: 10.1016/j.compind.2022.103661

Gitterman, A., & Knight, C. (2019). *Evidence-guided practice: Integrating the science and art of social work*. Oxford University Press.

Goldsmith, R. E., Moore, M. A., & Beaudoin, P. (1999). Fashion innovativeness and self-concept: A replication. *Journal of Product and Brand Management*, *8*(1), 7–18. DOI: 10.1108/10610429910257904

Graetz, G., & Michaels, G. (2018). Robots at work. *The Review of Economics and Statistics*, *100*(5), 753–768. DOI: 10.1162/rest_a_00754

Greene, R. R. (2017). *Human behavior theory and social work practice* (4th ed.). Routledge.

Gren, A., Jamarillo, B., Kiev, V., Shabrov, N., & Vasiliev, D. (2018). Development of digital simulation on the basis of technologies of virtual and augmented reality. *SHS Web of Conferences*. *44*, 00036. https://doi.org/DOI: 10.1051/shsconf/20184400036

Guha, S., & Kumar, S. (2018). Emergence of Big Data Research in Operations Management, Information Systems, and Healthcare: Past Contributions and Future Roadmap. *Production and Operations Management*, *27*(9), 1724–1735. DOI: 10.1111/poms.12833

Guiné, R. P. F., Florença, S. G., Barroca, M. J., & Anjos, O. (2020). The Link between the Consumer and the Innovations in Food Product Development. *Foods*, *9*(9), 1317. DOI: 10.3390/foods9091317 PMID: 32962007

Guisan, M. C., & Exposito, P. (2015). Manufacturing and development in countries and areas of Asia-Pacific, 2000-2010. *Regional and Sectoral Economic Studies*, *15*(2), 203–208.

Gupta, C. P., & Goyal, K. K. (2020). *Computer Concepts and Management Information Systems*. USA. *Mercury Learning and Information.*, *ISBN-10*, 1683925866.

Hackwood, S., & Wang, J. (1988). The engineering of cellular robotic systems. In *Proceedings 1988 IEEE International Symposium on Intelligent Control*. IEEE Computer Society.

Hamdan, A., Ibekwe, K. I., Ilojianya, V. I., Sonko, S., & Etukudoh, E. A. (2024). AI in renewable energy: A review of predictive maintenance and energy optimization. *International Journal of Science and Research Archive*, *11*(1), 718–729. DOI: 10.30574/ijsra.2024.11.1.0112

Hansen, E. B., & Bøgh, S. (2021). Artificial intelligence and internet of things in small and medium-sized enterprises: A survey. *Journal of Manufacturing Systems*, *58*, 362–372. DOI: 10.1016/j.jmsy.2020.08.009

Hanson, K. (2019). Policy advocacy in the digital age: Strategies for social workers. *Social Work Policy Review*, *13*(2), 45–62.

Haque, R., Bajwa, A., Siddiqui, N. A., & Ahmed, I. (2024). Predictive Maintenance In Industrial Automation: A Systematic Review Of IOT Sensor Technologies And AI Algorithms. *American Journal of Interdisciplinary Studies, 5*(01), 01-30.

Harris, C. E., Starnes, J. H.Jr, & Shuart, M. J. (2002). Design and manufacturing of aerospace composite structures, state-of-the-art assessment. *Journal of Aircraft*, *39*(4), 545–560. DOI: 10.2514/2.2992

Hassan, R., Qamar, F., Hasan, M. K., Aman, A. H. M., & Ahmed, A. S. (2020). Internet of Things and Its Applications: A Comprehensive Survey. *Symmetry*, *12*(10), 1674. DOI: 10.3390/sym12101674

Heckman, J. J., & Kautz, T. (2012). Hard evidence on soft skills. *Labour Economics*, *19*(4), 451–464. DOI: 10.1016/j.labeco.2012.05.014 PMID: 23559694

Heshmati, A. (2003). Productivity growth, efficiency and outsourcing in manufacturing and service industries. *Journal of Economic Surveys*, *17*(1), 79–112. DOI: 10.1111/1467-6419.00189

Hinsch, C., Felix, R., & Rauschnabel, P. A. (2020). Nostalgia beats the wow-effect: Inspiration, awe and meaningful associations in augmented reality marketing. *Journal of Retailing and Consumer Services*, *53*, 101987. DOI: 10.1016/j.jretconser.2019.101987

Hoffmann, S., & Mai, R. (2022). Consumer behavior in augmented shopping reality. A review, synthesis, and research agenda. *Frontiers in Virtual Reality*, *3*, 961236. DOI: 10.3389/frvir.2022.961236

Hollnagel, E. (2014). *Safety-I and Safety-II: The past and future of safety management*. Ashgate Publishing.

Holmström, J., & Partanen, J. (2014). Digital manufacturing-driven transformations of service supply chains for complex products. *Supply Chain Management, 19*(4), 421–430. DOI: 10.1108/SCM-10-2013-0387

Ho, P. T., Albajez, J. A., Santolaria, J., & Yagüe-Fabra, J. A. (2022). Study of Augmented Reality Based Manufacturing for Further Integration of Quality Control 4.0: A Systematic Literature Review. *Applied Sciences (Basel, Switzerland), 12*(4), 1961. DOI: 10.3390/app12041961

Hrnjica, B., & Softic, S. (2020, August). Explainable AI in manufacturing: a predictive maintenance case study. In *IFIP international conference on advances in production management systems* (pp. 66–73). Springer International Publishing. DOI: 10.1007/978-3-030-57997-5_8

Huang, M.-H., & Rust, R. T. (2021). A strategic framework for artificial intelligence in marketing. *Journal of the Academy of Marketing Science, 49*(1), 30–50. DOI: 10.1007/s11747-020-00749-9

Huber, R., Oberländer, A. M., Faisst, U., & Röglinger, M. (2022). Disentangling Capabilities for Industry 4.0 - an Information Systems Capability Perspective. *Information systems frontiers: a journal of research and innovation*, 1–29. Advance online publication. https://doi.org/DOI: 10.1007/s10796-022-10260-x

Hudson, P. (2016). Social work and technological change: Historical perspectives and future directions. *Social Work History Review, 28*(3), 123–145.

Hussain, Z., Akhunzada, A., Iqbal, J., Bibi, I., & Gani, A. (2021). Secure IIoT-enabled industry 4.0. *Sustainability (Basel), 13*(22), 12384. Advance online publication. DOI: 10.3390/su132212384

Hwa, L. S., & Te Chuan, L. (2024). A brief review of artificial intelligence robotic in food industry. *Procedia Computer Science, 232*, 1694–1700. DOI: 10.1016/j.procs.2024.01.167

Imaginovation. (n.d.). *IoT in aviation: How connected aircraft re-define flying*. https://imaginovation.net/blog/iot-in-aviation/

International Federation of Social Workers. (2014). *Global definition of social work*. IFSW Press.

Ismagilova, E., Slade, E., Rana, N. P., & Dwivedi, Y. K. (2020). The effect of characteristics of source credibility on consumer behaviour: A meta-analysis. *Journal of Retailing and Consumer Services*, *53*, 101736. DOI: 10.1016/j.jretconser.2019.01.005

Jackson, K., Efthymiou, K., & Borton, J. (2016). Digital manufacturing and flexible assembly technologies for reconfigurable aerospace production systems. *Procedia CIRP*, *52*, 274–279. DOI: 10.1016/j.procir.2016.07.054

Jamwal, A., Agrawal, R., Sharma, M., Kumar, A., Kumar, V., & Garza-Reyes, J. A. A. (2022). Machine learning applications for sustainable manufacturing: A bibliometric-based review for future research. *Journal of Enterprise Information Management*, *35*(2), 566–596. DOI: 10.1108/JEIM-09-2020-0361

Jandyal, A., Chaturvedi, I., Wazir, I., Raina, A., & Ul Haq, M. I. (2022). 3D printing - A review of processes, materials and applications in industry 4.0. *Sustainable Operations and Computers*, *3*, 33–42. DOI: 10.1016/j.susoc.2021.09.004

Javaid, M., Haleem, A., Singh, R. P., Suman, R., & Gonzalez, E. S. (2022). Understanding the adoption of Industry 4.0 technologies in improving environmental sustainability. *Sustainable Operations and Computers*, *3*, 203–217. DOI: 10.1016/j.susoc.2022.01.008

Jayawardhana, A., & Colombage, S. (2020). *Does Blockchain Technology Drive Sustainability? An Exploratory Review*. Emerald Publications. DOI: 10.1108/S2043-052320200000015002

Johnson, A., & Lee, C. (2023). Digital literacy and technology adaptation in manufacturing settings. *Technology and Learning Quarterly*, *35*(4), 67–82.

Johnson, M., Brown, A., & Wilson, K. (2022). Digital transformation and social inequality: A person-in-environment analysis. *Social Work Research*, *46*(3), 187–201.

Johnson, R., Martinez, S., & Davis, L. (2024). Inclusive community planning for technological transition. *Community Development Journal: An International Forum*, *59*(2), 234–251.

Jomo, K. S. (2003). *Manufacturing competitiveness in Asia*. Taylor & Francis.

Joshi, N. (n.d.). *Revolutionizing aviation with IoT*. Allerin Blog. https://www.allerin.com/blog/revolutionizing-aviation-with-iot

Judd, R., & Knasinski, A. (1987). A technique to calibrate industrial robots with experimental verification. In *Proceedings. 1987 IEEE International Conference on Robotics and Automation* (Vol. 4). IEEE. DOI: 10.1109/ROBOT.1987.1088010

Kagermann, H., Helbig, J., Hellinger, A., & Wahlster, W. (2013). *Recommendations for implementing the strategic initiative INDUSTRIE 4.0: Securing the future of German manufacturing industry*. National Academy of Science and Engineering.

Kalpakjian, S. (2001). *Manufacturing engineering and technology*. Pearson Education India.

Kane, G. C., Phillips, A. N., Copulsky, J. R., & Andrus, G. R. (2019). *The technology fallacy: How people are the real key to digital transformation*. MIT Press. DOI: 10.7551/mitpress/11661.001.0001

Kang, J., & Park-Poaps, H. (2010). Hedonic and utilitarian shopping motivations of fashion leadership. *Journal of Fashion Marketing and Management*, *14*(2), 312–328. DOI: 10.1108/13612021011046138

Kapoor, K., Bigdeli, A. Z., Dwivedi, Y. K., & Raman, R. (2024). How is COVID-19 altering the manufacturing landscape? A literature review of imminent challenges and management interventions. *Annals of Operations Research*, *335*(3), 1567–1599. DOI: 10.1007/s10479-021-04397-2 PMID: 34803204

Kapp, J. (2018). The fourth industrial revolution: assessing the intelligences of engineers in the South African automotive industry. Master Thesis: Faculty of Economic and Bussiness, Nelson Mandela University

Katz, L. F., & Margo, R. A. (2014). Technical change and the relative demand for skilled labor: The United States in historical perspective. In Boustan, L., Frydman, C., & Margo, R. (Eds.), *Human capital in history: The American record* (pp. 15–57). University of Chicago Press.

Kegyes, T., Süle, Z., & Abonyi, J. (2021). The applicability of reinforcement learning methods in the development of industry 4.0 applications. *Complexity*, *2021*(1), 7179374. DOI: 10.1155/2021/7179374

Keleko, A. T., Kamsu-Foguem, B., Ngouna, R. H., & Tongne, A. (2022). Artificial intelligence and real-time predictive maintenance in industry 4.0: A bibliometric analysis. *AI and Ethics*, *2*(4), 553–577. DOI: 10.1007/s43681-021-00132-6

Khalek, I., Chalhoub, J., & Ayer, S. (2019). Augmented Reality for Identifying Maintainability Concerns during Design. *Advances in Civil Engineering*, *2019*(1), 1–12. DOI: 10.1155/2019/8547928

Khan, Z. U., Nasim, B., & Rasheed, Z. (2025). Generative AI-based predictive maintenance in aviation: A systematic literature review. *CEAS Aeronautical Journal*, *16*(2), 1–19. DOI: 10.1007/s13272-025-00818-1

Kiefer, D., Grimm, F., Straub, T., Bitsch, G., & Van Dinther, C. (2024). Enhancing power skiving tool longevity: The synergy of AI and robotics in manufacturing automation. *International Journal of Mechatronics and Manufacturing Systems, 17*(2), 201–224. DOI: 10.1504/IJMMS.2024.143059

Kilari, S. D. (2025). The impact of artificial intelligence on automated manufacturing and supply chain systems: Advancements, challenges, and future prospects. *Challenges, and Future Prospects*. DOI: 10.2139/ssrn.4799112

Kim, H., Fiore, A. M., Niehm, L. S., & Jeong, M. (2010). Psychographic characteristics affecting behavioral intentions towards pop-up retail. *International Journal of Retail & Distribution Management, 38*(2), 133–154. DOI: 10.1108/09590551011020138

Kim, W. B., Kim, C., & Kurata, K. (2025). Relationships among consumer innovativeness, learning, and global product purchases: Store manager perspectives in retailing. *Journal of Retailing and Consumer Services, 82*, 104084. DOI: 10.1016/j.jretconser.2024.104084

Korfiatis, N., García-Bariocanal, E., & Sánchez-Alonso, S. (2012). Evaluating content quality and helpfulness of online product reviews: The interplay of review helpfulness vs. review content. *Electronic Commerce Research and Applications, 11*(3), 205–217. DOI: 10.1016/j.elerap.2011.10.003

KPMG International. (2016). *Global aerospace and defense outlook: The dawn of a new day*.

Krajčovič, M., Matys, M., Gabajová, G., & Komačka, D. (2024). Utilization of Immersive Virtual Reality as an Interactive Method of Assignment Presentation. *Electronics (Basel), 13*(8), 1430. DOI: 10.3390/electronics13081430

Krishnamurthy, V., Nezafati, K., Stayton, E., & Singh, V. (2020). Explainable AI framework for imaging-based predictive maintenance for automotive applications and beyond. *Data-Enabled Discovery and Applications, 4*(1), 7. DOI: 10.1007/s41688-020-00042-2

Krol, F., Saeed, M. A., & Kersten, W. (2020). A holistic digitalization KPI framework for the aerospace industry. In *Data Science and Innovation in Supply Chain Management: How Data Transforms the Value Chain. Proceedings of the Hamburg International Conference of Logistics (HICL),* Vol. 29 (pp. 797-847). Berlin: epubli GmbH.

Krugman, P. (2013). *End this depression now!* W. W. Norton & Company.

Krupitzer, C., Müller, S., Lesch, V., Züfle, M., Edinger, J., Lemken, A., Schäfer, D., Kounev, S., & Becker, C. (2020). A Survey on Human Machine Interaction in Industry 4.0. *ArXiv*. https://arxiv.org/abs/2002.01025

Kubik, C., Hofmann, J., Molitor, D. A., Becker, M., & Groche, P. (2025). Reliable machine learning models for manufacturing processes. *International Journal of Advanced Manufacturing Technology*, *138*(11-12), 5913–5927. Advance online publication. DOI: 10.1007/s00170-025-15785-4

Kumar, S., Patel, R., & Anderson, M. (2022). COVID-19 and digital transformation acceleration in manufacturing. *Industrial Technology Review*, *45*(6), 123–140.

Kumar, V., Thompson, J., & Wilson, S. (2023). Organizational culture and digital transformation success: A manufacturing perspective. *Organizational Dynamics*, *52*(3), 234–248.

Kumar, V., & Vidhyalakshmi, P. (2012). Cloud Computing for Business Sustainability. *Asia-Pacific Journal of Management Research and Innovation*, *8*(4), 461–474. DOI: 10.1177/2319510X13481905

Lagomarsino, M., Lorenzini, M., De Momi, E., & Ajoudani, A. (2021). *An Online Framework for Cognitive Load Assessment in Assembly Tasks*. ArXiv., DOI: 10.1016/j.rcim.2022.102380

Lahane, R. (n.d.). *Smart MRO: The digital revolution in aviation, aerospace & defense*. Ramco Blog. https://www.ramco.com/blog/aviation/tech-revolution-in-aviation-and-defense-mro

Laudon, K. C., & Laudon, J. P. (2020). *Management Information Systems. Managing the Digital Firm* (16th ed.). Pearson Education Limited.

Lazarus, M. A., Blume, M. A. U., Brown, K. B., Caid, W. R., Dunning, T. E., Peranich, L. S., Russell, G. R., & Sitze, K. L. (2004). *Predictive modeling of customer financial behavior* (European Union Patent No. EP1050833A3). https://patents.google.com/patent/EP1050833A3/en

Lee, J. (2001). *The empowerment approach to social work practice: Building the beloved community* (2nd ed.). Columbia University Press.

Lee, S., & Zhang, W. (2024). Positive outcomes of digital transformation: Worker perspectives from manufacturing. *Work and Occupations*, *51*(2), 178–203.

Lee, W. J., Wu, H., Yun, H., Kim, H., Jun, M. B., & Sutherland, J. W. (2019). Predictive maintenance of machine tool systems using artificial intelligence techniques applied to machine condition data. *Procedia CIRP*, *80*, 506–511. DOI: 10.1016/j.procir.2018.12.019

Lei, K., Liu, L. M., Chen, X., Feng, Q., Yang, L., & Guo, L. (2022). Navigation and robotics improved alignment compared with PSI and conventional instrument, while clinical outcomes were similar in TKA: A network meta-analysis. *Knee Surgery, Sports Traumatology, Arthroscopy : Official Journal of the ESSKA*, *30*(2), 721–733. DOI: 10.1007/s00167-021-06436-8 PMID: 33492410

Levy, F., & Murnane, R. J. (2013). *Dancing with robots: Human skills for computerized work*. Third Way.

Liang, Y.-Ch., Shiu, S.-Ch., & Liu, Ch.-W. (2025). Enhancing yield and process efficiency through dual internet of things and augmented reality for AI-driven human-machine interaction in centering mass production. *Journal of Manufacturing Systems*, *82*, 26–41. DOI: 10.1016/j.jmsy.2025.06.003

Libert, K., Mosconi, E., & Cadieux, N. (2020). Human-Machine Interaction and Human Resource Management Perspective for Collaborative Robotics Implementation and Adoption. *Hawaii International Conference on System Sciences*. Proceedings of the 53rd Hawaii International Conference on System Sciences. https://hdl.handle.net/10125/63805

Li-Hua, R., Wilson, J., Aouad, G., & Li, X. (2011). Strategic aspects of innovation and internationalization in higher education: The Salford PMI2 experience. Journal of Chinese Entrepreneurship .

Li, L., Aslam, S., Wileman, A., & Perinpanayagam, S. (2021). Digital twin in aerospace industry: A gentle introduction. *IEEE Access : Practical Innovations, Open Solutions*, *10*, 9543–9562. DOI: 10.1109/ACCESS.2021.3136458

Lim, C. H., & Tan, S. Y. (2021). *Ergonomic work system implementation in electronics manufacturing: A Malaysian case study*. Ministry of Human Resources Malaysia.

Lin, K.-Y., & Huang, T. K. (2024). Shopping in the digital world: How augmented reality mobile applications trigger customer engagement. *Technology in Society*, *77*, 102540. DOI: 10.1016/j.techsoc.2024.102540

Liu, L., Guo, F., Zou, Z., & Duffy, V. G. (2024). Application, development and future opportunities of collaborative robots (cobots) in manufacturing: A literature review. *International Journal of Human-Computer Interaction*, *40*(4), 915–932. DOI: 10.1080/10447318.2022.2041907

Li, Y., Jin, X., & Xue, H. (2025). Avoid following individual shopping time habit to promote new products: Evidence from a randomized field experiment. *Journal of Retailing and Consumer Services*, *82*, 104058. DOI: 10.1016/j.jretconser.2024.104058

Li, Y., Tian, X., Feng, Y., Yuan, G., Guo, X., Wang, D., He, B., Hu, L., & Jiang, G. (2025). Assembling in 3D printing: Functions and applications. *International Journal of Advanced Manufacturing Technology*, *138*(11-12), 5123–5135. Advance online publication. DOI: 10.1007/s00170-025-15860-w

Lozano-Perez, T., Mason, M. T., & Taylor, R. H. (1984). Automatic synthesis of fine-motion strategies for robots. *The International Journal of Robotics Research*, *3*(1), 3–24. DOI: 10.1177/027836498400300101

Lyu, Y., Zhang, L., & Wang, D. (2023). The impact of digital transformation on low-carbon development of manufacturing. *Frontiers in Environmental Science*, *11*, 1134882. DOI: 10.3389/fenvs.2023.1134882

Mahmood, A., Arshad Ali, A., Nazam, M., & Nazim, M. (2021). Developing an interplay among the psychological barriers for the adoption of industry 4.0 phenomenon. *PLoS One*, *16*(8), e0255115. DOI: 10.1371/journal.pone.0255115 PMID: 34339448

Makinen, G. (2002). The economic effects of 9/11: A retrospective assessment. United States: Congress: Congressional Research Service (CRS).

Makridis, C. A., & Han, J. H. (2021). Future of work and employee empowerment and satisfaction: Evidence from a decade of technological change. *Technological Forecasting and Social Change*, *173*, 121162. DOI: 10.1016/j.techfore.2021.121162

Mamun, A. A., Sohel, M., Mohammad, N., Haque Sunny, M., Dipta, D. R., & Hossain, E. (2020). A Comprehensive Review of the Load Forecasting Techniques Using Single and Hybrid Predictive Models. *IEEE Access : Practical Innovations, Open Solutions*, *8*, 134911–134939. DOI: 10.1109/ACCESS.2020.3010702

Manyika, J., Chui, M., Miremadi, M., Bughin, J., George, K., Willmott, P., & Dewhurst, M. (2017). *A future that works: Automation, employment, and productivity*. McKinsey Global Institute.

Martinez, A., Johnson, P., & Brown, L. (2023). Worker well-being and digital transformation outcomes in manufacturing. *Human Relations*, *76*(8), 1234–1259.

Martinez, C., & Garcia, R. (2024). Family resilience building during technological transition. *Family Process*, *63*(2), 145–162.

Massa, E., & Ladhari, R. (2023). Augmented reality in marketing: Conceptualization and systematic review. *International Journal of Consumer Studies*, *47*(6), 2335–2366. DOI: 10.1111/ijcs.12930

Matzka, S. (2020, September). Explainable artificial intelligence for predictive maintenance applications. In *2020 third international conference on artificial intelligence for industries (ai4i)* (pp. 69-74). IEEE. DOI: 10.1109/AI4I49448.2020.00023

Mayer, A., Schneider, K., & Braun, D. (2020). Reducing complexity in operator dashboards: A case study. *International Journal of Industrial Ergonomics*, *76*, 102935. DOI: 10.1016/j.ergon.2020.102935

McClamroch, N. H., & Wang, D. (1988). Feedback stabilization and tracking of constrained robots. *IEEE Transactions on Automatic Control*, *33*(5), 419–426. DOI: 10.1109/9.1220

McNally, P., & Inayatullah, S. (1988). The rights of robots: Technology, culture and law in the 21st century. *Futures*, *20*(2), 119–136. DOI: 10.1016/0016-3287(88)90019-5

Mehra, M. (2023). Smart materials and their applications in aerospace engineering. *Journal of Air Transport Management*, *114*, 102437. Advance online publication. DOI: 10.1016/j.jairtraman.2023.102437

Meindl, B., & Mendonça, J. (2021). Mapping Industry 4.0 Technologies: From Cyber-Physical Systems to Artificial Intelligence. *ArXiv*. https://arxiv.org/abs/2111.14168

Mekonnen, A. (2024). Augmented Reality (AR) in Retailing: Customers' Experience in Luxury Fashion. In Ozuem, W., Ranfagni, S., & Willis, M. (Eds.), *Digital Transformation for Fashion and Luxury Brands*. Palgrave Macmillan., DOI: 10.1007/978-3-031-35589-9_5

Milanovic, B. (2016). *Global inequality: A new approach for the age of globalization*. Harvard University Press.

Miller, K., Davis, S., & Wilson, J. (2022). Trauma-informed approaches to workplace change in manufacturing. *Social Work in Mental Health*, *20*(4), 267–284.

Minh Trieu, N., & Truong Thinh, N. (2023). A comprehensive review: Interaction of appearance and behavior, artificial skin, and humanoid robot. *Journal of Robotics*, *2023*(1), 5589845. DOI: 10.1155/2023/5589845

Modgil, S., Singh, R. K., & Hannibal, C. (2022). Artificial intelligence for supply chain resilience: Learning from Covid-19. *International Journal of Logistics Management*, *33*(4), 1246–1268. DOI: 10.1108/IJLM-02-2021-0094

Molina Vargas, D. G., Vijayan, K. K., & Mork, O. J. (2020). Augmented Reality for Future Research Opportunities and Challenges in the Shipbuilding Industry: A Literature Review. *Procedia Manufacturing*, *45*, 497–503. DOI: 10.1016/j.promfg.2020.04.063

Morales Méndez, G., & del Cerro Velázquez, F. (2024). Augmented Reality in Industry 4.0 Assistance and Training Areas: A Systematic Literature Review and Bibliometric Analysis. *Electronics (Basel)*, *13*(6), 1147. DOI: 10.3390/electronics13061147

Morales Méndez, G., & del Cerro Velázquez, F. (2024). Impact of Augmented Reality on Assistance and Training in Industry 4.0: Qualitative Evaluation and Meta-Analysis. *Applied Sciences (Basel, Switzerland)*, *14*(11), 4564. DOI: 10.3390/app14114564

Muntean, M. (2018). *Business Intelligence Issues for Sustainability Projects. Knowledge Management, Innovation and Big Data: Implications for Sustainability*. Policy Making and Competitiveness.

Mustak, M., Salminen, J., Plé, L., & Wirtz, J. (2021). Artificial intelligence in marketing: Topic modeling, scientometric analysis, and research agenda. *Journal of Business Research*, *124*, 389–404. DOI: 10.1016/j.jbusres.2020.10.044

Nacchia, M., Fruggiero, F., Lambiase, A., & Bruton, K. (2021). A systematic mapping of the advancing use of machine learning techniques for predictive maintenance in the manufacturing sector. *Applied Sciences (Basel, Switzerland)*, *11*(6), 2546. Advance online publication. DOI: 10.3390/app11062546

National Aeronautics and Space Administration. (n.d.). *EAP mission overview*. https://www.nasa.gov/mission/eap/

National Association of Social Workers. (2021). *Code of ethics of the National Association of Social Workers*. NASW Press.

Nelson, P. J. B., Biddle, J., & Shapira, P. (2023). Applications and Societal Implications of Artificial Intelligence in Manufacturing: A Systematic Review, *ArXiv, abs/2308.02025*. https://doi.org//arXiv.2308.02025DOI: 10.48550

Nishant, R., Kennedy, M., & Corbett, J. (2020). Artificial intelligence for sustainability: Challenges, opportunities, and a research agenda. *International Journal of Information Management*, *53*, 102104. DOI: 10.1016/j.ijinfomgt.2020.102104

Nižetić, S., Šolić, P., González-de-Artaza, D. L.-I., & Patrono, L. (2020). Internet of Things (IoT): Opportunities, issues and challenges towards a smart and sustainable future. *Journal of Cleaner Production*. PMID: 32834567

Norman, D. A. (2013). *The Design of Everyday Things (Revised and Expanded Edition)*. Basic Books.

Nuhu, A. A., Zeeshan, Q., Safaei, B., & Shahzad, M. A. (2023). Machine learning-based techniques for fault diagnosis in the semiconductor manufacturing process: A comparative study. *The Journal of Supercomputing*, *79*(2), 2031–2081. DOI: 10.1007/s11227-022-04730-x

Ohalete, N. C., Aderibigbe, A. O., Ani, E. C., Ohenhen, P. E., & Akinoso, A. (2023). Advancements in predictive maintenance in the oil and gas industry: A review of AI and data science applications. *World J. Adv. Res. Rev*, *20*(3), 167–181. DOI: 10.30574/wjarr.2023.20.3.2432

Okano, M. T., Antunes, S. N., & Fernandes, M. E. (2021). Digital transformation in the manufacturing industry under the optics of digital platforms and ecosystems. *Independent Journal of Management & Production*, *12*(4), 1139–1159. DOI: 10.14807/ijmp.v12i4.1375

Okorie, O., Subramoniam, R., Charnley, F., Patsavellas, J., Widdifield, D., & Salonitis, K. (2020). Manufacturing in the time of COVID-19: An assessment of barriers and enablers. *IEEE Engineering Management Review*, *48*(3), 167–175. DOI: 10.1109/EMR.2020.3012112

Olshavsky, R. W., & Spreng, R. A. (1996). An Exploratory Study of the Innovation Evaluation Process. *Journal of Product Innovation Management*, *13*(6), 512–529. DOI: 10.1111/1540-5885.1360512

Oriji, O., & Joel, O. S. (2024). Integrating accounting models with supply chain management in the aerospace industry: A strategic approach to enhancing efficiency and reducing costs in the US. *World Journal of Advanced Research and Reviews*, *21*(3), 1476–1489. DOI: 10.30574/wjarr.2024.21.3.0873

Ortiz-Ruiz, E., Bermejo, J. R., Sicilia, J. A., & Bermejo, J. (2024). Machine Learning Techniques for Cyberattack Prevention in IoT Systems: A Comparative Perspective of Cybersecurity and Cyberdefense in Colombia. *Electronics (Basel)*, *13*(5), 824. Advance online publication. DOI: 10.3390/electronics13050824

Pacifico, M. G., Marchiano, G., De Medici, S., & Novellino, A. (2024, June). Application of Dynamic and AI Approaches for Predictive Maintenance. In *2024 9th International Conference on Smart and Sustainable Technologies (SpliTech)* (pp. 1-6). IEEE. DOI: 10.23919/SpliTech61897.2024.10612544

Parasuraman, R., Sheridan, T. B., & Wickens, C. D. (2008). Situation awareness, mental workload, and trust in automation: Viable, empirically supported cognitive engineering principles. *Theoretical Issues in Ergonomics Science*, *8*(4), 285–301. DOI: 10.1080/14639220701754741

Parekh, P., Patel, S., Patel, N., & Shah, M. (2020). Systematic review and meta-analysis of augmented reality in medicine, retail, and games. *Visual Computing for Industry, Biomedicine, and Art*, *3*(1), 21. DOI: 10.1186/s42492-020-00057-7 PMID: 32954214

Park, C., & Jun, J. (2003). A cross-cultural comparison of Internet buying behavior: Effects of Internet usage, perceived risks, and innovativeness. *International Marketing Review*, *20*(5), 534–553. DOI: 10.1108/02651330310498771

Parmentola, A., Petrillo, A., Tutore, I., & Felice, F. D. (2021). Is blockchain able to enhance environmental sustainability? A systematic review and research agenda from the perspective of Sustainable Development Goals (SDGs). *Business Strategy and the Environment.*

Parrott, L. (2014). *Values and ethics in social work practice* (2nd ed.). Learning Matters.

Paszkiewicz, A., Bolanowski, M., Budzik, G., Przeszłowski, Ł., & Oleksy, M. (2020). Process of Creating an Integrated Design and Manufacturing Environment as Part of the Structure of Industry 4.0. *Processes (Basel, Switzerland)*, *8*(9), 1019. DOI: 10.3390/pr8091019

Patalas-Maliszewska, J., & Łosyk, H. (2022). An approach to maintenance sustainability level assessment integrated with Industry 4.0 technologies using Fuzzy-TOPSIS: A real case study. *Advances in Production Engineering & Management*, *17*(4), 455–468. DOI: 10.14743/apem2022.4.448

Payne, M. (2014). *Modern social work theory* (4th ed.). Palgrave Macmillan. DOI: 10.1007/978-1-137-40603-3

Penumuru, D. P., Muthuswamy, S., & Karumbu, P. (2020). Identification and classification of materials using machine vision and machine learning in the context of industry 4.0. *Journal of Intelligent Manufacturing*, *31*(5), 1229–1241. DOI: 10.1007/s10845-019-01508-6

Pereira, A. C., Alves, A. C., & Arezes, P. (2023). Augmented Reality in a Lean Workplace at Smart Factories: A Case Study. *Applied Sciences (Basel, Switzerland)*, *13*(16), 9120. DOI: 10.3390/app13169120

Peterson, J., & Liu, M. (2024). Business innovation and strengths-based workforce development. *Harvard Business Review, 102*(3), 78–87.

Pham, T.-T. T., & Ho, J. C. (2015). The effects of product-related, personal-related factors and attractiveness of alternatives on consumer adoption of NFC-based mobile payments. *Technology in Society, 43*, 159–172. DOI: 10.1016/j.techsoc.2015.05.004

Phanden, R. K., Sharma, P., & Dubey, A. (2021). A review on simulation in digital twin for aerospace, manufacturing and robotics. *Materials Today: Proceedings, 38*, 174–178. DOI: 10.1016/j.matpr.2020.06.446

Piketty, T. (2014). *Capital in the twenty-first century.* Harvard University Press. DOI: 10.4159/9780674369542

Pingulkar, C., & Joshi, A. (2021). AI and Data Analytics for Predictive Maintenance in Solar Power Plants. [IJPREMS]. *International Journal of Progressive Research in Engineering Management and Science, 1*, 52–69.

Pinzone, M., Albueque, F., Orlandelli, D., Giovanna Barletta, I., Berlin, C., Johansson, B., & Taisch, M. (2020). A framework for operative and social sustainability functionalities in Human-Centric Cyber-Physical Production Systems. *Computers & Industrial Engineering, 139*, 105132. DOI: 10.1016/j.cie.2018.03.028

Plakas, G., Ponis, S. T., Agalianos, K., Aretoulaki, E., & Gayialis, S. P. (2020). Augmented Reality in Manufacturing and Logistics: Lessons Learnt from a Real-Life Industrial Application. *Procedia Manufacturing, 51*, 1629–1635. DOI: 10.1016/j.promfg.2020.10.227

Pookkuttath, S., Rajesh Elara, M., Sivanantham, V., & Ramalingam, B. (2021). AI-enabled predictive maintenance framework for autonomous mobile cleaning robots. *Sensors (Basel), 22*(1), 13. DOI: 10.3390/s22010013 PMID: 35009556

Pop, G. I., Titu, A. M., & Pop, A. B. (2023). Enhancing aerospace industry efficiency and sustainability: Process integration and quality management in the context of Industry 4.0. *Sustainability (Basel), 15*(23), 16206. DOI: 10.3390/su152316206

Popple, P. R., & Leighninger, L. (2017). *The policy-based profession: An introduction to social welfare policy analysis for social workers* (6th ed.). Pearson.

Presidential Communications Office. (n.d.). *PBBM gives all-out support for RP-made MULA satellite.* https://pco.gov.ph/news_releases/pbbm-gives-all-out-support-for-rp-made-mula-satellite/

Pressman, R., & Maxim, B. (2020). *Software engineering. a practitioner's approach* (9th ed.). Edit. McGraw-Hill.

Qiu, H., Yu, Z., Zhao, T., Zhang, Q., Xu, M., Li, P., Li, T., Bao, W., Chai, Y., Chen, S., Chen, Y., Cheng, H.-M., Dai, D., Di, Z., Dong, Z., Duan, X., Feng, Y., Fu, Y., Guo, J., & Wang, X. (2024). Two-dimensional materials for future information technology: Status and prospects. *Science China. Information Sciences*, *67*(6), 160400. DOI: 10.1007/s11432-024-4033-8

Rai, R., Tiwari, M. K., Ivanov, D., & Dolgui, A. (2021). Machine learning in manufacturing and industry 4.0 applications. *International Journal of Production Research*. Taylor & Francis.

Ramos, R., Rita, P., & Vong, C. (2024). Mapping research in marketing: Trends, influential papers and agenda for future research. *Spanish Journal of Marketing - ESIC*, *28*(2), 187–206. DOI: 10.1108/SJME-10-2022-0221

Ramzi, B., Ahmad, H., & Zakaria, N. (2019). A Conceptual Model on People Approach and Smart Manufacturing. 8(4), 1102-1106. https://doi.org/DOI: 10.59160/ijscm.v8i4.3583

Rasmussen, J. (1983). Skills, rules, and knowledge; signals, signs, and symbols, and other distinctions in human performance models. *IEEE Transactions on Systems, Man, and Cybernetics*, *13*(3), 257–266. DOI: 10.1109/TSMC.1983.6313160

Reamer, F. G. (2018). *Social work values and ethics* (4th ed.). Columbia University Press. DOI: 10.7312/ream18828

Reljić, V., Milenković, I., Dudić, S., Šulc, J., & Bajči, B. (2021). Augmented Reality Applications in Industry 4.0 *Environment. Applied Sciences (Basel, Switzerland)*, *11*(12), 5592. DOI: 10.3390/app11125592

Republic Act No. 7916. (1995). *Special Economic Zone Act of 1995 (Philippines)*.

RevFine. (n.d.). *Aerospace companies: List of aircraft and spacecraft manufacturers*. https://www.revfine.com/aviation-industry/

Riar, M., Korbel, J. J., Xi, N., Meywirth, S., Zarnekow, R., & Hamari, J. (2023). Augmented Reality in Interactive Marketing: The State-Of-The-Art and Emerging Trends. In Wang, C. L. (Ed.), *The Palgrave Handbook of Interactive Marketing*. Palgrave Macmillan., DOI: 10.1007/978-3-031-14961-0_14

Roach, G. (2009). Consumer perceptions of mobile phone marketing: A direct marketing innovation. *Direct Marketing (Bradford)*, *3*(2), 124–138. DOI: 10.1108/17505930910964786

Roberto, R., Newnes, L., & Shokrani, A. (2024). Challenges to Deploy Augmented Reality in Manufacturing. *Procedia CIRP*, *130*, 381–386. DOI: 10.1016/j.procir.2024.10.104

Roberts, M., & Davis, K. (2023). Digital financial planning tools for families in transition. *Family Economics Review*, *34*(2), 89–104.

Roberts, S., Johnson, M., & Thompson, K. (2023). Digital fatigue and surveillance anxiety in modern manufacturing. *Work and Stress*, *37*(4), 456–473.

Rocha-Jácome, C., Carvajal, R. G., Chavero, F. M., Guevara-Cabezas, E., & Hidalgo Fort, E. (2022). Industry 4.0: A Proposal of Paradigm Organization Schemes from a Systematic Literature Review. *Sensors (Basel)*, *22*(1), 66. DOI: 10.3390/s22010066 PMID: 35009609

Rodriguez, A., & Kim, J. (2023). Augmented reality and digital twins in manufacturing: Workforce implications. *Technological Forecasting and Social Change*, *189*, 122–135.

Rodriguez, L., & Thompson, M. (2024). Artificial intelligence and worker displacement: Social work perspectives. *Journal of Technology in Human Services*, *42*(1), 23–41.

Ronell Leo, J. A. (2018). A model for smart factories in the automotive sector. Thesis. Faculty of Business and Economics Sciences, Nelson Mandela University.

Roschli, A., Gaul, K. T., Boulger, A. M., Post, B. K., Chesser, P. C., Love, L. J., Blue, F., & Borish, M. (2019). Designing for Big Area Additive Manufacturing. *Additive Manufacturing*, *25*, 275–285. DOI: 10.1016/j.addma.2018.11.006

Ros, F., & Guillaume, S. (2020). *Sampling Techniques for Supervised or Unsupervised Tasks*. Springer Nature., DOI: 10.1007/978-3-030-29349-9

Rothwell, W. J. (2013). *Career planning and succession management: Developing your organization's talent—for today and tomorrow*. Praeger.

Rubio, L. R., Martin-Mariscal, A., & Peralta, E. (2022, July). First approach to a theoretical framework for carbon footprint management in the aerospace manufacturing industry. In *IFIP International Conference on Product Lifecycle Management* (pp. 620-629). Cham: Springer Nature Switzerland.

Rupenyan, A., & Balta, E. C. (2024). Robotics and manufacturing automation. In *The Impact of Automatic Control Research on Industrial Innovation: Enabling a Sustainable Future* (pp. 169–197). DOI: 10.1002/9781119983644.ch6

Saberi, S., Kouhizadeh, M., Sarkis, J., & Shen, L. (2019). Blockchain technology and its relationships to sustainable supply chain management. *International Journal of Production Research*, *57*(7), 2117–2135. DOI: 10.1080/00207543.2018.1533261

Saha, P. K. (2016). *Aerospace manufacturing processes*. CRC Press. DOI: 10.1201/9781315367965

Sahu, A. K., Padhy, R. K., & Dhir, A. (2020). Envisioning the Future of Behavioral Decision-Making: A Systematic Literature Review of Behavioral Reasoning Theory. *Australasian Marketing Journal*, *28*(4), 145–159. DOI: 10.1016/j.ausmj.2020.05.001

Sakr, A., & Abdullah, T. (2024). Virtual, augmented reality and learning analytics impact on learners, and educators: A systematic review. *Education and Information Technologies*, *29*(15), 19913–19962. DOI: 10.1007/s10639-024-12602-5

Saleebey, D. (2012). *The strengths perspective in social work practice* (6th ed.). Pearson.

Samatas, G. G., Moumgiakmas, S. S., & Papakostas, G. A. (2021, May). Predictive maintenance-bridging artificial intelligence and iot. In *2021 IEEE World AI IoT Congress (AIIoT)* (pp. 0413-0419). IEEE. DOI: 10.1109/AIIoT52608.2021.9454173

Santi, G. M., Ceruti, A., Liverani, A., & Osti, F. (2021). Augmented Reality in Industry 4.0 and Future Innovation Programs. *Technologies*, *9*(2), 33. DOI: 10.3390/technologies9020033

Sarter, N. B., & Woods, D. D. (1991). Situation awareness: A critical but ill-defined phenomenon. *The International Journal of Aviation Psychology*, *1*(1), 45–57. DOI: 10.1207/s15327108ijap0101_4

Savastano, M., Amendola, C., & D'Ascenzo, F. (2018). How digital transformation is reshaping the manufacturing industry value chain: The new digital manufacturing ecosystem applied to a case study from the food industry. In *Network, Smart and Open: Three Keywords for Information Systems Innovation* (pp. 127–142). Springer International Publishing. DOI: 10.1007/978-3-319-62636-9_9

Scaife, A. D. (2024). Improve predictive maintenance through the application of artificial intelligence: A systematic review. *Results in Engineering*, *21*, 101645. DOI: 10.1016/j.rineng.2023.101645

Schifferstein, H. N. J., Lemke, M., & de Boer, A. (2022). An exploratory study using graphic design to communicate consumer benefits on food packaging. *Food Quality and Preference*, *97*, 104458. DOI: 10.1016/j.foodqual.2021.104458

Schumann, M., Fuchs, C., Kollatsch, C., & Klimant, P. (2021). Evaluation of augmented reality supported approaches for product design and production processes. *Procedia CIRP*, *97*, 160–165. DOI: 10.1016/j.procir.2020.05.219

Schwab, K. (2016). *The fourth industrial revolution*. World Economic Forum.

Scuotto, V., Del Giudice, M., & Carayannis, E. G. (2017). The effect of social networking sites and absorptive capacity on SMES' innovation performance. *The Journal of Technology Transfer*, *42*(2), 409–424. DOI: 10.1007/s10961-016-9517-0

Sharma, A., Aggarwal, K. K., & Malik, M. (2023). IoT integration with Big Data and Cloud Computing: A Multidisciplinary Approach. *2023 International Conference on Communication, Security and Artificial Intelligence, ICCSAI 2023*, 139–144. DOI: 10.1109/ICCSAI59793.2023.10420987

Sharma, R., & Villányi, B. (2022). Evaluation of corporate requirements for smart manufacturing systems using predictive analytics. *Internet of Things : Engineering Cyber Physical Human Systems*, *19*, 100554. DOI: 10.1016/j.iot.2022.100554

Shierholz, H. (2019). The consequences of the decline in unions for the working class. *Economic Policy Institute Briefing Paper*, 522.

Shyr, W.-J., Tsai, C.-J., Lin, C.-M., & Liau, H.-M. (2022). Development and Assessment of Augmented Reality Technology for Using in an Equipment Maintenance and Diagnostic System. *Sustainability (Basel)*, *14*(19), 12154. DOI: 10.3390/su141912154

Siegel, E. (Ed.). (2015). *Predictive Analytics: The Power to Predict Who Will Click, Buy, Lie, or Die, Revised and Updated* (1st ed.). Wiley., DOI: 10.1002/9781119172536

Simion, D., Postolache, F., Fleacă, B., & Fleacă, E. (2024). Ai-driven predictive maintenance in modern maritime transport—Enhancing operational efficiency and reliability. *Applied Sciences (Basel, Switzerland)*, *14*(20), 9439. DOI: 10.3390/app14209439

Siriborvornratanakul, T. (2018). Enhancing User Experiences of Mobile-Based Augmented Reality via Spatial Augmented Reality: Designs and Architectures of Projector-Camera Devices. *Advances in Multimedia*, *17*, 1–17. Advance online publication. DOI: 10.1155/2018/8194726

Smith, J. K. (2021). *Robotic persons: our future with social robots*. Westbow Press.

Smith, R., Johnson, A., & Davis, M. (2023). Digital counseling platforms for workplace transition support. *Clinical Social Work Journal*, *51*(3), 234–248.

Snyder, W. E. (1985). *Industrial Robots*. Prentice-Hall International.

Song, Y., Koeck, R., & Luo, S. (2021). Review and analysis of augmented reality (AR) literature for digital fabrication in architecture. *Automation in Construction*, *128*, 103762. DOI: 10.1016/j.autcon.2021.103762

Sorescu, A. B., & Spanjol, J. (2008). Innovation's Effect on Firm Value and Risk: Insights from Consumer Packaged Goods. *Journal of Marketing*, *72*(2), 114–132. DOI: 10.1509/jmkg.72.2.114

Stadnicka, D., Sęp, J., Amadio, R., Mazzei, D., Tyrovolas, M., Stylios, C., Carreras-Coch, A., Merino, J. A., Żabiński, T., & Navarro, J. (2022). Industrial Needs in the Fields of Artificial Intelligence, Internet of Things and Edge Computing. *Sensors (Basel)*, *22*(12), 4501. DOI: 10.3390/s22124501 PMID: 35746287

Stohr, A., Ollig, P., Keller, R., & Rieger, A. (2024). Generative mechanisms of AI implementation: A critical realist perspective on predictive maintenance. *Information and Organization*, *34*(2), 100503. DOI: 10.1016/j.infoandorg.2024.100503

Strube, G., Eloot, K., Griessmann, N., Dhawan, R., & Ramaswamy, S. (2016). Trends in the commercial aerospace industry. In *Supply Chain Integration Challenges in Commercial Aerospace: A Comprehensive Perspective on the Aviation Value Chain* (pp. 141–159). Springer International Publishing.

Sun, Y., Shahzad, M., Ali, A., & Razzaq, A. (2025). Synergistic effect of green knowledge sharing and green creative climate for circular economy practices: role of artificial intelligence information quality. *Journal of Manufacturing Technology Management*, Vol. ahead-of-print No. ahead-of-print. https://doi.org/DOI: 10.1108/JMTM-09-2024-0503

Szejka, A. L., Junior, O. C., & Mas, F. (2024). Knowledge-based expert system to drive an informationally interoperable manufacturing system: An experimental application in the Aerospace Industry. *Journal of Industrial Information Integration*, *41*, 100661. DOI: 10.1016/j.jii.2024.100661

Székely, F., & Knirsch, M. (2005). Responsible Leadership and Corporate Social Responsibility: Metrics for Sustainable Performance. *European Management Journal*, *23*(6), 628–647. DOI: 10.1016/j.emj.2005.10.009

Szmigin, I., & Foxall, G. (1998). Three forms of innovation resistance: The case of retail payment methods. *Technovation*, *18*(6–7), 459–468. DOI: 10.1016/S0166-4972(98)00030-3

Taber, K. S. (2018). The Use of Cronbach's Alpha When Developing and Reporting Research Instruments in Science Education. *Research in Science Education*, *48*(6), 1273–1296. DOI: 10.1007/s11165-016-9602-2

Tan, Y.-C., Chandukala, S. R., & Reddy, S. K. (2022). Augmented Reality in Retail and Its Impact on Sales. *Journal of Marketing*, *86*(1), 48–66. DOI: 10.1177/0022242921995449

Tao, F., Ma, X., Liu, W., & Zhang, C. (2024). Digital Engineering: State-of-the-art and perspectives. *Digital Engineering*, *1*, 100007. DOI: 10.1016/j.dte.2024.100007

Taylor, A. T., Berrueta, T. A., & Murphey, T. D. (2021). Active learning in robotics: A review of control principles. *Mechatronics*, *77*, 102576. DOI: 10.1016/j.mechatronics.2021.102576

Taylor, J., & Johnson, P. (2024). Change management capabilities and digital transformation success. *Change Management Review*, *28*(2), 145–162.

Taylor, M., Wilson, S., & Brown, A. (2023). Regional collaboration in rural workforce development. *Rural Sociology*, *88*(3), 567–589.

Teater, B. (2019). *An introduction to applying social work theories and methods* (2nd ed.). Open University Press.

Tegethoff, T., Santa, R., Bucheli, J. M., Cabrera, B., & Scavarda, A. (2025). Navigating Industry 4.0: Leveraging additive technologies for competitive advantage in Colombian aerospace and manufacturing industries. *PLoS One*, *20*(2), e0318339. DOI: 10.1371/journal.pone.0318339 PMID: 39928702

Thang, T. Q., & Kim, J. (2025). Applications of 3D Printing in Paper-Based Devices for Biochemical and Environmental Analyses. *Chemosensors (Basel, Switzerland)*, *13*(3), 89. DOI: 10.3390/chemosensors13030089

Thompson, K. (2015). Systems approaches to organizational change in manufacturing. *Systems Research and Behavioral Science*, *32*(4), 456–468.

Thompson, L., & Rivera, M. (2023). Trauma-informed organizational change in manufacturing settings. *Administration in Social Work*, *47*(2), 123–140.

Timmer, M. (2000). The dynamics of Asian manufacturing: a comparative perspective in the late twentieth century. In *The Dynamics of Asian Manufacturing*. Edward Elgar Publishing. DOI: 10.4337/9781035304103

Tiwari, S. P. (2008). Information and communication technology initiatives for knowledge sharing in agriculture.

Tiwari, S. P. (2023, November). A transformative and action plan-oriented approach to manufacturing using artificial intelligence and digital technologies. In *BISTIC business innovation sustainability and technology international conference (BISTIC 2023)* (pp. 22-31). Atlantis Press. DOI: 10.2991/978-94-6463-302-3_4

Tiwari, S. P. (2025). The implications of artificial intelligence in management consulting: A risk and barrier assessment. *Multidisciplinary Reviews*, *8*(8), 2025244–2025244. DOI: 10.31893/multirev.2025244

Tiwari, S. P., Kostenko, O., & Yekhanurov, Y. (2025). *Understanding technology in the context of national development: Critical reflections*. SciFormat Publishing Inc. DOI: 10.69635/978-1-0690482-3-3

Toffler, A. (1980). *The third wave*. William Morrow and Company.

Tohanean, D., Buzatu, A.-I., Baba, C. A., & Georgescu, B. (2020). Business Model Innovation Through the Use of Digital Technologies: Managing Risks and Creating Sustainability. *Amfiteatru Economic*, •••, 758–774.

Toubia, O., & Florès, L. (2007). Adaptive Idea Screening Using Consumers. *Marketing Science*, *26*(3), 342–360. DOI: 10.1287/mksc.1070.0273

Troisi, O., Visvizi, A., & Grimaldi, M. (2024). The different shades of innovation emergence in smart service systems: The case of Italian cluster for aerospace technology. *Journal of Business and Industrial Marketing*, *39*(6), 1105–1129. DOI: 10.1108/JBIM-02-2020-0091

Trstenjak, M., Opetuk, T., Cajner, H., & Hegedić, M. (2022). Industry 4.0 Readiness Calculation-Transitional Strategy Definition by Decision Support Systems. *Sensors (Basel)*, *22*(3), 1185. DOI: 10.3390/s22031185 PMID: 35161929

Turner, F. J. (2017). *Social work treatment: Interlocking theoretical approaches* (6th ed.). Oxford University Press.

Turovets, Y., Vishnevskiy, K., Tokareva, M., & Kukushkin, K. (2019). Technology foresight for digital manufacturing: Russian case. *IOP Conference Series. Materials Science and Engineering*, *497*, 012062. https://iopscience.iop.org/article/10.1088/1757-899X/497/1/012062. DOI: 10.1088/1757-899X/497/1/012062

U.S. Department of Labor. (2023). *Workforce development policies and programs*. Government Printing Office.

Ucar, A., Karakose, M., & Kırımça, N. (2024). Artificial intelligence for predictive maintenance applications: Key components, trustworthiness, and future trends. *Applied Sciences (Basel, Switzerland)*, *14*(2), 898. DOI: 10.3390/app14020898

Ünlü, R., & Söylemez, İ. (2024). AI-Driven Predictive Maintenance. In *Engineering Applications of AI and Swarm Intelligence* (pp. 207–233). Springer Nature Singapore.

Usuga Cadavid, J. P., Lamouri, S., Grabot, B., Pellerin, R., & Fortin, A. (2020). Machine learning applied in production planning and control: A state-of-the-art in the era of industry 4.0. *Journal of Intelligent Manufacturing, 31*(6), 1531–1558. DOI: 10.1007/s10845-019-01531-7

Van Wormer, K., & Besthorn, F. H. (2018). *Human behavior and the social environment, macro level: Groups, communities, and organizations* (3rd ed.). Oxford University Press.

Vicente, K. J. (2006). *The human factor: Revolutionizing the way people live with technology*. Routledge.

Vilar, E., Filgueiras, E., & Rebelo, F. (2022). *Virtual and Augmented Reality for Architecture and Design*. CRC Press. DOI: 10.1201/9781003051381

Voinea, G.-D., Gîrbacia, F., Duguleană, M., Boboc, R. G., & Gheorghe, C. (2023). Mapping the Emergent Trends in Industrial Augmented Reality. *Electronics (Basel), 12*(7), 1719. DOI: 10.3390/electronics12071719

Vollrath, M. D., & Villegas, S. G. (2022). Avoiding digital marketing analytics myopia: Revisiting the customer decision journey as a strategic marketing framework. *Journal of Marketing Analytics, 10*(2), 106–113.

Von Leipzig, T., Gamp, M., Manz, D., Schottle, K., Ohlhausen, P., Oosthuizen, G., Palm, D., & Von Leipzig, K. (2017). Initialising customer-orientated digital transformation in enterprises. *Procedia Manufacturing, 8*, 517–524. DOI: 10.1016/j.promfg.2017.02.066

Wacquant, L. (2009). *Punishing the poor: The neoliberal government of social insecurity*. Duke University Press.

Walker, C. M., Agarwal, V., Lin, L., Hall, A. C., Hill, R. A., Mortenson, T. J., & Lybeck, N. J. (2023). *Explainable artificial intelligence technology for predictive maintenance* (No. INL/RPT-23-74159-Rev000). Idaho National Laboratory (INL), Idaho Falls, ID (United States).

Wallace, P. (2021). *Introduction to Information Systems* (4th ed.). Pearson.

Walsh, F. (2019). *Strengthening family resilience* (3rd ed.). Guilford Press.

Wang, J., Shi, E., Hu, H., Ma, C., Liu, Y., Wang, X., ... & Zhang, S. (2024). Large language models for robotics: Opportunities, challenges, and perspectives. Journal of Automation and Intelligence.

Wang, K., & Wang, Y. (2017, September). How AI affects the future predictive maintenance: a primer of deep learning. In *International workshop of advanced manufacturing and automation* (pp. 1-9). Singapore: Springer Singapore.

Wang, B. (2018). The future of manufacturing: A new perspective. *Engineering (Beijing)*, *4*(5), 722–728. DOI: 10.1016/j.eng.2018.07.020

Wang, M. (2009). Manufacturing FDI and economic growth: Evidence from Asian economies. *Applied Economics*, *41*(8), 991–1002. DOI: 10.1080/00036840601019059

Wang, W., Guo, Q., Yang, Z., Jiang, Y., & Xu, J. (2023). A state-of-the-art review on robotic milling of complex parts with high efficiency and precision. *Robotics and Computer-integrated Manufacturing*, *79*, 102436. DOI: 10.1016/j.rcim.2022.102436

Wang, Y., & Mason, M. (1987). Modeling impact dynamics for robotic operations. In *Proceedings. 1987 IEEE International Conference on Robotics and Automation* (Vol. 4). IEEE. DOI: 10.1109/ROBOT.1987.1087851

Westkämper, E. (2007). Digital Manufacturing in the global Era. In *Digital enterprise technology: Perspectives and future challenges* (pp. 3–14). Springer US. DOI: 10.1007/978-0-387-49864-5_1

Wheelwright, S. C., & Hayes, R. H. (1985). *Competing Through Manufacturing*. Harvard Business Review Case Services.

Whysall, Z., Owtram, M., & Brittain, S. (2019). The new talent management challenges of Industry 4.0. *Journal of Management Development*, *38*(2), 118–129. DOI: 10.1108/JMD-06-2018-0181

Wickens, C. D. (2008). Multiple resources and mental workload. *Human Factors*, *50*(3), 449–455. DOI: 10.1518/001872008X288394 PMID: 18689052

Williams, A. (2006). Tourism and hospitality marketing: Fantasy, feeling and fun. *International Journal of Contemporary Hospitality Management*, *18*(6), 482–495. DOI: 10.1108/09596110610681520

Williams, S. J., & Glover, K. (1985). Robot dynamic control instabilities. In *1985 American Control Conference*. IEEE.

Williams, T., & Davis, R. (2023). Demographic factors in digital transformation adaptation. *Work, Aging and Retirement*, *9*(2), 134–151.

Wilson, J., & Whitmore, E. (2019). Community resilience and economic transition. *Community Development Journal: An International Forum*, *54*(3), 445–462.

Wilson, P., Johnson, K., & Martinez, L. (2022). Intergenerational effects of technological change in manufacturing families. *Journal of Family Issues*, *43*(7), 1567–1589.

World Economic Forum. (2020). *The future of jobs report 2020*. WEF Publications.

Wu, Ch., & Thompson, M. E. (2020). *Sampling Theory and Practice*. Springer Nature Switzerland., DOI: 10.1007/978-3-030-44246-0

Wu, D., Rosen, D. W., Wang, L., & Schaefer, D. (2015). Cloud-based design and manufacturing: A new paradigm in digital manufacturing and design innovation. *Computer Aided Design*, *59*, 1–14. DOI: 10.1016/j.cad.2014.07.006

Xiong, J., Hsiang, E. L., He, Z., Zhan, T., & Wu, S.-T. (2021). Augmented reality and virtual reality displays: Emerging technologies and future perspectives. *Light, Science & Applications*, *10*(1), 216. DOI: 10.1038/s41377-021-00658-8 PMID: 34697292

Xu, Z., Feng, Z., Jelodar, M. B., & Guo, B. H. W. (2024). Augmented reality applications in construction productivity: A systematic literature review. *Advanced Engineering Informatics*, *62*(Part C), 102798. DOI: 10.1016/j.aei.2024.102798

Ye, Y., Yong, Z., & Han, D. (2020). Research on key technology of industrial artificial intelligence and its application in predictive maintenance. *Acta Automatica Sinica*, *46*(10), 2013–2030.

Yilmaz, A. A. (Ed.). (2024). *Harnessing digital innovation for air transportation*. IGI Global. DOI: 10.4018/979-8-3693-0732-8

Young, P., Byrne, G., & Cotterell, M. (1997). Manufacturing and the environment. *International Journal of Advanced Manufacturing Technology*, *13*(7), 488–493. DOI: 10.1007/BF01624609

Zastrow, C., Kirst-Ashman, K. K., & Hessenauer, S. L. (2019). *Understanding human behavior and the social environment* (10th ed.). Cengage Learning.

Zekić-Sušac, M., Has, A., & Knežević, M. (2021). Predicting energy cost of public buildings by artificial neural networks, CART, and random forest. *Neurocomputing*, *439*, 223–233. DOI: 10.1016/j.neucom.2020.01.124

Zemskov, A. D., Fu, Y., Li, R., Wang, X., Karkaria, V., Tsai, Y. K., . . . Li, P. (2024). Security and privacy of digital twins for advanced manufacturing: A survey. *arXiv preprint arXiv:2412.13939*.

Zhang, L., Kumar, S., & Patel, R. (2023). Machine learning and predictive analytics in digital manufacturing. *Manufacturing Science and Engineering*, 145(8), 081-095.

Zhang, C., Xu, Q., Yu, Y., Zhou, G., Zeng, K., Chang, F., & Ding, K. (2025). A survey on potentials, pathways and challenges of large language models in new-generation intelligent manufacturing. *Robotics and Computer-integrated Manufacturing*, *92*, 102883. DOI: 10.1016/j.rcim.2024.102883

Zheng, H., Paiva, A. R., & Gurciullo, C. S. "Advancing from predictive maintenance to intelligent maintenance with ai and iiot." *arXiv preprint arXiv:2009.00351* (2020).

Zheng, C., Soosay, C., & Hyland, P. (2007). Manufacturing to Asia: Who will win the emerging battle for talent between Dragons and Tigers? *Journal of Manufacturing Technology Management*, *19*(1), 52–72. DOI: 10.1108/17410380810843453

Zheng, X., Hu, X., Arista, R., Lu, J., Sorvari, J., Lentes, J., Ubis, F., & Kiritsis, D. (2024). A semantic-driven tradespace framework to accelerate aircraft manufacturing system design. *Journal of Intelligent Manufacturing*, *35*(1), 175–198. DOI: 10.1007/s10845-022-02043-7

Zhou, W. (2024). Competition and development in the aviation industry: An analysis of strategic adaptability and challenges.

Zhu, M., Liang, C., Yeung, A. C., & Zhou, H. (2024). The impact of intelligent manufacturing on labor productivity: An empirical analysis of Chinese listed manufacturing companies. *International Journal of Production Economics*, *267*, 109070. DOI: 10.1016/j.ijpe.2023.109070

Zonta, T., da Costa, C. A., da Rosa Righi, R., de Lima, M. J., da Trindade, E. S., & Li, G. P. (2020). Predictive maintenance in the Industry 4.0: A systematic literature review. *Computers & Industrial Engineering*, *150*, 106889. Advance online publication. DOI: 10.1016/j.cie.2020.106889

About the Contributors

Siddhartha Paul Tiwari, PhD, FRAS, FRAI is an award-winning academic, technologist, and researcher known for his significant contributions to educational innovation, mobile technology, digital strategies, and ICT. He presently serves as the Head of Policy and Planning for Android and Mobile Technologies across Google Asia Pacific, based in Singapore, a key role in shaping the region's mobile technology landscape. He holds a doctorate from the prestigious Southeast Asia Interdisciplinary Development Institute (SAIDI) in the Philippines, underscoring his profound expertise and commitment to interdisciplinary development in the tech field.

Eliseo A. Aurellado is the President of Southeast Asia Interdisciplinary Development Institute (SAIDI) Graduate School of OD. He is a CPA, management practitioner, academician and entrepreneur with more than 40 years\' experience in public accounting, controllership, general management, management consultancy services, and business practice. He holds degrees in BBA, Accountancy, magna cum laude (University of the East), MBA, Beta Gamma Sigma (University of California, Los Angeles (UCLA), and PhD (Southeast Asia Interdisciplinary Development Institute). He is one of the peer reviewers of the Horizon Publishing Company in the US that issues professional journals. He started his career with SGV & Co., CPAS, worked with Deloitte Haskins & Sells, CPAs in New York, became First Vice President of PCIBank, President and CEO of PCIB Property Care, Inc., Audit Partner of R.S. Bernaldo & Associates (CPAs), and Audit Director of ML Aguirre & Co, CPAs. He is former Cluster Chair of Finance Faculty at the Ateneo Graduate School of Business, a position he held for 13 years. He still teaches accounting, finance and strategic management courses in Ateneo\'s MBA programs. He is also a lecturer at Ateneo\'s Center for Continuing Education (CCE) and part-time faculty of CCE\'s Master in Entrepreneurship program. He is Vice Chairman of the Board and COO of Metro Stonerich Corporation, a Triple A, ISO-certified construction company catering to big-time developers, Chairman of the Board of Metrorich Realty and President of Metrorich Builders Corporation. He is a member of CPA professional organizations such as PICPA, ACPAPP and ACPACI. He is a former member of the Professional Regulations Commission Board of Accountancy that regulates the accountancy profession and administers licensure examinations for CPAs. He was also a member of CHED's Technical Committee on Accountancy that created three new Bachelor\'s degree programs in accountancy. He is Past President of the Rotary Club of Downtown Manila and former director of AMCI Mountaineering Club. He is a commissioned officer of the Armed Forces of the Philippines Reserve Command with the rank of Lt. Colonel.

Adi Fahrudin was born on January 12, 1966. He is a Professor in Social Work at the Ministry of Education, Culture, Research and Technology, the Republic of Indonesia under the High Education Service Agency Region III Jakarta seconded to the Faculty of Psychology, Universitas Bhayangkara Jakarta Raya, Indonesia (from 1 March 2022 until present). Start 26 June 2022, Prof Adi was appointed as Head of the Institute for Research, Community Service and Publication, and since 10 August 2023, he has also been appointed as Acting Dean of the Faculty of Psychology, Universitas Bhayangkara Jakarta Raya til 10 August 2024. Before that, he was appointed as Professor in Social Work seconded to the Faculty of Social and Political Sciences, at the University of Muhammadiyah Jakarta (1 January 2012-28 February 2022) and the Bandung Polytechnic of Social Welfare (1 March 1993 – December 2011). From April 2018 until 31 March 2025, he was appointed as a Visiting Professor at the Japan University College of Social Work, Tokyo Japan. From 2015-2023, he was appointed as a Visiting Professor and Associate Research Fellow at The Psychology and Social Health Research Unit, Faculty of Psychology and Education, Universiti Malaysia Sabah. Prof Adi is also an Associate Research Scholar at the Asian Research Institute for International Social Work (ARIISW), Shukutoku University, Chiba, Jepang (2015 until now). Previously he was appointed as a Visiting Professor in the Faculty of Applied Social Sciences, Universiti Sultan Zainal Abidin (UnisZa), Malaysia (2014-2015), and Visiting Associate Professor at Universiti Sains Malaysia (1997-1998).

***

Kapil Kumar Aggarwal is Professor of Finance at the University School of Business, Chandigarh University. He has earned PhD, M. Phil, MFC, MBA & M. Com. He has more than 16 years of teaching and research experience. His research experience is inclusive of 22 research papers published in National journals, International Journals, Edited Books, ABDC-listed journals, Scopus Indexed,

Web of Sciences and Peer Reviewed journals, 32 research papers presented in National/International conferences and two books published in his account. His teaching interest is primarily focused on finance, management sciences, quantitative techniques, business mathematics, and business statistics. He is the reviewer of the International Journal of Educational Development Journal, indexed in Scopus Elsevier, Social Sciences Citation Index and Educational and Information Technologies Journal indexed in SCOPUS, Social Science Citation Index, and UGC-CARE List (India) and Editorial Board member of International Journal of Accounting, Finance and Risk Management (IJAFRM) of He is also a pioneer in the field of SPSS and R and has delivered multiple lectures and workshops on the same.

Saket Bansod, HoD & Assistant Professor, Dr.Ambedkar Institute of Management Studies and Research, Nagpur.

Nihar Suresh Dahake, Assistant Professor, G H Raisoni College of Engineering, Nagpur.

Parihar Suresh Dahake is working as an Assistant Professor at Ramdeobaba University,Nagpur. He received his BTech, MBA, and Ph.D. from the Rashtrasant Tukdoji Maharaj Nagpur University and PGDIBO, M.Com from Indira Gandhi National Open University (IGNOU). Dr.Parihar dahake has Qualified SET(Commerce) and SET (Management). He has 14 Years of Teaching Experience and 13 years of Research Experience. His areas of expertise are Consumer behavior, Organizational Behaviour, Marketing management, and IMC.

Fahmi Ilman Fahrudin, PhD, is a lecturer at the Faculty of Science and Technology, Universitas Muhammadiyah Bandung. He graduated with a Bachelor of Food Technology from Universitas Diponegoro, Semarang, Central Java, Indonesia; a Master of Food Technology from Universiti Putra Malaysia, Serdang, Malaysia; and a PhD in Food Science and Technology from the Faculty of Agroindustry, Chiang Mai University, Thailand.

G Prasad is currently working as faculty in the department of Aerospace Engineering Chandigarh University, India. He has 9 years of experience in Teaching and research. He received his Bachelor of Engineering and Master of Engineering degrees with first class and distinction and Doctor of Philosophy in the field of Aeronautical Engineering from the Anna University, Chennai, India. Prof. Prasad G is the author of over 30 technical publications, 1 Patent Published and 1 Book published. His research interests include Unmanned Aerial Vehicles, Aircraft Structures, Computational Fluid Dynamics and Interdisciplinary research. He has completed three funded project sponsored by The Institution of Engineers (India) and Tamilnadu State Council for Science and Technology. Awarded Indian National Science Academy (INSA) Visiting Scientist Programme 2019 and Awarded Science Academies' Summer Research Fellowship Programme (SRPF) 2019. Further he is SPOC of NPTEL- IIT Madras, SPOC of ISRO IIRS outreach coordinator and SPOC for NASA International SpaceApps Challenge. Further key positions such as coordinator of Board of Studies, Outcome Based Education, Project Coordinator, Research Coordinator

Rashid Hashmi is a seasoned senior journalist with a rich and diverse experience spanning over 20 years in Broadcast, Digital, and Print Media. Prior to joining Sharda University, he held the position of Executive Editor and Senior Anchor at the renowned national news channel, 'India News'. Throughout his illustrious journalistic career, Prof. Rashid Hashmi has served in various capacities across prominent media platforms including Dainik Jagran Newspaper, Zee News, ETV, Total TV, FM News, and India News. His expertise encompasses anchoring, debating, and ground zero reporting of numerous parliamentary and assembly elections, providing insightful coverage of pivotal political events. Prof. Rashid Hashmi is renowned for his engaging interviews with prominent personalities from the worlds of politics, sports, and Bollywood. His interviews have featured distinguished celebrities, political leaders, and sports icons, offering unique perspectives and valuable insights to viewers and readers alike. With a keen eye for detail and a commitment to journalistic integrity, Rashid Hashmi has earned widespread acclaim for his contribution to media discourse in India. His dynamic approach to storytelling and extensive experience across multiple media formats underline his versatility and depth as a journalist. Prof. Rashid Hashmi's tenure in the media industry reflects a dedication to delivering compelling and informative content while upholding the values of responsible journalism. His role in shaping public discourse through impactful reporting and incisive interviews underscores his standing as a respected figure in Indian journalism.

Mohd Dahlan A. Malek is the Dean of the Faculty of Psychology and Social Work at Universiti Malaysia Sabah (UMS), where he also serves as a senior academic in the field of Industrial and Organizational Psychology. With over two decades of experience in higher education and psychological research, he has made significant contributions to the advancement of occupational well-being, psychological assessment, and leadership development in both academic and applied settings. Professor Dahlan holds a Ph.D. in Industrial and Organizational Psychology and has led numerous research projects focused on workplace mental health, employee engagement, and digital interventions for psychological well-being. He is actively involved in cross-disciplinary collaborations and international academic networks, particularly across Southeast Asia, including Malaysia, Indonesia, and the Philippines. He has published extensively in peer-reviewed journals, academic books, and conference proceedings, and is often invited as a keynote speaker, thesis examiner, and consultant for government and private-sector initiatives related to mental health, human capital development, and inclusive leadership. Professor Dahlan is committed to translating psychological theories into practical solutions that enhance individual and organizational performance. His leadership at UMS reflects his passion for integrating research, education, and community engagement in building a psychologically resilient and innovative society.

Ida Shafinaz Mohamed Kamil is a senior lecturer at the Faculty of Business, Economics & Accountancy, Universiti Malaysia Sabah. Ida received her bachelor degree in law (LL.B.) from the International Islamic University Malaysia (IIUM), Malaysia. She obtained her master degree in law (LL.M.) (Business Law) from the University of Wales, Aberystwyth, Wales, United Kingdom. Ida taught courses such as Malaysian business law, international business law, financial law and regulation, and law for accountants. Her major research interests include business law, intellectual property law, law and psychology, law and society, and native laws.

Prachita Patil Assistant Professor, Dr.Ambedkar Institute of Management Studies and Research, Nagpur.

Kus Hanna Rahmi, M.Psi., Psychologist, is a Lecturer at the Faculty of Psychology, Bhayangkara University Jakarta Raya. Born in Malang on October 26, 1977, he completed his Bachelor's degree in Psychology (S1), then continued his education with a Master's degree in Professional Psychology (S2) and a Doctorate in Psychology (S3) at Universitas Persada Indonesia YAI, Jakarta. As a practicing psychologist, Dr. Hanna actively provides clinical services at Muhammadiyah Taman Puring Hospital, South Jakarta, with a primary focus on psychological interventions for adults with various psychological issues. With over 10 years of experience in clinical practice, she is committed to developing evidence-based intervention approaches. His research and teaching are largely related to adult clinical interventions and therapeutic approaches. Currently, Dr. Hanna is developing a book titled "Trauma and Violence: A Psychological Review" as a contribution to deepening the understanding and handling of trauma faced by individuals in the context of violence. Dr. Hanna is also active in providing training and seminars in various institutions and communities, as well as writing scientific articles in the field of clinical psychology. With a high dedication to the profession and the science of psychology, he continues to strive to advance the field of adult clinical psychology in Indonesia.

Gerardo Reyes Ruiz is an Actuary by training from the Faculty of Sciences-UNAM; he studied a Specialty in Econometrics in the Postgraduate Studies Division of the Faculty of Economics-UNAM; he obtained a Master's and Doctorate (with a scholarship from CONACYT and the Carolina Foundation) in Business Studies (actuarial profile) from the Faculty of Economics and Business of the University of Barcelona, Spain. In the Doctorate he graduated with the highest honors awarded to a thesis of this nature, that is, he obtained the qualification of Excellent Cum Laude. Later he did a postdoctoral stay at the Center for Economic, Administrative, and Social Studies (CIECAS) of the National Polytechnic Institute (IPN). His work experience has been in public education institutions (UNAM, UAEM, IPN, Centro de Estudios Superiores Navales-CESNAV) where he has had the opportunity to teach classes in Bachelor's, Specialty, Master's, and Doctorate.

Binastya Anggara Sekti has been a Lecturer at the faculty of Computer Science, at Esa Unggul University, Jakarta since 2016 and is active in the Artificial Intelligence and Information Security Study Group. He is a graduate of Informatics Engineering from the Bandung Institute of Technology and achieved a Master's Degree from Trisakti University, Jakarta. Before as a Lecture, he was an IT practitioner at several companies in Jakarta for more than 25 years. Some of the projects that have been completed are building Data Centers at Internet Service Providers . He also finished many other IT Projects.

Sonal Sisodia, currently the Principal of a prestigious Business School in Indore, is an alumna of Sophia College Ajmer and holds an MBA (Gold Medalist), PhD in Management and a post graduate diploma in Human Resource Management. Specializing in Marketing, Dr Sisodia has over 20 years of experience in mentoring graduate and post graduate students, engaging in quality research, conducting and handling MDPs and handling consultancy assignments. She is a Professional Certified Marketer (PCM) trainer of American Management Association and contributes to the Smart Girl program as a trainer. Her prolific research career includes winning prestigious awards such as Manu Bhai Shah Memorial Research Gold Medal, Best Research award at ISTD Bhubaneshwar and Best Case study Award by Emerald Publication. Dr Sisodia has presented and published over sixty papers in national and international journals, including Scopus. Recognized for her excellence, she has received Best Teacher awards solidifying her status as a distinguished in the field of education.

Husmiati Yusuf is a Senior Researcher at the National Research and Innovation Regency (BRIN). She completed her Bachelor's degree in social work at the Polytechnic of Social Welfare, Bandung, Indonesia, and completed her Master of

Social Science (Social Work) and Doctor of Philosophy in Social Work at University Sains Malaysia (USM), Penang, Malaysia. She is interested in studies on social welfare issues, especially vulnerable groups such as the disabled, elderly, child abused, families, and indigenous social workers.

Index

A

C

D

H

I

M

N

P

R

S

T

W

www.ingramcontent.com/pod-product-compliance
Lightning Source LLC
LaVergne TN
LVHW080909300926
849162LV00003B/612
* 9 7 9 8 3 3 7 3 1 0 8 2 4 *